Distributed Computing Through Combinatorial Topology

This page is intentionally left blank

Distributed Computing Through Combinatorial Topology

Maurice Herlihy

Dmitry Kozlov

Sergio Rajsbaum

AMSTERDAM • BOSTON • HEIDELBERG • LONDON
NEW YORK • OXFORD • PARIS • SAN DIEGO
SAN FRANCISCO • SINGAPORE • SYDNEY • TOKYO

Morgan Kaufmann is an imprint of Elsevier

Acquiring Editor: *Todd Green*
Editorial Project Manager: *Lindsay Lawrence*
Project Manager: *Punithavathy Govindaradjane*
Designer: *Maria Inês Cruz*

Morgan Kaufmann is an imprint of Elsevier
225 Wyman Street, Waltham, MA 02451, USA

Library of Congress Cataloging-in-Publication Data
Herlihy, Maurice.
 Distributed computing through combinatorial topology / Maurice Herlihy, Dmitry Kozlov, Sergio Rajsbaum.
 pages cm
 Includes bibliographical references and index.
 ISBN 978-0-12-404578-1 (alk. paper)
1. Electronic data processing–Distributed processing–Mathematics.
2. Combinatorial topology. I. Kozlov, D. N. (Dmitrii Nikolaevich) II. Rajsbaum, Sergio. III. Title.
 QA76.9.D5H473 2013
 004'.36–dc23

                              2013038781

British Library Cataloguing-in-Publication Data
A catalogue record for this book is available from the British Library

ISBN: 978-0-12-404578-1

For my parents, David and Patricia Herlihy, and for Liuba, David, and Anna.
To Esther, David, Judith, and Eva-Maria.
Dedicated to the memory of my grandparents, Itke and David, Anga and Sigmund,
and to the memory of my Ph.D. advisor, Shimon Even.

This page is intentionally left blank

CONTENTS

Acknowledgments

We thank all the students, colleagues, and friends who helped improve this book: Hagit Attiya, Irina Calciu, Armando Castañeda, Lisbeth Fajstrup, Eli Gafni, Eric Goubault, Rachid Guerraoui, Damien Imbs, Petr Kuznetsov, Hammurabi Mendez, Yoram Moses, Martin Raussen, Michel Raynal, David Rosenblueth, Ami Paz, Vikram Seraph, Nir Shavit, Christine Tasson, Corentin Travers, and Mark R. Tuttle. We apologize for any names inadvertently omitted.

Special thanks to Eli Gafni for his many insights on the algorithmic aspects of this book.

This page is intentionally left blank

Preface

This book is intended to serve as a textbook for an undergraduate or graduate course in theoretical distributed computing or as a reference for researchers who are, or want to become, active in this area. Previously, the material covered here was scattered across a collection of conference and journal publications, often terse and using different notations and terminology. Here we have assembled a self-contained explanation of the mathematics for computer science readers and of the computer science for mathematics readers.

Each of these chapters includes exercises. We think it is essential for readers to spend time solving these problems. Readers should have some familiarity with basic discrete mathematics, including induction, sets, graphs, and continuous maps. We have also included mathematical notes addressed to readers who want to explore the deeper mathematical structures behind this material.

The first three chapters cover the *fundamentals* of combinatorial topology and how it helps us understand distributed computing. Although the mathematical notions underlying our computational models are elementary, some notions of combinatorial topology, such as simplices, simplicial complexes, and levels of connectivity, may be unfamiliar to readers with a background in computer science. We explain these notions from first principles, starting in Chapter 1, where we provide an intuitive introduction to the new approach developed in the book. In Chapter 2 we describe the approach in more detail for the case of a system consisting of two processes only. Elementary graph theory, which is well-known to both computer scientists and mathematicians, is the only mathematics needed.

The graph theoretic notions of Chapter 2 are essentially one-dimensional simplicial complexes, and they provide a smooth introduction to Chapter 3, where most of the topological notions used in the book are presented. Though similar material can be found in many topology texts, our treatment here is different. In most texts, the notions needed to model computation are typically intermingled with a substantial body of other material, and it can be difficult for beginners to extract relevant notions from the rest. Readers with a background in combinatorial topology may want to skim this chapter to review concepts and notations.

The next four chapters are intended to form the core of an advanced undergraduate course in distributed computing. The mathematical framework is self-contained in the sense that all concepts used in this section are defined in the first three chapters.

In this part of the book we concentrate on the so-called *colorless* tasks, a large class of coordination problems that have received a great deal of attention in the research literature. In Chapter 4, we describe our basic operational and combinatorial models of computation. We define tasks and asynchronous, fault-tolerant, wait-free shared-memory protocols. This chapter explains how the mathematical language of combinatorial topology (such as simplicial complexes and maps) can be used to describe concurrent computation and to identify the colorless tasks that can be solved by these protocols. In Chapter 5, we apply these mathematical tools to study colorless task solvability by more powerful protocols. We first consider computational models in which processes fail by *crashing* (unexpectedly halting). We give necessary and sufficient conditions for solving colorless tasks in a range of different computational models, encompassing different crash-failure models and different forms of communication. In Chapter 6, we show how the same mathematical notions can be extended to deal with *Byzantine* failures, where faulty processes, instead of crashing, can display arbitrary behavior. In Chapter 7, we show how to use *reductions* to transform results about one model of computation to results about others.

Chapters 8–11 are intended to form the core of a graduate course. Here, too, the mathematical framework is self-contained, although we expect a slightly higher level of mathematical sophistication. In this part, we turn our attention to *general tasks*, a broader class of problems than the colorless tasks covered earlier. In Chapter 8, we describe how the mathematical framework previously used to model colorless tasks can be generalized, and in Chapter 9 we consider manifold tasks, a subclass of tasks with a particularly nice geometric structure. We state and prove Sperner's lemma for manifolds and use this to derive a separation result showing that some problems are inherently "harder" than others. In Chapter 10, we focus on how computation affects *connectivity*, informally described as the question of whether the combinatorial structures that model computations have "holes." We treat connectivity in an axiomatic way, avoiding the need to make explicit mention of homology or homotopy groups. In Chapter 11, we put these pieces together to give necessary and sufficient conditions for solving general tasks in various models of computation. Here notions from elementary point-set topology, such as open covers and compactness are used.

The final part of the book provides an opportunity to delve into more advanced topics of distributed computing by using further notions from topology. These chapters can be read in any order, mostly after having studied Chapter 8. Chapter 12 examines the renaming task, and uses combinatorial theorems such as the Index Lemma to derive lower bounds on this task. Chapter 13 uses the notion of shellability to show that a number of models of computation that appear to be quite distinct can be analyzed with the same formal tools. Chapter 14 examines simulations and reductions for general tasks, showing that the shared-memory models used interchangeably in this book really are equivalent. Chapter 15 draws a connection between a certain class of tasks and the Word Problem for finitely-presented groups, giving a hint of the richness of the universe of tasks that are studied in distributed computing. Finally, Chapter 16 uses Schlegel diagrams to prove basic topological properties about our core models of computation.

Maurice Herlihy was supported by NSF grant 000830491.

Sergio Rajsbaum by UNAM PAPIIT and PAPIME Grants.

Dmitry Kozlov was supported by the University of Bremen and the German Science Foundation.

Companion Site

This book offers complete code for all the examples, as well as slides, updates, and other useful tools on its companion web page at:

https://store.elsevier.com/product.jsp?isbn=9780124045781&pagename=search.

Fundamentals

This page intentionally left blank

Introduction

1

Concurrency is confusing. Most people who find it easy to follow sequential procedures, such as preparing an omelette from a recipe, find it much harder to pursue concurrent activities, such as preparing a 10-course meal with limited pots and pans while speaking to a friend on the telephone. Our difficulties in reasoning about concurrent activities are not merely psychological; there are simply too many ways in which such activities can interact. Small disruptions and uncertainties can compound and cascade, and we are often ill-prepared to foresee the consequences. A new approach, based on topology, helps us understand concurrency.

1.1 Concurrency everywhere

Modern computer systems are becoming more and more concurrent. Nearly every activity in our society depends on the Internet, where distributed databases communicate with one another and with human beings. Even seemingly simple everyday tasks require sophisticated distributed algorithms. When a customer asks to withdraw money from an automatic teller machine, the banking system must either both

Distributed Computing Through Combinatorial Topology. http://dx.doi.org/10.1016/B978-0-12-404578-1.00001-2

provide the money and debit that account or do neither, all in the presence of failures and unpredictable communication delays.

Concurrency is not limited to wide-area networks. As transistor sizes shrink, processors become harder and harder to physically cool. Higher clock speeds produce greater heat, so processor manufacturers have essentially given up trying to make processors significantly faster. Instead, they have focused on making processors more parallel. Today's laptops typically contain *multicore* processors that encompass several processing units (cores) that communicate via a shared memory. Each core is itself likely to be *multithreaded*, meaning that the hardware internally divides its resources among multiple concurrent activities. Laptops may also rely on specialized, internally parallel *graphics processing units* (GPUs) and may communicate over a network with a "cloud" of other machines for services such as file storage or electronic mail. Like it or not, our world is full of concurrency.

This book is about the theoretical foundations of concurrency. For us, a *distributed system*[1] is a collection of sequential computing entities, called *processes*, that cooperate to solve a problem, called a *task*. The processes may communicate by message passing, shared memory, or any other mechanism. Each process runs a program that defines how and when it communicates with other processes. Collectively these programs define a *distributed algorithm* or *protocol*. It is a challenge to design efficient distributed algorithms in the presence of failures, unpredictable communication, and unpredictable scheduling delays. Understanding when a distributed algorithm exists to solve a task, and why, or how efficient such an algorithm can be is the aim of the book.

1.1.1 Distributed computing and topology

In the past decade, exciting new techniques have emerged for analyzing distributed algorithms. These techniques are based on notions adapted from *topology*, a field of mathematics concerned with properties of objects that are *innate*, in the sense of being preserved by continuous deformations such as stretching or twisting, although not by discontinuous operations such as tearing or gluing. For a topologist, a cup and a torus are the same object; Figure 1.1 shows how one can be continuously deformed into the other. In particular, we use ideas adapted from *combinatorial topology*, a branch of topology that focuses on discrete constructions. For example, a sphere can be approximated by a figure made out of flat triangles, as illustrated in Figure 1.2.

Although computer science itself is based on discrete mathematics, combinatorial topology and its applications may still be unfamiliar to many computer scientists. For this reason, we provide a self-contained, elementary introduction to the combinatorial topology concepts needed to analyze distributed computing. Conversely, although the systems and models used here are standard in computer science, they may be unfamiliar to readers with a background in applied mathematics. For this reason, we also provide a self-contained, elementary description of standard notions of distributed computing.

Distributed computing encompasses a wide range of systems and models. At one extreme, there are tiny GPUs and specialized devices, in which large arrays of simple processors work in lock-step. In the middle, desktops and servers contain many multithreaded, multicore processors, which use shared memory communication to work on common tasks. At the other extreme, "cloud" computing and

---

[1]The term *distributed system* is often used for a concurrent system in which the participants are geographically far apart. We do not emphasize this distinction, so we use the terms *distributed computing* and *concurrent computing* more or less interchangeably.

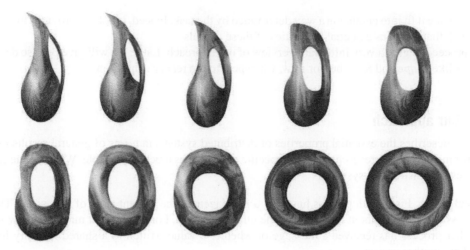

FIGURE 1.1

Topologically identical objects.

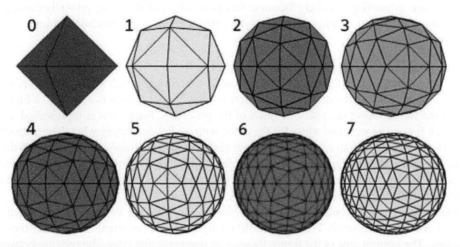

FIGURE 1.2

Starting with a shape constructed from two pyramids, we successively subdivide each triangle into smaller triangles. The finer the degree of triangulation, the closer this structure approximates a sphere.

peer-to-peer systems may encompass thousands of machines that span every continent. These systems appear to have little in common besides the common concern with complexity, failures, and timing. Yet the aim of this book is to reveal the astonishing fact that they do have much in common, more specifically, that computing in a distributed system is essentially a form of stretching one geometric object to make it fit into another in a way determined by the task. Indeed, topology provides the common framework that explains essential properties of these models.

We proceed to give a very informal overview of our approach. Later, we will give precise definitions for terms like *shape* and *hole*, but for now, we appeal to the reader's intuition.

1.1.2 Our approach

The book describes the essential properties of distributed systems in terms of general results that hold for all (or most) models, restricting model-specific reasoning as much as possible. What are the essential properties of a distributed system?

- **Local views.** First, each process has only a *local view* of the current state of the world. That is, a process is uncertain about the views of the other processes. For example, it may not know whether another process has received a message or whether a value written to a shared memory has been read by another process.
- **Evolution of local views.** Second, processes communicate with one another. Each communication modifies local views. If they communicate everything they know and the communication is flawless and instantaneous, they end up with identical local views, eliminating uncertainty. The systems we study are interesting precisely because this is usually not the case, either because processes communicate only part of what they know (for efficiency) or communication is imperfect (due to delays or failures).

A process's local view is sometimes called its *local state*.

Figure 1.3 presents a simple example. There are two processes, each with a three-bit local view. Each process "knows" that the other's view differs by one bit, but it does not "know" which one. The left side of the figure shows the possible views of the processes. Each view is represented as a *vertex*, colored black for one process and white for the other, with a label describing what the process "knows." A pair of views is joined by an *edge* if those views can coexist. The *graph* consisting of all the vertices and edges outlines a cube. It represents in one single combinatorial object the initial local views of the processes and their uncertainties. Each vertex belongs to three edges, because the process corresponding to the vertex considers possible that the other process is in one of those three initial states.

Suppose now that each process then sends its view to the other via an unreliable medium that may lose at most one of the messages. The right side of the figure shows the graph of new possible views and uncertainties. The bottom area of the figure focuses on one particular edge, the relation between views 110 and 111. That edge splits into three edges, corresponding to the three possibilities: Black learns White's view but not vice versa; each learns the other's; and White learns Black's view but not vice versa. An innate property of this model is that, although unreliable communication adds new vertices to the graph, it does not change its overall shape, which is still a cube. Indeed, no matter how many times the processes communicate, the result is still a cube.

Figure 1.4 shows the same transformation except that the processes use a reliable communication medium that does not drop messages. At the end of a communication round, each process knows the other's view. Reliable communication changes the overall shape, transforming a cube into a set of disconnected edges.

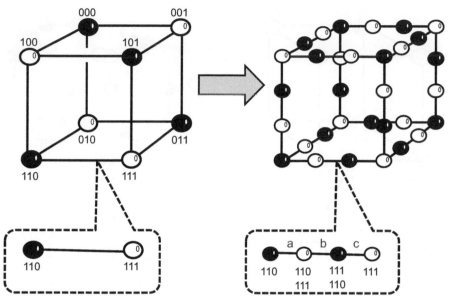

FIGURE 1.3

The left side of the figure shows the possible views of two processes, each with a three-bit local view, in black for one process and white for the other. A pair of views is joined by an edge if those views can coexist. Each process then sends its view to the other, but communication is unreliable and at most one message may be lost. The new set of possible views appears on the right. The bottom view focuses on the changing relation between views 110 and 111. After communicating, each process may or may not have learned the other's views. At edge *a*, White learns Black's view but not vice versa, whereas at edge *b*, each learns the other's view, and at edge *c*, Black learns White's view but not vice versa. Unreliable communication leaves the structure of the left and right sides essentially unchanged.

The key idea is that we represent all possible local views of processes at some time as a single, static, combinatorial geometric object, called a *simplicial complex*. For the case of two processes, the complex is just a graph. The complex is obtained by "freezing" all possible interleavings of operations and failure scenarios up to some point in time. Second, we analyze the model-specific evolution of a system by considering how communication changes this complex. Models differ in terms of their reliability and timing guarantees (processing and communication delays). These properties are often reflected as "holes" in the simplicial complex induced by communication: In our simple example, unreliable communication leaves the overall cubic shape unchanged, whereas reliable communication tears "holes" in the cube's edges. The model-dependent theorems specify when the holes are introduced (if at all) and their type. The model-independent theorems say which tasks can be solved (or how long it takes to solve them), solely in terms of the "holes" of the complex.

The appeal of this approach is that it reduces the difficult problem of reasoning about computations that unfold in time to the more tractable problem of reasoning about static combinatorial structures. Equally important, we can call upon a vast literature of results in combinatorial and algebraic topology.

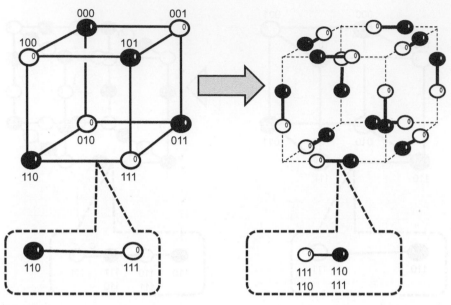

FIGURE 1.4

If we replace the unreliable communication of Figure 1.3 with perfectly reliable communication, the structure of the left and right sides looks quite different.

1.1.3 Two ways of thinking about concurrency

Consider a distributed system trying to solve a task. The initial views of the processes are just the possible inputs to the task and are described by an *input complex*, $\mathcal{X}$. The outputs that the processes are allowed to produce, as specified by the task, are described by the *output complex, $\mathcal{Y}$*. Distributed computation is simply a way to stretch, fold, and possibly tear $\mathcal{X}$ in ways that depend on the specific model, with the goal of transforming $\mathcal{X}$ into a form that can be mapped to $\mathcal{Y}$. We can think of this map as a continuous map from the space occupied by the transformed $\mathcal{X}$ into $\mathcal{Y}$, where the task's specification describes which parts of the transformed $\mathcal{X}$ must map to which parts of $\mathcal{Y}$. See Figure 1.5.

This approach is particularly well suited for impossibility results. Topology excels at using invariants to prove that two structures are fundamentally different in the sense that no continuous map from one to the other can preserve certain structures. For example, consider the task shown schematically in Figure 1.5. The input complex is represented by a two-dimensional disk, and the output complex is represented by an annulus (a two-dimensional disk with a hole). Assume that the task specification requires the boundary of the input disk to be mapped around the boundary of the output annulus. In a model where the input complex can be arbitrarily stretched but not torn, it is impossible to map the transformed input to the annulus without "covering" the hole, i.e., producing outputs in the hole, and hence is illegal. In such a model this task is not solvable.

Instead, as suggested by the figure, this task requires a more powerful computational model that tears holes in the input, but with no need to tear the complex into disconnected pieces. An even more

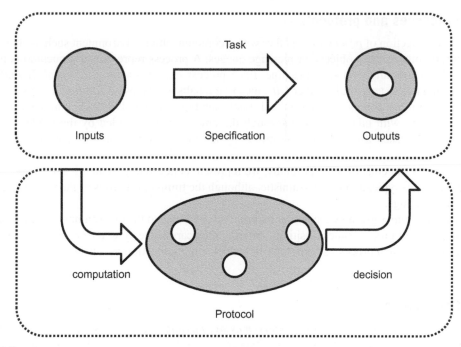

FIGURE 1.5

Geometric representation of a task specification and a protocol.

powerful model that tears the model into disjoint parts would also suffice but would be stronger than necessary.

The approach is so powerful because the preceding explanations go both ways. A task is solvable in a given model of computation if and only if the input complex can be arbitrarily stretched, adding "holes" as permitted by the model to map the transformed input to the output complex and sending regions of the transformed input complex to regions of the output complex, as specified by the task. Thus, we get two different ways of *thinking about* concurrency: operational and topological. With its powers of abstraction and vast armory of prior results, topology can abstract away from model-dependent detail to provide a concise mathematical framework unifying many classical models. Classic distributed computing techniques combine with topology to obtain a solid, powerful theoretical foundation for concurrency.

1.2 Distributed computing

The literature on distributed computing includes a bewildering array of different models, encompassing different choices of communication, failures, and timing. Not all possible combinations of these choices make sense, but many of them do. Here we briefly describe the various models. Our description here is informal and intended to motivate later formal models.

1.2.1 Processes and protocols

A *system* is a collection of *processes*, together with a communication environment such as shared read-write memory, other shared objects, or message queues. A process represents a sequential computing entity, modeled formally as a state machine. Each process executes a finite *protocol*. It starts in an initial state and takes steps until it either *fails*, meaning it halts and takes no additional steps, or it *halts*, usually because it has completed the protocol. Each step typically involves local computation as well as communicating with other processes through the environment provided by the model. Processes are deterministic: Each transition is determined by the process's current state and the state of the environment.

The processes run concurrently. Formally, we represent concurrency by interleaving process steps. This interleaving is typically nondeterministic, although the timing properties of the model can restrict possible interleavings.

The *protocol state* is given by the nonfaulty processes' views and the environment's state. An *execution* is a sequence of process state transitions. An execution carries the system from one state to another, as determined by which processes take steps and which communication events occur.

1.2.2 Communication

Perhaps the most basic communication model is *message passing*. Each process sends messages to other processes, receives messages sent to it by the other processes, performs some internal computation, and changes state. Usually we are interested in whether a task can be solved instead of how efficiently it can be solved; thus we assume processes follow a *full-information* protocol, which means that each process sends its entire local state to every process in every round.

In some systems, messages are delivered through communication channels that connect pairs of processes, and a graph describes the network of pairs of processes that share a channel. To send a message from one process to another that is not directly connected by a channel, a routing protocol must be designed. In this book we are not interested in the many issues raised by the network structure. Instead, we abstract away this layer and assume that processes communicate directly with each other so we can concentrate on task computability, which is not really affected by the network layer.

In *shared-memory* models, processes communicate by applying operations to objects in shared memory. The simplest kind of shared-memory object is *read-write* memory, where the processes share an array of memory locations. There are many models for read-write memory. Memory variables may encompass a single bit, a fixed number of bits, or an arbitrary number. A variable that may be written by a single process but read by all processes is called a *single-writer* variable. If the variable can be written by all processes, it is called *multiwriter*. Fortunately, all such models are equivalent in the sense that any one can be implemented from any other. From these variables, in turn, one can implement an *atomic snapshot memory*: an array in which each process writes its own array entry and can atomically read (take a *snapshot* of) the entire memory array.

In models that more accurately reflect today's multiprocessor architectures, read-write memory is augmented with shared objects such as stacks, queues, test-and-set variables, or objects of arbitrary abstract type. (Many such synchronization primitives cannot be implemented directly in read-write memory and must be provided by the underlying hardware.)

1.2.3 **Failures**

The theory of distributed computing is largely about what can be accomplished in the presence of timing uncertainty and failures. In some timing models, such failures can eventually be detected, whereas in other models, a failed process is indistinguishable from a slow process.

In the most basic model, the goal is to provide *wait-free* algorithms that solve particular tasks when any number of processes may fail. The wait-free failure model is very demanding, and sometimes we are willing to settle for less. A *t-resilient* algorithm is one that works correctly when the number of faulty processes does not exceed a value t. A wait-free algorithm for $n + 1$ processes is n-resilient.

A limitation of these classical models is that they implicitly assume that processes fail independently. In a distributed system, however, failures may be correlated for processes running on the same node, running in the same network partition, or managed by the same provider. In a multiprocessor system, failures may be correlated for processes running on the same core, the same process, or the same card. To model these situations, it is natural to introduce the notion of an *adversary* scheduler that can cause certain subsets of processes to fail.

In this book, we first consider *crash failures*, in which a faulty process simply halts and falls silent. We consider *Byzantine* failures, where a faulty process can display arbitrary (or malicious) behavior, in Chapter 6.

1.2.4 **Timing**

As we will see, the most basic timing model is *asynchronous*, whereby processes run at arbitrary, unpredictable speeds and there is no bound on process step time. In this case, a failed process cannot be distinguished from a slow process. In *synchronous* timing models, all nonfaulty processes take steps at the same time. In synchronous models, it is usually possible to detect process failures. In between there are *semisynchronous* models, whereby there is an upper bound on how long it takes for a nonfaulty process to communicate with another. In such models, a failed process can be detected following a (usually lengthy) timeout.

1.2.5 **Tasks**

The question of what it means for a function to be *computable* is one of the deepest questions addressed by computer science. In sequential systems, computability is understand through the *Church-Turing thesis*: Anything that *can* be computed can be computed by a Turing machine. The Church-Turing thesis led to the remarkable discovery that most functions from integers to integers are not computable.[2] Moreover, many specific functions, such as the famous "halting problem," are also known to be not computable.

In distributed computing, where computations require coordination among multiple participants, computability questions have a different flavor. Here, too, there are many problems that are not computable, but these computability failures reflect the difficulty of making decisions in the face of ambiguity and have little to do with the inherent computational power of individual participants. If the participants could reliably and instantaneously communicate with one another, then each one could learn

---

[2]The set of Turing machines is countable, and each one computes functions from integers to integers. However, the set of functions from integers to integers is not countable.

the complete system state and perform the entire computation by itself. In any realistic model of distributed computing, however, each participant initially knows only part of the global system state, and uncertainties caused by failures and unpredictable timing limit each participant to an incomplete picture.

In sequential computing, a function can be defined by an algorithm, or more precisely a Turing machine, that starts with a single input, computes for a finite duration, and halts with a single output. In sequential computing one often studies nondeterministic algorithms, in which there is more than one output allowed for each input. In this case, instead of functions, *relations* are considered.

In distributed computing, the analog of a function is called a *task*. An input to a task is distributed: Only part of the input is given to each process. The output from a task is also distributed: Only part of the output is computed by each process. The task specification states which outputs can be produced in response to each input. A *protocol* is a concurrent algorithm to solve a task; initially each process knows its own part of the input, but not the others'. Each process communicates with the others and eventually halts with its own output value. Collectively, the individual output values form the task's output. Unlike a function, which deterministically carries a single input value to a single output value, an interesting task specification is a non-deterministic relation that carries each input value assignment to multiple possible output value assignments.

1.3 Two classic distributed computing problems

Distributed algorithms are more challenging than their sequential counterparts because each process has only a limited view of the overall state of the computation. This uncertainty begins at the very beginning. Each process starts with its own private input, which could come from a person (such as a request to withdraw a sum from a cash machine) or from another application (such as a request to enqueue a message in a buffer). One process typically does not "know" the inputs of other processes, nor sometimes even "who" the other processes are. As the computation proceeds, the processes communicate with each other, but uncertainty may persist due to nondeterministic delays or failures. Eventually, despite such lingering uncertainty, each process must decide a value and halt in such a way that the collective decisions form a correct output for the given inputs for the task at hand.

To highlight these challenges, we now examine two classic distributed computing problems. These problems are simple and idealized, but each one is well known and illustrates principles that will recur throughout this book. For each problem, we consider two kinds of analysis. First, we look at the conventional, *operational* analysis, in which we reason about the computation as it unfolds in time. Second, we look at the new, *combinatorial* approach to analysis, in which all possible executions are captured in one or more static, topological structures. For now, our exposition is informal and sketchy; the intention is to motivate essential ideas, still quite simple, that are described in detail later on.

1.3.1 The muddy children problem

When reasoning about concurrency, we often end up talking about what a process "knows," what it knows about what other processes know, and so on. The following problem has been used to motivate the use of logic to reason formally about knowledge. We encounter a synchronous model in which the uncertainty begins at the very beginning and then diminishes in a simple way as the computation proceeds.

A group of children is playing in the garden, and some of them end up with mud on their foreheads. Each child can see the other children's foreheads but not his or her own. At noon, their teacher summons the children and says: "At least one of you has a muddy forehead. You are not allowed to communicate with one another about it in any manner. But whenever you become certain that you are dirty, you must announce it to everybody, exactly on the hour" The children resume playing normally, and nobody mentions the state of anyone's forehead. There are six muddy children, and at 6:00 they all announce themselves. How does this work?

The usual operational explanation is by induction on the number of children that are dirty, say, k. If $k = 1$, then, as soon as the teacher speaks, the unique muddy child knows she is muddy, since there are no other muddy children. At 1:00 she announces herself. If $k = 2$ and A and B are dirty, then at 1:00, A notices that B does not announce himself and reasons that B must see another dirty child, which can only be A. Of course, B follows the same reasoning, and they announce themselves at 2:00. A similar argument is done for any k.

By contrast, the combinatorial approach, which we now explore, provides a geometric representation of the problem's input values and evolving knowledge about those input values. In particular, it gives a striking answer to the following seeming paradox: The information conveyed by the teacher, that there is a muddy child, seems to add nothing to what everyone knows and yet is somehow essential to solving the problem.

A child's *input* is its initial state of knowledge. If there are $n + 1$ children, then we represent a child's input as an $(n + 1)$-element vector. The input for child i has 0 in position $j \neq i$ if child j is clean, and it has 1 if child j is dirty. Because child i does not know his own status, his input vector has $\perp$ in position i.

For three children, conveniently named Black, White, and Gray, the possible initial configurations are shown in Figure 1.6. Each vertex represents a child's possible input. Each vertex is labeled with an input vector and colored either, black, white, or gray, to identify the corresponding child. Each

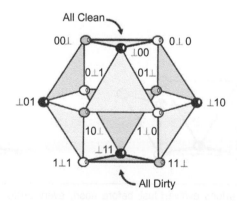

FIGURE 1.6

Input configurations for the Muddy Children problem. Each vertex is labeled with a child's name (color) and input vector, and every solid triangle represents a possible configuration (squares are holes). Every vertex lies in exactly two triangles, reflecting each child's degree of uncertainty about the actual situation.

possible configuration is represented as a solid triangle, linking *compatible* states for the three children, meaning that the children can be in these states simultaneously. The triangle at the very top represents the configurations where all three children are clean, the one at the bottom where they are all dirty, and the triangles in between represent configurations where some are clean and some are dirty.

Notice that in contrast to Figure 1.3, where we had a one-*dimensional complex* consisting of vertices and edges (i.e., a graph) representing the possible configurations for two processes, for three processes we use a two-dimensional complex, consisting of vertices, edges, and triangles.

Inspecting this figure reveals something important: Each vertex belongs to exactly two triangles. This geometric fact reflects each child's *uncertainty* about the actual situation: his or her own knowledge (represented by its vertex) is compatible with two possible situations—one where it is dirty, and one where it is clean.

Figure 1.7 shows how the children's uncertainty evolves over time. At 11:59 AM, no children can deduce their own status from their input. At noon, however, when the teacher announces that at least one child is dirty, the all-clean triangle at the top is eliminated. Now there are three vertices that belong to a

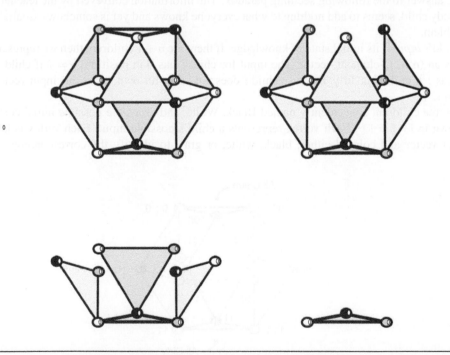

FIGURE 1.7

How Muddy Children configurations evolve. Just before noon, every child's vertex lies on two triangles, reflecting each child's uncertainty about its status. At noon, when the teacher announces that at least one child is dirty, the top triangle is eliminated, exposing three vertices where some child knows his status. At 1:00, if no child announces, the top tier of triangles is eliminated, exposing more such vertices. Every hour on the hour, if the configuration is not resolved, additional vertices are exposed.

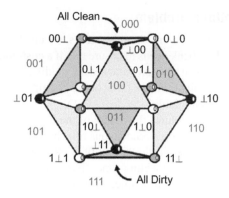

FIGURE 1.8

Output configurations for the Muddy Children problem. Each vertex is labeled with a child's name (color) and decision value indicating whether the child is clean or muddy. Every edge lies in exactly two triangles.

single triangle: $00\perp$, $0\perp0$, $\perp00$. Any child whose input matches one of those vertices, and only those, will announce itself at 1:00. Since every triangle contains at most one such vertex, exactly one child will make an announcement. If nobody says anything at 1:00, then the top tier of triangles is eliminated, and now there are more vertices that are included in exactly one triangle. Since every triangle containing such a vertex has exactly two of them, exactly two children will make an announcement, and so on.

The Muddy Children puzzle, like any distributed task, requires the participants to produce outputs. Each child must announce whether it is clean or dirty. In Figure 1.8 these decisions are represented as binary values, 0 or 1. In the triangle at the top, all three children announce they are clean; at the bottom, all three announce they are dirty.

The appeal of the combinatorial approach is that all possible behaviors are captured statically in a single structure, such as the one appearing in Figure 1.6. Inspecting this figure helps solve the mystery of why it is so important that the teacher announces that some child is dirty. Without that announcement, every vertex remains ambiguously linked to two triangles, and no evolution is possible. Interestingly, any announcement by the teacher that eliminates one or more triangles would allow each child eventually to determine his status.

Of course, the computational model implicit in this puzzle is highly idealized. Communication is synchronous: Every hour on the hour, every child knows the time and decides whether to speak. Communication never fails, so if nothing is heard from a child on the hour, it is because nothing was said, and if something was said, everybody hears it. No one cheats or naps, and the children reason perfectly.

In this situation, uncertainties come from the partial knowledge inherent in the inputs and from the limited form of communication (a child cannot tell other children what it sees). The system can evolve in only one way: Each triangle either survives until the next hour or it is eliminated. When we replace idealized children with concurrent processes, there may be many more possible ways for the system to evolve, reflecting nondeterministic interleavings and failures. In such a case, instead of either persisting or vanishing, a triangle may be replaced by several triangles, each representing a possible new configuration.

1.3.2 **The coordinated attack problem**

Here is another classic problem. In many distributed systems, we need to ensure that two things happen together or not at all. For example, a bank needs to ensure that if a customer tries to transfer money from one account to another, then either both account balances are modified or neither one is changed (and an error is reported). This kind of coordination task turns out to be impossible if either the communication or the participants are sufficiently unreliable.

The following idealized problem captures the nature of the difficulty. Simply put, it shows that it is impossible for two participants to agree on a rendezvous time by exchanging messages that may fail to arrive. As in the Muddy Children problem, the difficulty is inherent in the initial system state, where the participants have not yet agreed on a meeting time. (Naturally, if both had agreed earlier when to rendezvous, they could simply show up at that time, with no additional communication.)

> *Two army divisions, one commanded by General Alice and one by General Bob, are camped on two hilltops overlooking a valley. The enemy is camped in the valley. If both divisions attack simultaneously, they will win, but if only one division attacks by itself, it will be defeated. As a result, neither general will attack without a guarantee that the other will attack at the same time. In particular, neither general will attack without communication from the other.*
>
> *At the time the divisions are deployed on the hilltops, the generals had not agreed on whether or when to attack. Now Alice decides to schedule an attack. The generals can communicate only by messengers. Normally it takes a messenger exactly one hour to get from one encampment to the other. However, it is possible that he will get lost in the dark or, worse yet, be captured by the enemy. Fortunately, on this particular night all the messengers happen to arrive safely. How long will it take Alice and Bob to coordinate their attack?*

To rule out the trivial solution in which both generals simply refrain from attacking, we will require that if all messages are successfully delivered, then Alice and Bob must agree on a time to attack. If enough messages are lost, however, Alice and Bob may refrain from attacking, but both must do so.

The standard operational way of analyzing this problem goes as follows: Suppose Bob receives a message at 1:00 PM from Alice, saying, "Attack at dawn." Should Bob schedule an attack? Although her message was in fact delivered, Alice has no way of knowing that it would be. She must therefore consider it possible that Bob did not receive the message (in which case Bob would not plan to attack). Hence Alice cannot decide to attack given her current state of knowledge. Knowing this and not willing to risk attacking alone, Bob will not attack based solely on Alice's message.

Naturally, Bob reacts by sending an acknowledgment back to Alice, and it arrives at 2:00 PM. Will Alice plan to attack? Unfortunately, Alice's predicament is similar to Bob's predicament at 1:00 PM. This time it is Bob who does not know whether his acknowledgment was delivered. Since Bob knows that Alice will not attack without his acknowledgment, Bob cannot attack as long as Alice might not have received his acknowledgment. Therefore, Alice cannot yet decide to attack.

So Alice sends an acknowledgment back to Bob. Similar reasoning shows that this message, too, is not enough. It is not hard to see that no number of successfully delivered acknowledgments will be enough to ensure that the generals can attack safely. The key insight is that the difficulty is not caused by what *actually* happens (all messages do arrive) but by the uncertainty regarding what *might* have happened. In the scenario we just considered, communication is flawless, but coordination is still impossible.

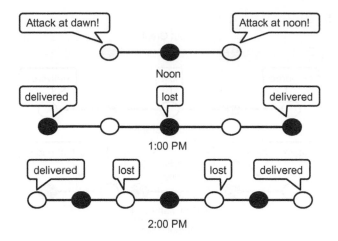

FIGURE 1.9

Evolution of the possible executions for the Coordinated Attack problem. (The 2:00 PM graph shows only a subset of the possible executions.)

Here is how to consider this problem using the combinatorial approach, encompassing all possible scenarios in a single geometric object—a graph.

Alice has two possible initial states: She intends to attack either at dawn or at noon the next day. The top structure on Figure 1.9 depicts each state as a white vertex. Bob has only one possible initial state: He awaits Alice's order. This state is the black vertex linking the two edges, representing Bob's uncertainty whether he is in a world where Alice intends to attack at dawn, as indicated on the left in the figure, or in a world where she intends to attack at noon, as shown on the right.

At noon Alice sends a message with her order. The second graph in Figure 1.9 shows the possible configurations one hour later, at 1:00 PM, in each of the possible worlds. Either her message arrives, or it does not. (We can ignore scenarios where the message arrives earlier or later, because if agreement is impossible even if messages always arrive on time, then it is impossible even if they do not. We will often rely on this style of argument.) The three black vertices represent Bob's possible states. On the left, Bob receives a message to attack at dawn, on the right, to attack at noon, and in the middle, he receives no message. Now Alice is the one who is uncertain whether Bob received her last message.

The bottom graph in Figure 1.9 shows a subset of the possible configurations an hour later, at 2:00 PM, when Bob's 1:00 PM acknowledgment may or may not have been received. We can continue this process for an arbitrary number of rounds. In each case, it is not hard to see that the graph of possible states forms a line. At time t, there will be $2t + 2$ edges. At one end, an initial "Attack at dawn" message is followed by successfully delivered acknowledgments, and at the other end, an initial "Attack at noon" message is followed by successfully delivered acknowledgments. In the states in the middle, however, messages were lost.

If it were possible to reach agreement after t message exchanges, we could then label each vertex with an attack order, either *dawn* or *noon*. Because the two generals must agree on the attack time, both

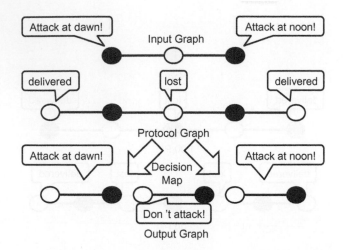

FIGURE 1.10

The input, output, and protocol graphs for the Coordinated Attack problem illustrate why the problem is not solvable.

vertices of each edge must be labeled with the same attack time. Here is the problem: The graph is *connected*; starting at one edge, where both generals agree to attack at dawn, we can follow a path of edges to the other edge, where both generals agree to attack at noon. One of the edges we traverse must switch from dawn to noon, representing a state where the two generals make incompatible decisions. This impossibility result holds no matter what rules the generals use to make their decisions and no matter how many messages they send.

This observation depends on a *topological* property of the graph, namely, that it is always connected. In Figure 1.10 this is more explicitly represented. At the top, the complex of possible inputs to the task is a connected graph consisting of two edges, whereas at the bottom the output complex is a disconnected graph consisting of two disjoint edges. In the middle, the protocol complex after sending one message is a larger connected graph, one that "stretches" the input graph by subdividing its edges into "smaller" edges. As in Figure 1.5, the task specification restricts which parts of the subdivided input graph can be mapped into which parts of the output graph: The endpoints of the subdivided graph should be mapped to different edges of the output graph, corresponding to the desired time of attack.

1.4 Chapter notes

The topological approach to distributed computing has its origins in the well-known paper by Fischer, Lynch, and Paterson [55], in which it was shown that there is no fault-tolerant message-passing protocol for the *consensus* task (which we will study in detail later on), a problem closely related to coordinated attack, even if only one process may crash. In that paper, process uncertainties (as described in this chapter) were identified as the main ingredient for the impossibility result. Furthermore, chains of uncertainties forming a graph and the connectivity of this graph were the basis of the proof. Another paper, by Loui and Abu-Amara [110], was required to extend the result to shared memory systems. Later on, Biran, Moran, and Zaks [19] used the graph connectivity arguments to provide a characterization of the tasks solvable in a message-passing system in which at most one process may crash, and even on the time needed to solve a task [21].

Three papers presented at the ACM Symposium on Theory of Computing in 1993 [23,90,134] (journal versions in [91,135]) realized that when more than one process may crash, a generalization of graph connectivity to higher-dimensional connectivity is needed. The discovery of the connection between distributed computing and topology was motivated by trying to prove that the *k-set agreement* task is not wait-free solvable, requiring processes to agree on at most k different values, a problem posed by Chaudhuri [38]. Originally it was known that processes cannot agree on a single value, even if only one process can crash. The techniques to prove this result need graph connectivity notions, namely one-dimensional connectivity only. In 1993 it was discovered that to prove that n processes cannot agree on $n - 1$ of their input values wait-free requires general topological connectivity notions.

Another connection between distributed computing and topology, based on homotopy theory, is due to Fajstrup, Raussen and Goubault [146].

The Muddy Children problem is also known as the Cheating Husbands problem as well as other names. Fagin, Halpern, Moses, and Vardi [52] use this problem to describe the notion of *common knowledge* and more generally the idea of using formal logic to reason about what processes know. Others who discuss this problem include Gamow and Stern [70] and Moses, Dolev, and Halpern [119].

The Coordinated Attack problem, also known as the Two Generals problem, was formally introduced by Jim Gray in [73] in the context of distributed databases. It appears often in introductory classes about computer networking (particularly with regard to the Transmission Control Protocol), database systems (with regard to commit/adopt protocols), and distributed systems. It is also an important concept in epistemic logic and knowledge theory, as discussed in Fagin, Halpern, Moses, and Vardi [52]. It is related to the more general Byzantine Generals problem [107].

1.5 Exercises

Exercise 1.1.   In the Muddy Children problem, describe the situation if the teacher announces at noon:

- Child number one is dirty.
- There are an odd number of dirty children.
- There are an even number of dirty children.

For three children, redraw the pictures in Figure 1.7.

Exercise 1.2.   In the original version of the puzzle, before the teacher's announcement, the children know there are at least 0 muddy children, and in general, after k rounds, they know there are at least k muddy children. Prove this invariant using the combinatorial approach, and do the same for the corresponding invariant in each of the variants of the puzzle listed in Exercise 1.1. Note that to prove such a statement, you need to define formally what *knowing* means, using the combinatorial approach.

Exercise 1.3.   Consider a binary variable initialized to 0. The *test-and-set* operation atomically sets that variable to 1 and returns its previous value. (So the first process to call *test-and-set* receives 0, and all later processes receive 1.) Draw the complex of possible results if two processes, A and B, each calling *test-and-set* on the same variable. Do the same for three processes A, B, and C.

Hint: Each vertex should be labeled with a process name and a binary value, and the result should look something like the first step of constructing a *Sierpinski triangle*.

Exercise 1.4.   Three processes, *A*, *B*, and *C*, are assigned distinct values from the set {0, 1, 2}. Draw the complex of all such possible assignments.

Hint: Each vertex should be labeled with a process name and an integer value, and your picture should look something like a star of David.

Exercise 1.5.   Three processes, *A*, *B*, and *C*, are assigned distinct values from the set {0, 1, 2, 3}. Draw the complex of all such possible assignments.

Hint: Each vertex should be labeled with a process name and an integer value, and your picture should be topologically equivalent to a torus.

Exercise 1.6.   Consider three processes, *A*, *B*, and *C*, which are assigned values from some set *S*. Find a set *S* and a way of assigning values to processes such that the resulting complex is a *Möbius strip*.

Two-Process Systems

CHAPTER OUTLINE HEAD

This chapter is an introduction to how techniques and models from combinatorial topology can be applied to distributed computing by focusing exclusively on two-process systems. It explores several distributed computing models, still somewhat informally, to illustrate the main ideas.

For two-process systems, the topological approach can be expressed in the language of graph theory. A protocol in a given model induces a graph. A *two-process* task is specified in terms of a pair of graphs: one for the processes' possible inputs and one for the legal decisions the processes can take. Whether a two-process protocol exists for a task can be completely characterized in terms of connectivity properties of these graphs. Moreover, if a protocol exists, that protocol is essentially a form of approximate agreement. In later chapters we will see that when the number of processes exceeds two and the number of failures exceeds one, higher-dimensional notions of connectivity are needed, and the language of graphs becomes inadequate.

Distributed Computing Through Combinatorial Topology. http://dx.doi.org/10.1016/B978-0-12-404578-1.00002-4

2.1 Elementary graph theory

It is remarkable that to obtain the characterization of two-process task solvability, we need only a few notions from graph theory, namely, maps between graphs and connectivity.

2.1.1 Graphs, vertices, edges, and colorings

We define graphs in a way that can be naturally generalized to higher dimensions in later chapters.

Definition 2.1.1.   A *graph* is a finite set S together with a collection $\mathcal{G}$ of subsets of S, such that

(1) If $X \in \mathcal{G}$, then $|X| \leq 2$;
(2) For all $s \in S$, we have $\{s\} \in \mathcal{G}$;
(3) If $X \in \mathcal{G}$ and $Y \subset X$, then $Y \in \mathcal{G}$.

We use $\mathcal{G}$ to denote the entire graph. An element of $\mathcal{G}$ is called a *simplex* (plural: *simplices*) of $\mathcal{G}$. We say that a simplex σ has *dimension* $|\sigma| - 1$. A zero-dimensional simplex $s \in S$ is called a *vertex* (plural: *vertices*) of $\mathcal{G}$, whereas a one-dimensional simplex is called an *edge* of $\mathcal{G}$. We denote the set of vertices of $\mathcal{G}$ by $V(\mathcal{G})$ (that is, $V(\mathcal{G}) := S$), and we denote the set of edges of $\mathcal{G}$ by $E(\mathcal{G})$.

We say that a vertex is *isolated* if it does not belong to any edge. A graph is called *pure* if either every vertex belongs to an edge or none does. In the first case, the graph is *pure of dimension* 1, whereas in the second it is *pure of dimension* 0.

Assume C is a set. A *coloring* of a graph $\mathcal{G}$ is a function $\chi : V(\mathcal{G}) \rightarrow C$, such that for each edge $\{s_0, s_1\}$ of $\mathcal{G}$, $\chi(s_0) \neq \chi(s_1)$. We say that a graph is *chromatic* or that it is *colored by C* if it is equipped with a coloring $\chi : V(\mathcal{G}) \rightarrow C$. Often we will color vertices with just two colors: $C = \{A, B\}$, where A and B are the names of the two processes.

More generally, given a set L, an *L-labeling* of $\mathcal{G}$ is defined as a function f that assigns to each vertex an element of L without any further conditions imposed by the existence of edges. We say the graph is *labeled* by L. A coloring is a labeling, but not vice versa.

We frequently consider graphs that simultaneously have a coloring (denoted by χ) and a vertex labeling (denoted by f). Figure 2.1 shows four such graphs: two "input" graphs at the top and two "output" graphs at the bottom. All are pure of dimension 1. In all figures, the coloring is shown as black or white, and the labeling is shown as a number in or near the vertex.

For distributed computing, if s is a vertex in a labeled chromatic graph, we denote by name(s) the value $\chi(s)$, and by view(s) the value $f(s)$. Moreover, we assume that each vertex in a chromatic labeled graph is uniquely identified by its values of name($\cdot$) and view($\cdot$). In Figure 2.1, for example, each graph has a unique black vertex labeled 1.

2.1.2 Simplicial maps and connectivity

Labeling functions can be more elegantly defined as maps, namely functions that satisfy some form of structure, from one graph to another. Given two graphs $\mathcal{G}$ and $\mathcal{H}$, a *vertex map* $\mu : V(\mathcal{G}) \rightarrow V(\mathcal{H})$ carries each vertex of $\mathcal{G}$ to a vertex of $\mathcal{H}$. However, for the map to preserve structure, it must also carry edges to edges. The vertex map μ is a *simplicial map* if it also carries simplices to simplices: that is, if $\{s_0, s_1\}$ is a simplex in $\mathcal{G}$, then $\{\mu(s_0), \mu(s_1)\}$ is a simplex of $\mathcal{H}$. Notice that $\mu(s_0)$ and $\mu(s_1)$ need not

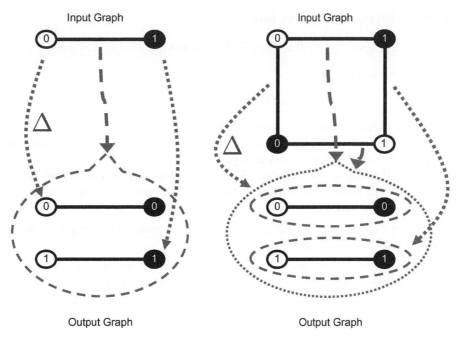

FIGURE 2.1

Graphs for fixed-input and binary consensus.

be distinct; the image of an edge may be a vertex. If, for $s_0 \neq s_i$, $\mu(s_0) \neq \mu(s_1)$, the map is said to be *rigid*. In the terminology of graph theory, a rigid simplicial map is called a *graph homomorphism*.

When $\mathcal{G}$ and $\mathcal{H}$ are chromatic, we usually assume that the simplicial map μ preserves names: name$(s) =$ name$(\mu(s))$. Thus, *chromatic simplicial* maps are rigid.

If s, t are vertices of a graph $\mathcal{G}$, then a *path* from s to t is a sequence of distinct edges $\sigma_0, \dots, \sigma_\ell$ linking those vertices: $s \in \sigma_0, \sigma_i \cap \sigma_{i+1} \neq \emptyset$, and $t \in \sigma_\ell$. A graph is *connected* if there is a path between every pair of vertices. The next claim is simple but important. Intuitively, simplicial maps are approximations of continuous maps.

Fact 2.1.2. The image of a connected graph $\mathcal{G}$ under a simplicial map is connected.

2.1.3 Carrier maps

Whereas a simplicial map carries simplices to simplices, it is also useful in the context of distributed computing to define maps that carry simplices to subgraphs.

Definition 2.1.3. Given two graphs $\mathcal{G}$ and $\mathcal{H}$, a *carrier map* $\Phi : \mathcal{G} \to 2^{\mathcal{H}}$ takes each simplex $\sigma \in \mathcal{G}$ to a subgraph $\Phi(\sigma)$ of $\mathcal{H}$ such that Φ satisfies the following *monotonicity* property: For all $\sigma, \tau \in \mathcal{G}$, if $\sigma \subseteq \tau$, then $\Phi(\sigma) \subseteq \Phi(\tau)$.

Notice that for arbitrary edges σ, τ, we have

$$\Phi(\sigma \cap \tau) \subseteq \Phi(\sigma) \cap \Phi(\tau). \tag{2.1.1}$$

The carrier map is *strict* if it satisfies

$$\Phi(\sigma \cap \tau) = \Phi(\sigma) \cap \Phi(\tau). \tag{2.1.2}$$

A carrier map Φ is called *rigid* if for every simplex σ in $\mathcal{G}$ of dimension d, the subgraph $\Phi(\sigma)$ is pure of dimension d. For vertex s, $\Phi(s)$ is a (non-empty) set of vertices, and if σ is an edge, then $\Phi(\sigma)$ is a graph where each vertex is contained in an edge.

We say that a carrier map is *connected* if it sends each vertex to a non-empty set of vertices and each edge to a connected graph. Carrier maps that are connected are rigid. Equation 2.1.1 implies the following property, reminiscent of Fact 2.1.2.

Fact 2.1.4. If Φ is a connected carrier map from a connected graph $\mathcal{G}$ to a graph $\mathcal{H}$, then the image of $\mathcal{G}$ under Φ, $\Phi(\mathcal{G})$, is a connected graph.

Definition 2.1.5. Assume we are given chromatic graphs $\mathcal{G}$ and $\mathcal{H}$ and a carrier map $\Phi : \mathcal{G} \to 2^{\mathcal{H}}$. We call Φ *chromatic* if it is rigid and for all $\sigma \in \mathcal{G}$ we have $\chi(\sigma) = \chi(\Phi(\sigma))$. Note that the image of $\mathcal{G}$ under Φ is simply the union of subgraphs $\Phi(\sigma)$ taken over all simplices σ of $\mathcal{G}$.

Here, for an arbitrary set S, we use the notation

$$\chi(S) = \{\chi(s) \,|\, s \in S\}.$$

When graphs are colored by process names (that is, by the function name$(\cdot)$), we say that Φ *preserves names* or that Φ is *name-preserving*.

2.1.4 Composition of maps

Simplicial maps and carrier maps compose. Let Φ be a carrier map from $\mathcal{G}$ to $\mathcal{H}$ and δ be a simplicial map from $\mathcal{H}$ to a graph $\mathcal{O}$. There is an induced carrier map $\delta(\Phi)$ from $\mathcal{G}$ to $\mathcal{O}$, defined in the natural way: $\delta(\Phi)$ sends a simplex σ of $\mathcal{G}$ to the subgraph $\delta(\Phi(\sigma))$.

Fact 2.1.6. Let Φ be a carrier map from $\mathcal{G}$ to $\mathcal{H}$, and let δ be a simplicial map from $\mathcal{H}$ to a graph $\mathcal{O}$. Consider the carrier map $\delta(\Phi)$ from $\mathcal{G}$ to $\mathcal{O}$. If Φ is chromatic and δ is chromatic, then so is $\delta(\Phi)$. If Φ is connected, then so is $\delta(\Phi)$.

We will be interested in the composition of chromatic carrier maps. Let Φ_0 be a chromatic carrier map from $\mathcal{G}$ to $\mathcal{H}_0$ and Φ_1 be a chromatic carrier map from $\mathcal{H}_0$ to $\mathcal{H}_1$. The induced chromatic carrier map Φ from $\mathcal{G}$ to $\mathcal{H}_1$ is defined in the natural way: $\Phi(\sigma)$ is the union of $\Phi_1(\tau)$ over all simplices $\tau \in \Phi_0(\sigma)$.

Fact 2.1.7. Let Φ_0 be a chromatic carrier map from $\mathcal{G}$ to $\mathcal{H}_0$ and Φ_1 be a chromatic carrier map from $\mathcal{H}_0$ to $\mathcal{H}_1$. The induced chromatic carrier map Φ from $\mathcal{G}$ to $\mathcal{H}_1$ is connected if both Φ_0 and Φ_1 are connected.

Proof. Let σ be a simplex of $\mathcal{G}$, then $\mathcal{K} = \Phi_0(\sigma)$ is connected, so it is enough to show that $\Phi_1(\mathcal{K})$ is connected whenever $\mathcal{K}$ is connected.

If $\mathcal{K}$ is just a vertex or if it has only one edge, we know that $\Phi_1(\mathcal{K})$ is connected, since Φ_1 is a connected carrier map. We can then use induction on the number of edges in $\mathcal{K}$. Given $\mathcal{K}$, if possible, pick an edge $e \in \mathcal{K}$ such that $\mathcal{K} \setminus \{e\}$ is still connected. Then $\Phi_1(\mathcal{K}) = \Phi_1(\mathcal{K} \setminus \{e\}) \cup \Phi_1(e)$, where both $\Phi_1(\mathcal{K} \setminus \{e\})$ and $\Phi_1(e)$ are connected. On the other hand, the intersection $\Phi_1(\mathcal{K} \setminus \{e\}) \cap \Phi_1(e)$ is nonempty; hence $\Phi_1(\mathcal{K})$ is connected as well.

If such an edge e does not exist, we know that the graph $\mathcal{K}$ does not have any cycles and is in fact a tree. In that case, we simply pick a leaf v and an edge e adjacent to v. We then repeat the argument above, representing $\Phi_1(\mathcal{K})$ as the union $\Phi_1(\mathcal{K} \setminus \{e, v\}) \cup \Phi_1(e)$. $\qquad\square$

2.2 Tasks

Let A, B be process names (sometimes *Alice* and *Bob*), V^{in} a domain of *input values*, and V^{out} a domain of *output values*. A *task* for these processes is a triple $(\mathcal{I}, \mathcal{O}, \Delta)$, where

- $\mathcal{I}$ is a pure chromatic *input graph* of dimension 1 colored by $\{A, B\}$ and labeled by V^{in};
- $\mathcal{O}$ is a pure chromatic *output graph* of dimension 1 colored by $\{A, B\}$ and labeled by V^{out};
- Δ is a name-preserving carrier map from $\mathcal{I}$ to $\mathcal{O}$.

The input graph defines all the possible ways the two processes can start the computation, the output graph defines all the possible ways they can end, and the carrier map defines which inputs can lead to which outputs. Each edge $\{(A, a), (B, b)\}$ in $\mathcal{I}$ defines a possible input configuration (initial system state) where A has input value $a \in V^{in}$ and B has input value $b \in V^{in}$. The processes communicate with one another, and each eventually decides on an output value and halts. If A decides x, and B decides y, then there is an output configuration (final system state) represented by an edge $\{(A, x), (B, y)\}$ in the output graph,

$$\{(A, x), (B, y)\} \in \Delta(\{(A, a), (B, b)\}).$$

Moreover, if A runs solo without ever hearing from B, it must decide a vertex (A, x) in $\Delta((A, a))$. Naturally, B is subject to the symmetric constraint.

The monotonicity condition of the carrier map Δ has a simple operational interpretation. Suppose A runs solo starting on vertex s_0, without hearing from B, and halts, deciding on a vertex t_0 in $\Delta(s_0)$. After A halts, B might start from any vertex s_1 such that $\{s_0, s_1\}$ is an edge of $\mathcal{I}$. Monotonicity ensures that there is a vertex t_1 in $\mathcal{O}$ for B to choose, such that $\{t_0, t_1\}$ is in $\Delta(\{s_0, s_1\})$.

2.2.1 Example: coordinated attack

Recall from Chapter 1 that in the coordinated attack task, Alice and Bob each commands an army camped on a hilltop overlooking a valley where the enemy army is camped. If they attack together, they will prevail, but if they attack separately, they may not. For simplicity, we suppose the only possible attack times are either dawn or noon.

The top of Figure 1.10 shows this task's input graph. Bob has only one possible input; he is indifferent to which time is chosen. Alice has two inputs: She might prefer either dawn or noon. As a result, the input graph is a path of two edges. The bottom of the figure shows the output graph. Alice and Bob each have two possible outputs: dawn or noon. Because they must agree, the output graph consists of two disjoint edges. The task specification Δ says that the decisions should agree.

Here is the formal specification of this task. We use 0 to denote *attack at dawn*, 1 to denote *attack at noon*, and $\perp$ to denote *do not attack*. The input graph contains three vertices: $(A, 0), (A, 1), (B, \perp)$. In the figure, Alice's vertices are shown as black and Bob's as white. Alice has two possible input values: 0 and 1, whereas Bob has only one: $\perp$. Similarly, the output graph has three edges, with vertices $(A, 0), (A, 1), (A, \perp), (B, 0), (B, 1), (B, \perp)$.

The carrier map Δ reflects the requirement that if Alice runs alone and never hears from Bob, she does not attack:

$$\Delta((A, 0)) = \Delta((A, 1)) = \{(A, \perp)\} \,.$$

If Bob runs alone and never hears from (Alice), then he does not attack:

$$\Delta((B, \perp)) = \{(B, \perp)\} \,.$$

Finally,

$$\Delta(\{(A, 0), (B, \perp)\}) = \{\{(A, 0), (B, 0)\}, \{(A, \perp), (B, \perp)\}, (A, 0), (B, 0), (A, \perp), (B, \perp)\}$$
$$\Delta(\{(A, 1), (B, \perp)\}) = \{\{(A, 1), (B, 1)\}, \{(A, \perp), (B, \perp)\}, (A, 1), (B, 1), (A, \perp), (B, \perp)\}$$

Note that the vertices on the left side of each equation are in $\mathcal{I}$ and the right side in $\mathcal{O}$.

Notice that this task specification does not rule out the trivial protocol whereby Alice and Bob always refrain from attacking. The requirement that they attack when no failures occur is not a property of the task specification, it is a property of any protocol we consider acceptable for this task. We saw in Chapter 1 that there is no nontrivial protocol for this task when processes communicate by taking turns sending unreliable messages. Later, however, we will see how to solve an approximate version of this task.

2.2.2 Example: consensus

In the *consensus* task, as in the coordinated attack problem, Alice and Bob must both decide one of their input values. In Figure 2.1 we see two versions of the consensus task. On the left side of the figure, surprisingly, the input graph consists of a single edge. This would seem to imply that a process has no initial uncertainty about the input of the other process. Indeed, there is only one possible input to each process. So why is the task nontrivial? Here we see the power that the task carrier map Δ has when defining the possible outputs for individual vertices: Intuitively, the uncertainty of a process is not on what input the other process has but on whether that process participates in the computation or not.

In this "fixed input" version of consensus, if one general deserts without communicating, the other decides on its own input. The input graph consists of a single edge: Alice's vertex is labeled with 0 and Bob's with 1. The output graph consists of two disjoint edges: One where Alice and Bob both decide 0, and another where they both decide 1. The carrier map Δ is similar to that of coordinated attack; if Alice runs solo, she must decide 0, if Bob runs solo, he must decide 1, and if both run, then they must agree on either 0 or 1.

On the right side of Figure 2.1 we see the "binary inputs" version of consensus, where each general can start with two possible inputs, either 0 or 1. To avoid cluttering the picture, the carrier map Δ is not shown on vertices. It sends each input vertex to the output vertex with the same process name and value. It sends each edge where both generals start with the same value to the edge where both generals decide that value, and it sends each edge with mixed values to both output edges.

Notice that the coordinated attack and consensus carrier maps satisfy monotonicity (Definition 2.1.3).

2.2.3 **Example: approximate agreement**

Let us consider a variation on the coordinated attack task. Alice and Bob have realized that they do not need to agree on an exact time to attack, because they will still prevail if their attack times are sufficiently close. In other words, they must choose values v_0 and v_1, between 0 and 1, such that $|v_0 - v_1| \leq \epsilon$, for some fixed $\epsilon > 0$. (Here, 0 means *dawn* and 1 means *noon*, so $\frac{1}{2}$ means the time halfway between dawn and noon.)

In this variant, for simplicity, we assume both Alice and Bob start with a preferred time, 0 or 1, and if either one runs alone without hearing from the other, that one decides his or her own preference.

Here is one way to capture the notion of agreement within ϵ as a discrete task. Given an odd positive integer k, the *k-approximate agreement* task for processes A, B has an input graph $\mathcal{I}$ consisting of a single edge, $I = \{(A, 0), (B, 1)\}$. The output graph $\mathcal{O}$ consists of a path of k edges, whose vertices are:

$$(A, 0), \left(B, \frac{1}{k}\right), \left(A, \frac{2}{k}\right), \ldots, \left(A, \frac{k-1}{k}\right), (B, 1).$$

The carrier map Δ is defined on vertices by

$$\Delta((A, 0)) = \{(A, 0)\} \text{ and } \Delta((B, 1)) = \{(B, 1)\}$$

and extends naturally to edges: $\Delta(\{(A, 0), (B, 1)\}) = \mathcal{O}$. Any protocol for k-approximate agreement causes the processes to decide values that lie within $1/k$ of each other. See Figure 2.2 for the case of $k = 5$. We can think of the path linking the output graph's end vertices as a kind of discrete approximation to a continuous curve between them. No matter how fine the approximation, meaning no matter how many edges we use, the two endpoints remain connected. Connectivity is an example of a topological property that is invariant under subdivision.

It is remarkable that approximate agreement turns out to be the essential building block of a solution to *every* task.

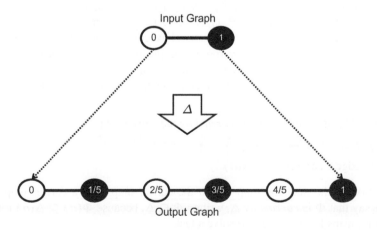

FIGURE 2.2

5-approximate agreement task for fixed inputs. Alice has input value 1 and Bob input value 0.

2.3 Models of computation

We now turn our attention from tasks, the problems we want to solve, to the models of computation with which we want to solve them. As noted earlier, there are many possible models of distributed computation. A model typically specifies how processes communicate, how they are scheduled, and how they may fail. Here we consider three simple models with different characteristics. Although these two-process models are idealized, they share some properties with the more realistic models introduced later.

We will see that the computational power of a model, that is, the set of tasks it can solve, is determined by the topological properties of a family of graphs, called *protocol graphs*, generated by the model.

2.3.1 The protocol graph

Let $(\mathcal{I}, \mathcal{O}, \Delta)$ be a task. Recall that $\mathcal{I}$ is the graph of all possible assignments of input values to processes, $\mathcal{O}$ is the graph of all possible assignments of output values, and the carrier map Δ specifies which outputs may be generated from which inputs.

Now consider a protocol execution in which the processes exchange information through the channels (message passing, read-write memory, or other) provided by the model. At the end of the execution, each process has its own view (final state). The set of all possible final views themselves forms a chromatic graph. Each vertex is a pair (P, p), where P is a process name, and p is P's view (final state) at the end of some execution. A pair of such vertices $\{(A, a), (B, b)\}$ is an edge if there is some execution where A halts with view a and B halts with view b. This graph is called the *protocol graph*.

There is a strict carrier map Ξ from $\mathcal{I}$ to $\mathcal{P}$, called the *execution carrier map*, that carries each input simplex to a subgraph of the protocol graph. Ξ carries each input vertex (P, v) to the solo execution in which P finishes the protocol without hearing from the other process. It carries each input edge $\{(A, a), (B, b)\}$ to the subgraph of executions where A starts with input a and B with b.

The protocol graph is related to the output graph by a *decision map* δ that sends each protocol graph vertex (P, p) to an output graph vertex (P, w), labeled with the same name. Operationally, this map should be understood as follows: If there is a protocol execution in which P finishes with view p and then chooses output w, then (P, p) is a vertex in the protocol graph, (P, w) a vertex in the output graph, and $\delta((P, p)) = (P, w)$. It is easy to see that δ is a simplicial map, carrying edges to edges, because any pair of mutually compatible final views yields a pair of mutually compatible decision values.

Definition 2.3.1. The decision map δ is *carried by* the carrier map Δ if

- For each input vertex s, $\delta(\Xi(s)) \subseteq \Delta(s)$, and
- For each input edge σ, $\delta(\Xi(\sigma)) \subseteq \Delta(\sigma)$.

The composition of the decision map δ with the carrier map Ξ is a carrier map $\Phi : \mathcal{I} \rightarrow 2^{\mathcal{O}}$, (Fact 2.1.6). We say that Φ is *carried by* Δ, written $\Phi \subseteq \Delta$, because $\Phi(\sigma) \subseteq \Delta(\sigma)$ for every $\sigma \in \mathcal{I}$.

Here is what it means for a protocol to solve a task.

Definition 2.3.2. The protocol $(\mathcal{I}, \mathcal{P}, \Xi)$ *solves* the task $(\mathcal{I}, \mathcal{O}, \Delta)$ if there is a simplicial *decision map* δ from $\mathcal{P}$ to $\mathcal{O}$ such that $\delta \circ \Xi$ is carried by Δ.

It follows that the computational power of a two-process model is entirely determined by the set of protocol graphs generated by that model. For example, we will see that some tasks require a disconnected protocol graph. These tasks cannot be solved in any model that permits only connected protocol graphs. More precisely,

Corollary 2.3.3. *Assume that every protocol graph $\mathcal{P}$ permitted by a particular model has the property that the associated strict carrier map $\Xi : \mathcal{I} \to 2^{\mathcal{P}}$ is connected. Then, the task $(\mathcal{I}, \mathcal{O}, \Delta)$ is solvable only if Δ contains a connected carrier map.*

This lemma, and its later higher-dimensional generalization, will be our principal tool for showing that tasks are not solvable. We will use model-specific reasoning to show that a particular model permits only connected protocol graphs, implying that certain tasks, such as the versions of consensus shown in Figure 2.1, are not solvable in that model.

2.3.2 The alternating message-passing model

The *alternating message-passing* model is a formalization of the model used implicitly in the discussion of the coordinated attack task. The model itself is not particularly interesting or realistic, but it provides a simple way to illustrate specific protocol graphs.

As usual, there are two processes, *A* (Alice) and *B* (Bob). Computation is *synchronous*: Alice and Bob take steps at exactly the same times. At step 0, Alice sends a message to Bob, which may or may not arrive. At step 1, if Bob receives a message from Alice, he changes his view to reflect the receipt and immediately sends that view to Alice in a reply message. This pattern continues for a fixed number of steps. Alice may send on even-numbered steps and Bob on odd-numbered steps. After step 0, a process sends a message only if it receives one.

Without loss of generality, we may restrict our attention to *full-information* protocols, whereby each process sends its entire current view (local state) in each message. For impossibility results and lower bounds, we do not care about message size. For specific protocol constructions, there are often task-specific optimizations that reduce message size.

Figure 2.3, shows protocol graphs for zero, one, and two-step protocols, starting with the same input graph as binary consensus. The white vertices are Alice's, and the black vertices Bob's. The protocol graph at step zero is just the input graph, and each process's view is its input value. The protocol graph at step one shows the possible views when Alice's initial message is or is not delivered to Bob. The one-step graph consists of a central copy of the input graph with two branches growing from each of Bob's vertices. The central copy of the input graph represents the processes' unchanged views if Alice's message is not delivered. The new branches reflect Bob's four possible views if Alice's message is delivered, combining Bob's possible inputs and Alice's. Similarly, the two-step graph consists of a copy of the one-step graph with four new branches reflecting Alice's possible views if Bob's message is delivered. Because each process falls silent if it fails to receive a message, subsequent protocol graphs grow only at the periphery, where all messages have been received.

The dotted lines in Figure 2.3 trace the evolution of $\Xi(\sigma)$ for an individual edge σ. It is not hard to see that each $\Xi(\sigma)$ is connected, and therefore the execution carrier map Ξ is connected. (We will see how to prove such claims later.) It follows from Corollary 2.3.3 that the alternating message-passing model cannot solve consensus. Of course, this model cannot solve much of anything, since it admits executions where Alice and Bob finish any protocol without ever communicating.

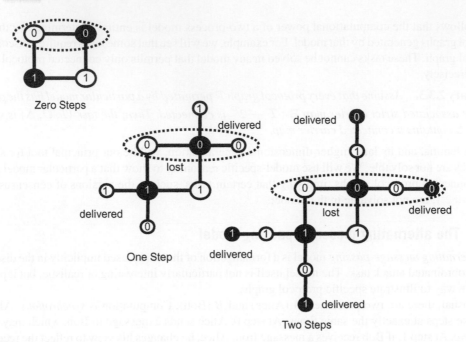

FIGURE 2.3

Alternating message-passing model: how the protocol graph evolves. The dotted lines trace the evolution of a single input edge.

2.3.3 The layered message-passing model

The *layered message-passing* model is stronger and more interesting than the alternating model. Here, too, computation is synchronous: Alice and Bob take steps at the same time. For reasons that will be apparent in later chapters, we will call each such step a *layer*. In each layer, Alice and Bob each sends his or her current view to the other in a message. In each layer, at most one message may fail to arrive, implying that either one or two messages will be received. A process may crash at any time, after which it sends no more messages.

Figure 2.4 shows two single-layer protocol graphs for this model. On the left, the input graph has fixed inputs, and on the right, the input graph has binary inputs. On the right side, each vertex in the input graph is labeled with a binary value, 0 for Alice (white) and 1 for Bob (black). Each vertex in the protocol graph is labeled with the pair of values received in messages, or $\perp$ if no message was received.

It is remarkable that the single-layer protocol graph in this model is the same as the input graph except that each input edge is subdivided into three. Moreover, each subsequent layer further subdivides the edges of the previous layer, and the topological invariant that the protocol graph remains a subdivision of the input graph is maintained. More precisely, consider an edge $\sigma \in \mathcal{I}, \sigma = \{(A, a), (B, b)\}$, where a and b are input values. The single-layer protocol graph σ is a path of three edges:

$$\{\{(A, a\perp), (B, ab)\}, \{(B, ab), (A, ab)\}, \{(A, ab), (B, \perp b)\}\},$$

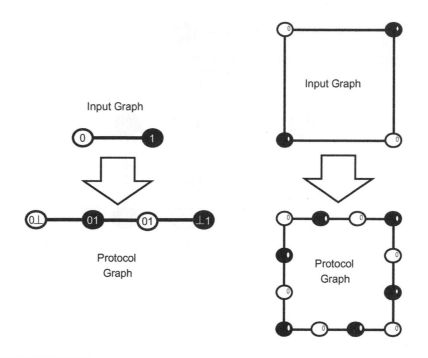

FIGURE 2.4

Layered message-passing model: single-layer protocol graphs, fixed-inputs left, binary inputs right.

where (X, yz) denotes a vertex colored with process name X, message y from A, and message z from B. Either message symbol can be $\bot$.

No matter how many layers we execute, the protocol graph will be a subdivision of the input graph. In particular, the image of an input edge is a subdivided edge, so the execution carrier map Ξ for any protocol in this model is connected. It follows from Corollary 2.3.3 that the consensus task has no protocol in the layered message-passing model. We will see later that it is possible, however, to solve any approximate agreement task. This example shows that the layered message-passing model is stronger than the alternating message model.

Small changes in the model can cause large changes in computational power. Suppose we change this model to guarantee that every message sent is eventually delivered, although processes may still crash. In this case, there is a simple one-layer consensus protocol, illustrated for fixed inputs in Figure 2.5. Each process sends its input to the other. If it does not receive a reply, it decides its own value. If it does receive a reply it decides the lesser of the two input values. Notice that the protocol graph for this model is not pure; the isolated vertices reflect configurations in which one process is certain the other has crashed.

2.3.4 The layered read-write model

We now turn our attention to shared read-write memory. This model, though still idealized, is closer to the programming model presented by modern multicore architectures. Here, computation is *asynchronous*:

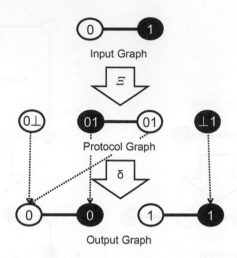

FIGURE 2.5

Reliable message delivery: a single-layer protocol for consensus.

There is no bound on processes' relative speeds. Alice and Bob communicate by reading and writing a shared memory. As before, computation is structured as a sequence of *layers*. In an L-layered protocol execution, the shared memory is organized as an $(L \times 2)$-element array mem[·][·]. At each layer ℓ, starting at 0 and halting at L, Alice writes her view to mem[ℓ][0] and reads mem[ℓ][1], while Bob writes his view to mem[ℓ][1] and reads mem[ℓ][0]. Because scheduling is asynchronous and because either Alice or Bob may crash, Alice reads each mem[ℓ][1] only once, and she may read mem[ℓ][1] before Bob writes to it. Bob's behavior is symmetric. Notice that at each level, at least one process observes the other's view.

Unlike the synchronous layered message-passing model, where a failure can be detected by the absence of a message, failures are undetectable in this model. If Alice does not hear from Bob for a while, she has no way of knowing whether Bob has crashed or whether he is just slow to respond. Because Alice can never wait for Bob to act, any such protocol is said to be *wait-free*.

Figure 2.6 shows a layered read-write protocol. Each process has a *view*, initially just its input value. At each layer $0 \le \ell \le L-1$, Alice, for example, writes her view mem[ℓ][0], reads Bob's view (possibly $\perp$) from mem[ℓ][1], and constructs a new view by joining them.

Note that this protocol, like most protocols considered in this book, is split into two parts. In the first, each process repeatedly writes its view to a shared memory and then constructs a new view by taking a snapshot of the memory. This part is *generic* in the sense that such a step could be part of any protocol for any task. The second part, however, is *task-specific*; each process applies its task-specific decision map to its new view to determine its decision value. This decision map depends on the task being solved. Any protocol can be structured in this way, isolating the task-specific logic in the final decision maps. The decision maps do not affect the protocol graph.

Fact 2.3.4. In the layered read-write model, the protocol graph for the L-layer normal-form protocol starting in a single edge is a path of length 3^L.

```
shared mem: array[0..L−1][0..1] of view  //  initially   all ⊥
private i: int                      // my name: 0 for Alice, 1 for Bob
private j: int := 1 − i             // other's name
protocol LayeredReadWrite (input: int)
  v: view := input                  //  initially  my view is my input
  for ℓ := 0 to L−1 do              // for L  layers
    mem[ℓ][i] := v                  // write my view to memory
    v := {v, mem[ℓ][j]}             // make new view
  decide δᵢ(v)                      // apply my decision map
```

FIGURE 2.6

Layered read-write model: an L-layer protocol.

Remarkably, this is exactly the same execution carrier map Ξ as in the layered message-passing model. Even though one model is synchronous and the other asynchronous, one model uses message passing and the other shared memory, they have exactly the same sets of protocol graphs and exactly the same computational power. In particular, the layered read-write model can solve approximate agreement but not consensus.

Corollary 2.3.5. *If $\mathcal{I}$ is an input graph, and $\Xi : \mathcal{I} \to 2^{\mathcal{O}}$ is an execution carrier map in the layered read-write model, then Ξ is a connected-carrier map.*

2.4 Approximate agreement

Topological methods can be used to establish when protocols exist as well as when they do not. The approximate agreement task of Section 2.2.3 plays a central role in protocol construction, as we shall see in Section 2.5. Here we consider approximate agreement protocols in the layered read-write model. Although k-approximate agreement can be defined for an arbitrary input graph, here we focus on a single-edge fixed-input graph consisting of a single edge, $I = \{(A, 0), (B, 1)\}$.

Recall that the k-approximate agreement task is specified by an odd positive integer k and output graph $\mathcal{O}$ consisting of a path of k edges whose i-th vertex, w_i, is $(A, i/k)$ if i is even and $(B, i/k)$ if i is odd.

The top part of Figure 2.7 shows the input, protocol, and output graphs for a three-approximate agreement protocol. Alice's vertices are white, Bob's are black, and each vertex is labeled with its view. Figure 2.8 shows an explicit single-layer protocol. The processes share a two-element array. Each process writes to its array element and reads from the other's. If the other has not written, the process decides its own value. Otherwise, the process switches to the middle of the range: If its input was 0, it decides 2/3, and if its input was 1, it decides 1/3.

Here is why this protocol works. Because each process writes to its own memory element before reading from the other's, some process must read the other's value. In particular, Alice cannot decide 0 if Bob decides 1. There are three possibilities, illustrated in Figure 2.9. At the top of the figure, Alice reads from Bob but not vice versa, and Alice moves to the middle, at $\frac{2}{3}$, while Bob stays at 1. At the

Input Graph

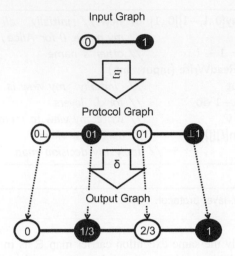

Protocol Graph

Output Graph

FIGURE 2.7

Input, protocol, and output graphs for a single-layer, 3-approximate agreement protocol.

```
shared mem: array[0..1] of real   // two-word shared memory, initially ⊥
private i: int                     // my name: 0 for Alice, 1 for Bob
private j: int := 1 − i            // other's name
protocol 3Approx (input: int)
  v: real := input                 //  initially  my view is my input
  mem[i] := v                      // write  my view to memory
  w: real := mem[j]                // read  other's value
  if w = ⊥ then      // if other is silent ..
    decide v         // decide my value
  else if v = 0 then      // otherwise jump to middle of range
    decide 2/3
  else decide 1/3
```

FIGURE 2.8

A single-layer protocol for 3-approximate agreement.

middle, each reads from the other, and Alice and Bob both move to the middle, at $\frac{2}{3}$ and $\frac{1}{3}$ respectively. At the bottom of the figure, Bob reads from Alice, but not vice versa, and Bob moves to the middle, at $\frac{1}{3}$, while Alice stays at 0. In all cases, each decision lies within $\frac{1}{3}$ of the other.

Figure 2.10 illustrates why there is no single-layer, 5-approximate agreement protocol: There are not enough vertices in the protocol graph to cover the output graph. With two layers, however, it is possible to generate enough vertices and edges to construct a correct decision map (Figure 2.11).

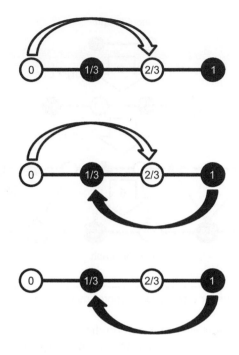

FIGURE 2.9

A single-layer, 3-approximate agreement protocol.

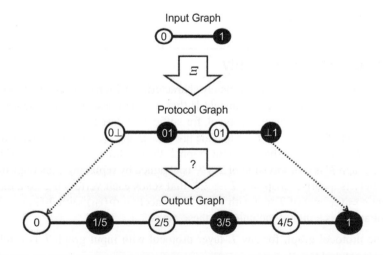

FIGURE 2.10

There is no single-layer, 5-approximate agreement protocol.

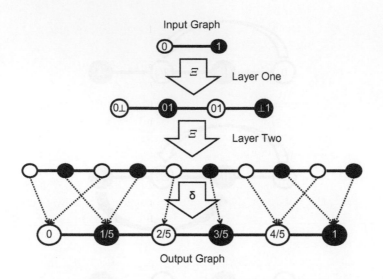

FIGURE 2.11

Input, protocol, and output graphs for a two-layer, 5-approximate agreement protocol.

Using k levels of recursion, it is easy to transform the protocol of Figure 2.8 to a 3^k-approximate agreement protocol. We leave it as an exercise to transform an explicit 3^k-approximate protocol into a K-approximate agreement protocol for $3^{k-1} < K \le 3^k$.

Fact 2.4.1. In the layered read-write model, the K-approximate agreement has a $\left(\lceil \log_3 K \rceil \right)$-layer protocol.

2.5 Two-process task solvability

We are now ready to give a theorem that completely characterizes which two-process tasks have protocols in the layered read-write model. The key insight is that we can construct a protocol for any solvable task from the k-approximate agreement protocol, for sufficiently large k.

For a single-edge input, Fact 2.3.4 states that the protocol graph for an L-layer read-write protocol is a path of length 3^L. Applied to an arbitrary input graph $\mathcal{I}$, the resulting protocol graph $\mathcal{P}$ is a *subdivision* of $\mathcal{I}$. In general, a graph $\mathcal{P}$ is a subdivision of $\mathcal{I}$ if $\mathcal{P}$ is obtained by replacing each edge of $\mathcal{I}$ with a path. More formally, there is a carrier map $\Phi : \mathcal{I} \to 2^{\mathcal{P}}$ that sends each vertex of $\mathcal{I}$ to a distinct vertex of $\mathcal{P}$, and each edge $e = (v_0, v_1)$ of $\mathcal{I}$ to a path P_e of $\mathcal{P}$ connecting $\Phi(v_0)$ with $\Phi(v_1)$ such that different paths are disjoint and $\mathcal{P}$ is equal to the union of these paths.

Fact 2.5.1. The protocol graph for any L-layer protocol with input graph $\mathcal{I}$ is a subdivision of $\mathcal{I}$, where each edge is subdivided 3^L times.

Recall that a task $(\mathcal{I}, \mathcal{O}, \Delta)$ is solvable in the layered read-write model if there exists an L-layer protocol $(\mathcal{I}, \mathcal{P}, \Xi)$, in the form of Figure 2.6, for some $L > 0$, and a simplicial decision map $\delta : \mathcal{P} \to \mathcal{O}$ carried by Δ.

We are now ready to give a complete characterization of the tasks solvable by two asynchronous processes that communicate by layered read-write memory.

Theorem 2.5.2. *The two-process task* $(\mathcal{I}, \mathcal{O}, \Delta)$ *is solvable in the layered read-write model if and only if there exists a connected carrier map* $\Phi : \mathcal{I} \to 2^{\mathcal{O}}$ *carried by* Δ.

Recall that a carrier map Φ is connected (Section 2.1.3) if $\Phi(\sigma)$ is a connected graph, for every $\sigma \in \mathcal{I}$. That is, for every vertex s in $\mathcal{I}$, $\Phi(s)$ is a vertex in $\Delta(s)$, and for every edge σ, $\Phi(\sigma)$ is a connected subgraph of $\Delta(\sigma)$. Finally, because $\Phi(\cdot)$ is a carrier map, if $s \subseteq \sigma \cap \tau$, then $\Phi(s) \subseteq \Phi(\sigma) \cap \Phi(\tau)$.

Here is a simple informal justification for the *if* part. We are given a carrier map Φ carried by Δ. For each vertex v in $\mathcal{I}$, $\Phi(v)$ is a single vertex. Let $\{\sigma_i | i \in I\}$ be the edges of $\mathcal{I}$, where I is an index set. For any edge $\sigma_i = \{s_i, t_i\}$ of $\mathcal{I}$, there is a path linking $\Phi(s_i)$ and $\Phi(t_i)$ in $\Phi(\sigma)$ of length ℓ_i. Let $\ell = \max_{i \in I} \ell_i$. By Fact 2.4.1, there is a protocol to solve approximate agreement on this path that takes $L = \lceil \log_3 \ell \rceil$ layers. The approximate agreement protocols for two intersecting edges $\{s, t\}$ and $\{s, u\}$ agree on their intersection, the solo execution starting at s, so these protocols can be "glued together" on $\mathcal{I}$ to yield a protocol for the entire task.

The informal justification for the *only if* direction is also straightforward. We are given a protocol with decision map δ that solves the task. Its protocol graph $\mathcal{P}$ is a subdivision of $\mathcal{I}$, and δ is a simplicial map from $\mathcal{P}$ to $\mathcal{O}$. The composition of Ξ and δ, $\Phi = \Xi \circ \delta$, is a carrier map from $\mathcal{I}$ to $\mathcal{O}$. By Fact 2.1.2, for every input edge σ, $\Phi(\sigma)$ is connected. Moreover, each input vertex is mapped to a single vertex of the protocol graph (by the solo execution), and from there to a single vertex of $\mathcal{O}$, by δ (the deterministic decision).

Theorem 2.5.2 has two immediate applications. Because the input complex for consensus is connected (an edge) but the output complex is disconnected (two vertices),

Corollary 2.5.3. *The consensus task has no layered read-write protocol.*

By contrast, the input complex $\mathcal{I}$ for approximate agreement is connected (an edge), but so is the output complex (a subdivided edge),

Corollary 2.5.4. *The approximate agreement task does have a layered read-write protocol.*

2.6 Chapter notes

Fischer, Lynch, and Paterson [55] proved that there is no message-passing protocol for the consensus task that tolerates even a single process failure. Later on, Biran, Moran, and Zaks [18] showed how to extend this style of impossibility proof to arbitrary tasks. Moreover, for the tasks that *can* be solved, they derived an approximate agreement-based protocol to solve them, expressing a task solvability characterization in terms of graph connectivity. Our characterization of the tasks solvable by two processes is based on these earlier papers. There are several reasons that our treatment is simpler: We consider only two processes, we use shared-memory communication, and we use a layer-by-layer model whereby each memory location is written only once.

Loui and Abu-Amara [110] showed that consensus is impossible in read-write memory, and Herlihy [78] extended this analysis to shared objects such as stacks, queues, and compare-and-swap variables. Borowsky and Gafni [26] use a round-by-round snapshot model for the wait-free model.

The results in this chapter can all be expressed in the language of graph theory. When at most one process can fail, graph theory is sufficient, even if the system consists of more than two processes. Indeed, graph theory is used in the work of Biran, Moran, and Zaks [18] to analyze task solvability and round complexity in message-passing models [21]. To analyze consensus specifically, graph theory is sufficient even if more processes can fail, as was shown in the work of Moses and Rajsbaum [120] in various models.

In synchronous message-passing models, graph theory is sufficient to analyze consensus. Santoro and Widmayer [136], introduced a model similar to layered message passing, which was further investigated by Charron-Bost and Schiper [36] and by Schmid *et al.* [138]. Santoro and Widmayer [137] investigate the model for arbitrary network interconnection.

The *t*-faulty model, where up to $t \leq n$ processes can fail, was studied by many researchers, including Dwork and Moses [50] and in a recent book of Raynal [133].

The first successful attempts to go beyond graph theory are due to Borowsky and Gafni [23], Herlihy and Shavit [91], and Zaks and Zaharoglou [134]. Higher-dimensional graphs, called *simplicial complexes*, are required to study general tasks in models in which more than one process can fail.

The approximate agreement task was first studied by Dolev *et al.* [47] and later by Abraham *et al.* [1] as a way to circumvent the impossibility of consensus in asynchronous models whereby a single process may crash. They presented algorithms to reach approximate agreement in both synchronous and asynchronous systems. Their algorithms work by successive approximation, with a convergence rate that depends on the ratio between the number of faulty processes and the total number of processes. They also proved lower bounds on this rate.

The two-cover task of Exercise 2.9 is from Fraigniaud *et al.* [58], where many other covering tasks can be found.

2.7 Exercises

Exercise 2.1.   Consider a simplicial map μ from a graph $\mathcal{G}$ to a graph $\mathcal{H}$. Prove that the image $\mu(\mathcal{G})$ is a subgraph of $\mathcal{H}$. Similarly, consider a carrier map Φ from a graph $\mathcal{G}$ to a graph $\mathcal{H}$. Prove that the image $\Phi(\mathcal{G})$ is a subgraph of $\mathcal{H}$. Also, if μ is a simplicial map from $\mathcal{H}$ to another graph, then $\mu(\Phi(\mathcal{G}))$ is a subgraph of that graph.

Exercise 2.2.   Following the previous exercise, prove that if $\mathcal{G}$ is a connected graph, so is the subgraph $\mu(\mathcal{G})$. Prove that it is not true that if $\mathcal{G}$ is connected, then $\Phi(\mathcal{G})$ is connected. However, if $\Phi(\sigma)$ is connected for each edge σ of $\mathcal{G}$, then $\Phi(\mathcal{G})$ is connected. Notice that in this case, $\mu(\Phi(\mathcal{G}))$ is also connected for any simplicial map μ from $\Phi(\mathcal{G})$.

Exercise 2.3.   Prove that a chromatic graph is connected if and only if there exists a (rigid) chromatic simplicial map to the graph consisting of one edge.

Exercise 2.4.   Prove that the composition of two simplicial maps is a simplicial map. Prove that if both are rigid, so is their composition.

Exercise 2.5.   Define the composition of two carrier maps. Prove that the composition satisfies the monotonicity requirement of a carrier map.

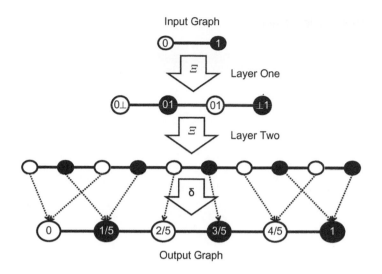

Input Graph

Layer One

Layer Two

Output Graph

FIGURE 2.12

A two-cover task.

Exercise 2.6.   Define the composition of a carrier map followed by a simplicial map. Prove that the composition is a carrier map. Moreover, if both are chromatic, their composition is chromatic.

Exercise 2.7.   In the model of Chapter 1 where Alice and Bob communicate by sending messages to each other in turn, describe the protocol graph and show that it is connected.

Exercise 2.8.   Consider the approximate coordinated attack task of Section 2.2.1. Prove that if Alice and Bob exchange messages in turn, the task is not solvable.

Exercise 2.9.   Consider the *two-cover task* of Figure 2.12. The inputs are binary, and the outputs are in the set $\{0, 1, 2, 3\}$. If a process starts with input b and runs solo, it outputs b or $b + 2$. When the processes start with an edge labeled ℓ in the input graph, they decide on any of the two edges labeled ℓ in the output graph. Prove that this task has no wait-free protocol in the layered read-write model.

Exercise 2.10.   Modify the code of Figure 2.8 to solve 3^k-approximate agreement. (Hint: Use recursion.)

Exercise 2.11.   Given a protocol for 3^k-approximate agreement modify it to solve K-approximate agreement for $3^{k-1} < K \leq 3^k$. Be sure to define the decision maps.

Exercise 2.12.   Consider a solvable task $\mathcal{T} = (\mathcal{I}, \mathcal{O}, \Delta)$, where $\mathcal{I}$ is a graph without cycles. In the layered read-write model, is there a protocol that solves $\mathcal{T}$ with carrier map $\mathcal{P}$, and decision map δ, such that $\delta(\mathcal{P}(\mathcal{I}))$ is an acyclic subgraph of $\mathcal{O}$?

This page is intentionally left blank

Elements of Combinatorial Topology

This chapter defines the basic notions of topology needed to formulate the language we use to describe distributed computation.

Topology is a branch of geometry devoted to drawing the distinction between the *essential* and *inessential* properties of spaces. For example, whether two edges intersect in a vertex is considered essential because it remains the same, no matter how the graph is drawn. By contrast, the length of the

Distributed Computing Through Combinatorial Topology. http://dx.doi.org/10.1016/B978-0-12-404578-1.00003-6

edge linking two vertices is not considered essential, because drawing the same graph in different ways changes that length.

Essential properties are those that endure when the space is subjected to continuous transformations. For example, a connected graph remains connected even if the graph is redrawn or an edge is subdivided into multiple edges that take up the same space (a discrete version of a continuous transformation).

The various branches of topology differ somewhat in the way of representing spaces and in the continuous transformations that preserve essential properties. The branch of topology that concerns us is *combinatorial topology*, because we are interested in spaces made up of simple pieces for which essential properties can be characterized by counting, such as the sum of the degrees of the nodes in a graph. Sometimes the counting can be subtle, and sometimes we will need to call on powerful mathematical tools for help, but in the end, it is all just counting.

3.1 Basic concepts

A distributed system can have a large and complex set of possible executions. We will describe these executions by breaking them into discrete pieces called *simplices*. The structure of this decomposition, that is, how the simplices fit together, is given by a structure called a *complex*. As the name suggests, a complex can be quite complicated, and we will need tools provided by combinatorial topology to cut through the confusing and inessential properties to perceive the simple, underlying essential properties.

We start with an informal geometric example. Perhaps the simplest figure is the disk. It consists of a single piece, without holes, and any attempt to divide it into more than one piece requires cutting or tearing. A 0-dimensional disk is a point; a 1-dimensional disk is a line segment; a 2-dimensional disk is, well, a disk; a 3-dimensional disk is a solid ball, and so on. A d-dimensional disk has a $(d-1)$-dimensional sphere as its boundary. A *cell* of dimension d is a convex polyhedron homeomorphic[1] to a disk of dimension d. We can "glue" cells together along their boundaries to construct a *cell complex*.

As noted, we are primarily interested in properties of complexes that can be expressed in terms of counting. To illustrate this style of argument, we review a classical result: Euler's formula for polyhedrons and some of its applications. This particular result is unrelated to the specific topics covered by this book, but it serves as a gentle informal introduction to the style and substance of the arguments used later. We use a similar approach for Sperner's Lemma in Chapter 9.

A *polyhedron* is a two-dimensional cell complex that is homeomorphic to a sphere. Figure 3.1 shows three such complexes: a tetrahedron, a cube, and an octahedron. Each is made up of a number of vertices, V, a number of edges, E, and a number of faces, F. (Vertices, edges, and faces are all cells of respective dimensions 0, 1, and 2.) Perhaps the earliest discovery of combinatorial topology is Euler's formula:

$$F - E + V = 2.$$

This formula says that the alternating sum of the numbers of faces, edges, and vertices (called the *Euler number*) for any complex homeomorphic to a sphere is always 2. The actual shape of the faces, whether triangles, squares, or other, is irrelevant.

---

[1] Two topological spaces X and Y are *homeomorphic* if there is a bijection $f : X \to Y$ such that both f and its inverse are continuous.

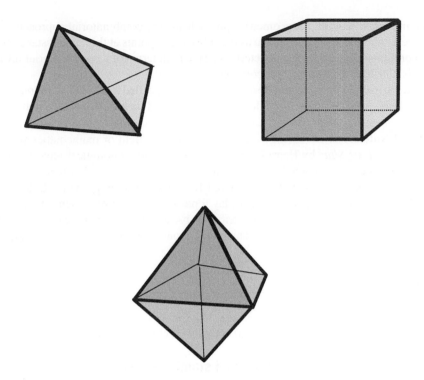

FIGURE 3.1

Three Platonic solids: a tetrahedron, a cube, and an octahedron.

The ancient Greeks discovered that, in addition to the three polyhedrons shown in the figure, there are only two more *Platonic solids:* the dodecahedron and the icosahedron. A *Platonic solid* is a regular polyhedron in which all faces have the same number of edges, and the same number of faces meet at each vertex. The proof that there are only five such polyhedrons is a simple example of the power of combinatorial topology, based on the Euler characteristic and a style of counting we will use later. Let a be the number of edges of each face and let b be the number of edges meeting at each vertex. The number aF counts all the edges, by face, so each edge is counted twice, once for each face to which it belongs. It follows that $aF = 2E$. Similarly, each edge has two vertices, so $bV = 2E$. We can now rewrite Euler's formula as

$$\frac{2E}{a} - E + \frac{2E}{b} = 2$$

or

$$\frac{1}{a} + \frac{1}{b} - \frac{1}{2} = \frac{1}{E}$$

It turns out that there are only five solutions to this equation, because a and b most be integers greater than 2 and less than 6. For example, for the tetrahedron $a = b = 3$; for the cube $a = 4$, $b = 3$; and for the octahedron $a = 3, b = 4$.

Notice the interplay between the geometric approach and the combinatorial approach. In geometry, we characterize a sphere in Euclidean space as the subspace of points at the same distance from a point, whereas the combinatorial approach characterizes a sphere in terms of a combinatorial invariant in the way a sphere is constructed from simpler components.

In this book, we use a more structured form of cell complex, called a *simplicial complex*, in which the cells consist only of vertices, edges, triangles, tetrahedrons, and their higher-dimensional extensions.

> **Mathematical Note 3.1.1.**   Topology emerged as a distinct field of mathematics with the 1895 publication of *Analysis Situs* by Henri Poincaré, although many topological ideas existed before. The work that is usually considered the beginning of topology is due to Leonhard Euler, in 1736, in which he describes a solution to the celebrated Königsberg bridge problem (Euler's work also is cited as the beginning of graph theory). Today topological ideas are present in almost all areas of mathematics and can be highly sophisticated and abstract. Topological ideas are also present in many application areas, including physics, chemistry, economics, biology, and, of course, computer science.

3.2 Simplicial complexes

There are three distinct ways to view simplicial complexes: combinatorial, geometric, and topological.

3.2.1 Abstract simplicial complexes and simplicial maps

We start with the combinatorial view, since it is the most basic and the more closely related to distributed computing. Abstract simplicial complexes and maps between them are the central objects of the combinatorial topology.

Definition 3.2.1.   Given a set S and a family $\mathcal{A}$ of finite subsets of S, we say that $\mathcal{A}$ is an *abstract simplicial complex* on S if the following are satisfied:

(1) If $X \in \mathcal{A}$, and $Y \subseteq X$, then $Y \in \mathcal{A}$; and
(2) $\{v\} \in \mathcal{A}$ for all $v \in S$.

An element of S is a called a *vertex* (plural: *vertices*), and an element of $\mathcal{A}$ is called a *simplex* (plural: *simplices*). The set of all vertices of $\mathcal{A}$ is denoted by $V(\mathcal{A})$. A simplex $\sigma \in \mathcal{A}$ is said to have *dimension* $|\sigma| - 1$. In particular, vertices are 0-dimensional simplices. We sometimes mark a simplex's dimension with a superscript: σ^n. A simplex of dimension n is sometimes called an *n-simplex*. We often say *complex* for brevity when no confusion arises with geometric complex, defined below.

We usually use lowercase Latin letters to denote vertices ($x, y, z, \ldots$), lowercase Greek letters to denote simplices ($\sigma, \tau, \ldots$), and calligraphic font to denote simplicial complexes ($\mathcal{A}, \mathcal{B}, \ldots$).

A simplex τ is a *face* of σ if $\tau \subseteq \sigma$, and it is a *proper face* if $\tau \subset \sigma$. If τ has dimension k, then τ is a *k-face* of σ. Clearly, the 0-faces of σ and vertices of σ are the same objects, so, for $v \in S$, we may write $\{v\} \subseteq \sigma$ or $v \in \sigma$, depending on which aspect of relation between v and σ we want to emphasize. Let $\sigma = \{s_0, \ldots, s_n\}$ be an *n*-simplex. Define $\mathrm{Face}_i\sigma$, the ith *face* of σ, to be the $(n-1)$-simplex $\{s_0, \ldots, \hat{s}_i, \ldots, s_n\}$, where the circumflex denotes omission.

A simplex σ in a complex $\mathcal{A}$ is a *facet* if it is not a proper face of any other simplex in $\mathcal{A}$. The *dimension* of a complex $\mathcal{A}$ is the maximum dimension of any of its facets. A complex is *pure* if all facets have the same dimension. A complex $\mathcal{B}$ is a *subcomplex* of $\mathcal{A}$ if every simplex of $\mathcal{B}$ is also a simplex of $\mathcal{A}$. If $\mathcal{A}$ is a pure complex, the *codimension* $\mathrm{codim}(\sigma, \mathcal{A})$ of $\sigma \in \mathcal{A}$ is $\dim \mathcal{A} - \dim \sigma$, in particular, any facet has codimension 0. When $\mathcal{A}$ is clear from context, we denote the codimension simply by $\mathrm{codim}\, \sigma$.

Let $\mathcal{C}$ be an abstract simplicial complex and ℓ a nonnegative integer. The set of simplices of $\mathcal{C}$ of dimension at most ℓ is a subcomplex of $\mathcal{C}$, called the *ℓ-skeleton*, denoted $\mathrm{skel}^{\ell}(\mathcal{C})$. In particular, the 0-skeleton of a complex is simply its set of vertices.

For an n-dimensional simplex σ, we sometimes denote by 2^{σ} the complex containing σ and all its faces and by $\partial 2^{\sigma}$ the complex of faces of σ of dimension at most $n - 1$. (When there is no ambiguity, we sometimes denote these complexes simply as σ and $\partial \sigma$.) If σ is an n-simplex, its *boundary complex*, $\partial 2^{\sigma}$, or $\mathrm{skel}^{n-1}\sigma$, is its set of proper faces.

Given two complexes $\mathcal{A}$ and $\mathcal{B}$, a *vertex map* $\mu : V(\mathcal{A}) \to V(\mathcal{B})$ carries each vertex of $\mathcal{A}$ to a vertex of $\mathcal{B}$. In topology, however, we are interested in maps that preserve structure.

Definition 3.2.2.   For two simplicial complexes $\mathcal{A}$ and $\mathcal{B}$, a vertex map μ is called a *simplicial map* if it carries simplices to simplices; that is, if $\{s_0, \ldots, s_n\}$ is a simplex of $\mathcal{A}$, then $\{\mu(s_0), \ldots, \mu(s_n)\}$ is a simplex of $\mathcal{B}$.

Note that $\mu(\sigma)$ may have a smaller dimension than σ.

Definition 3.2.3.   Two simplicial complexes $\mathcal{A}$ and $\mathcal{B}$ are *isomorphic*, written $\mathcal{A} \cong \mathcal{B}$, if there are simplicial maps $\phi : \mathcal{A} \to \mathcal{B}$ and $\psi : \mathcal{B} \to \mathcal{A}$ such that for every vertex $a \in \mathcal{A}$, $a = \psi(\phi(a))$, and for every vertex $b \in \mathcal{B}$, $b = \phi(\psi(b))$.

Isomorphic complexes have identical structures.

Definition 3.2.4.   Given: Two abstract simplicial complexes, $\mathcal{A}$ and $\mathcal{B}$. A simplicial map $\varphi : \mathcal{A} \to \mathcal{B}$ is *rigid* if the image of each simplex σ has the same dimension as σ, i.e., $|\varphi(\sigma)| = |\sigma|$.

Rigid maps are rarer than simplicial maps. There are many possible simplicial maps between any two abstract complexes (for example, one could map every vertex of the first complex to any vertex of the second), but there may be no rigid maps. For example, there is no rigid simplicial map from the boundary complex of a triangle to a single edge.

We note that a composition of simplicial maps is a simplicial map, and if the maps are rigid, so is their composition.

3.2.2 The geometric view

We next switch to geometry. Let $\mathbb{R}^d$ denote d-dimensional Euclidean space. In the geometric view, we embed a complex in $\mathbb{R}^d$ and forget about how the complex is partitioned into simplices, considering only the underlying space occupied by the complex.

We use $[m : n]$, where $n \geq m$, as shorthand for $\{m, m + 1, \ldots, n\}$, and we write $[n]$ as shorthand for $[0 : n]$. A point y in $\mathbb{R}^d$ is the *affine combination* of a finite set of points $X = \{x_0, \ldots, x_n\}$ in $\mathbb{R}^d$ if it can be expressed as the weighted sum

$$y = \sum_{i=0}^{n} t_i \cdot x_i, \tag{3.2.1}$$

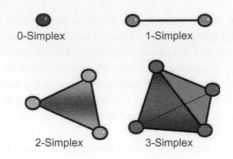

FIGURE 3.2

Simplices of various dimensions.

where the coefficients t_i sum to 1. These coefficients are called the *barycentric coordinates* of y with respect to X. If, in addition, all barycentric coordinates are positive, y is said to be a *convex combination* of the x_i. The *convex hull* of X, conv X, is the set of convex combinations whereby for each coefficient $t_i, 0 \leq t_i \leq 1$. (The convex hull is also the minimal convex set containing X.) The set X is *affinely independent* if no point in the set can be expressed as an affine combination of the others.

The *standard n-simplex* Δ^n is the convex hull of the $n + 1$ points in $\mathbb{R}^{n+1}$ with coordinates $(1, 0, \ldots, 0), (0, 1, 0, \ldots, 0), \ldots, (0, \ldots, 0, 1)$. More generally, a *geometric n-simplex*, or a *geometric simplex of dimension n*, is the convex hull of any set of $n + 1$ affinely independent points in $\mathbb{R}^d$ (in particular, we must have $d \geq n$). As illustrated in Figure 3.2, a 0-dimensional simplex is a point, a 1-simplex is an edge linking two points, a 2-simplex is a solid triangle, a 3-simplex is a solid tetrahedron, and so on.

In direct analogy with the combinatorial framework, we use the following terminology: When $v_0, \ldots, v_n \in \mathbb{R}^d$ are affinely independent, we call them *vertices* of the n-simplex $\sigma = \text{conv}\{v_0, \ldots, v_n\}$. In this case, for any $S \subseteq [n]$, the $(|S| - 1)$-simplex $\tau = \text{conv}\{v_s \mid s \in S\}$ is called a *face*, or an $(|S| - 1)$-face of σ; it is called a *proper face* if, in addition, $S \neq [n]$. We set $\text{Face}_i \sigma := \text{conv}\{v_0, \ldots, \hat{v}_i, \ldots, v_n\}$. Gluing geometric simplices together along their faces yields the geometric analog of Definition 3.2.1.

Definition 3.2.5.   A *geometric simplicial complex* $\mathcal{K}$ in $\mathbb{R}^d$ is a collection of geometric simplices, such that

(1) Any face of a $\sigma \in \mathcal{K}$ is also in $\mathcal{K}$;
(2) For all $\sigma, \tau \in \mathcal{K}$, their intersection $\sigma \cap \tau$ is a face of each of them.

For each geometric n-simplex $\sigma = \text{conv}(v_0, \ldots, v_n)$ with a fixed order on its set of vertices, we have a unique affine map $\varphi : \Delta^n \to \sigma$ taking the ith vertex of Δ^n to v_i. This map φ is called the *characteristic map* of σ.

Given a geometric simplicial complex $\mathcal{K}$, we can define the underlying abstract simplicial complex $\mathcal{C}(\mathcal{K})$ as follows: Take the union of all the sets of vertices of the simplices of $\mathcal{K}$ as the vertices of $\mathcal{C}(\mathcal{K})$; then, for each simplex $\sigma = \text{conv}\{v_0, \ldots, v_n\}$ of $\mathcal{K}$, take the set $\{v_0, \ldots, v_n\}$ to be a simplex of $\mathcal{C}(\mathcal{K})$. In the opposite direction, given an abstract simplicial complex $\mathcal{A}$ with finitely many vertices, there exist many geometric simplicial complexes $\mathcal{K}$, such that $\mathcal{C}(\mathcal{K}) = \mathcal{A}$. The simplest construction is as follows: Assume $\mathcal{A}$ has d vertices; take the standard simplex σ in $\mathbb{R}^d$, and take the subcomplex of σ consisting of the geometric simplices that correspond to the sets in the set family $\mathcal{A}$. Usually one can find $\mathcal{K}$ of a much lower dimension than d, but then the construction could be quite a bit more complicated.

We will see that many of the notions defined for abstract simplicial complexes generalize in a straightforward way to geometric complexes. For now, we remark that there is a standard way in which a simplicial map $\mu : \mathcal{A} \to \mathcal{B}$ induces a locally affine map between the associated geometric complexes: Simply take the map μ on the vertices and linearly extend it to each simplex, using the barycentric coordinate representation from (3.2.1), cf. (3.2.2).

3.2.3 **The topological view**

Finally, we proceed to the topological framework. Given a geometric simplicial complex $\mathcal{K}$ in $\mathbb{R}^d$, we let $|\mathcal{K}|$ denote the union of its simplices, called its *polyhedron*. This space has the usual topology as the subspace of $\mathbb{R}^d$. Somewhat confusingly, the space $|\mathcal{K}|$ is called the *geometric realization* of $\mathcal{K}$. If $\mathcal{A}$ is an abstract simplicial complex, we can first construct $\mathcal{K}$, such that $\mathcal{C}(\mathcal{K}) = \mathcal{A}$, and then let $|\mathcal{A}| = |\mathcal{K}|$. This construction does not depend on the choice of $\mathcal{K}$, only the choice of $\mathcal{A}$. One can also construct $|\mathcal{A}|$ by starting with a set of disjoint simplices, then gluing them together along their boundaries using the combinatorial data as the gluing schema.

Let us now look at the maps between the objects we just described. Let $\mathcal{A}$ and $\mathcal{B}$ be abstract simplicial complexes. Recall that a vertex map $\mu : V(\mathcal{A}) \to V(\mathcal{B})$ maps each vertex of $\mathcal{A}$ to a vertex of $\mathcal{B}$ and that μ is a simplicial map if it also carries simplices to simplices. A vertex map $\mu : V(\mathcal{A}) \to V(\mathcal{B})$ need not induce a continuous map between the geometric realizations $|\mathcal{A}|$ and $|\mathcal{B}|$. For example, if both $\mathcal{A}$ and $\mathcal{B}$ have the vertex set $\{0, 1\}$, and the edge $\{0, 1\}$ is a simplex of $\mathcal{A}$ but not of $\mathcal{B}$, then the identity map id $: \{1, 2\} \to \{1, 2\}$ is a vertex map, but there is no continuous map from an edge to its endpoints that is the identity on the endpoints. However, any simplicial map μ induces a continuous map $|\mu|$ between geometric realizations. For each n-simplex $\sigma = \{s_0, \ldots, s_n\}$ in $\mathcal{A}$, $|\mu|$ is defined on points of $|\sigma|$ by extending barycentric coordinates:

$$|\mu| \left(\sum_{i=0}^{n} t_i s_i \right) = \sum_{i=0}^{n} t_i \mu(s_i). \tag{3.2.2}$$

Before proceeding with constructions, we would like to mention that in standard use in algebraic topology the word *simplex* is overloaded. It is used to denote the abstract simplicial complex consisting of *all* subsets of a certain finite set, but it is also used to refer to individual elements of the family of sets constituting and abstract simplicial complex. There is a relation here: With a simplex in the second sense one can associate a subcomplex of the considered abstract simplicial complex, which is a simplex in the first sense. We will use simplex in both of these meanings. In some texts, *simplex* is also used to denote the geometric realization of that abstract simplicial complex; here we say *geometric simplex* instead.

3.3 **Standard constructions**

There are two standard constructions that characterize the neighborhood of a vertex or simplex: the star and the link (Figure 3.3).

Assume that σ is a simplex of a simplicial complex $\mathcal{C}$. We describe the constructions in the abstract simplicial case. The case of geometric simplicial complexes is completely analogous.

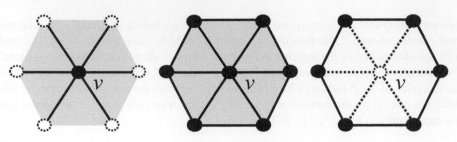

FIGURE 3.3

The open star St°(v), the star St(v), and the link Lk(v) of the vertex v.

3.3.1 Star

The *star* of a simplex $\sigma \in C$, written St(σ, C), or St(σ) when C is clear from context, is the subcomplex of C whose facets are the simplices of C that contain σ. The complex St(σ, C) consists of all the simplices τ which contain σ and, furthermore, all the simplices contained in such a simplex τ. The geometric realization of St(σ, C) is also called the star of σ. Using our previous notations, we write $|$St(σ, C)$|$.

The *open star*, denoted St°(σ), is the union of the interiors of the simplices that contain σ:

$$St°(\sigma) = \bigcup_{\tau \supseteq \sigma} \operatorname{Int}\tau.$$

Note that St°(σ) is not an abstract or geometric simplicial complex but just a topological space, which is open in C. The open sets (St°(v))$_{v \in V(C)}$ provide an open covering of $|C|$.

We have St°(σ) = $\cap_{v \in V(\sigma)}$St°(v), i.e., the open star of a simplex is the intersection of the open stars of its vertices. Here the interior of a vertex is taken to be the vertex itself, and the interior of a higher-dimensional simplex is the topological interior of the corresponding topological space. To distinguish the two notions, the geometric realization of a star is also sometimes called the *closed star*.

3.3.2 Link

The *link* of $\sigma \in C$, written Lk(σ, C) (or Lk σ), is the subcomplex of C consisting of all simplices in St(σ, C) that do not have common vertices with σ. The geometric realization of Lk(σ, C) is also called the link of σ.

Examples of the link of a vertex and of an edge are shown in Figure 3.4.

3.3.3 Join

Given two abstract simplicial complexes A and B with disjoint sets of vertices $V(A)$ and $V(B)$, their *join*, $A * B$, is the abstract simplicial complex with the set of vertices $V(A) \cup V(B)$ whose simplices are all the unions $\alpha \cup \beta$, where $\alpha \in A$, and $\beta \in B$. Note that either α or β can be be empty sets. In particular, both A and B are subcomplexes of $A * B$. For example, as shown in Figure 3.5, the join of two intervals is a tetrahedron. The join operation is commutative and associative.

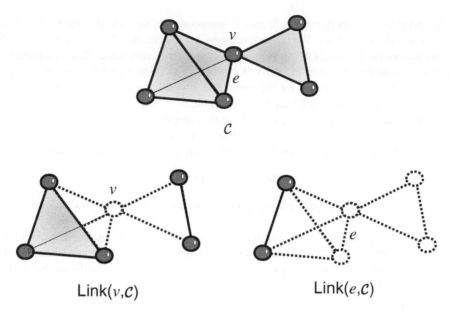

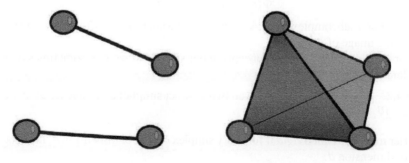

FIGURE 3.4

The link of a vertex and of an edge.

Assume furthermore, that $\mathcal{K}$ is a geometric simplicial complex in $\mathbb{R}^m$, such that $\mathcal{C}(\mathcal{K}) = \mathcal{A}$, and $\mathcal{L}$ is a geometric simplicial complex in $\mathbb{R}^n$, such that $\mathcal{C}(\mathcal{L}) = \mathcal{B}$. Then there is a standard way to construct a geometric simplicial complex in $\mathbb{R}^{m+n+1}$ whose underlying abstract simplicial complex is $\mathcal{A} * \mathcal{B}$. Consider the following embeddings: $\varphi : \mathbb{R}^m \to \mathbb{R}^{m+n+1}$, given by

$$\varphi(x_1, \ldots, x_m) = (x_1, \ldots, x_m, 0, \ldots, 0),$$

and $\psi : \mathbb{R}^n \to \mathbb{R}^{m+n+1}$ given by

$$\psi(y_1, \ldots, y_n) = (0, \ldots, 0, y_1, \ldots, y_n, 1).$$

FIGURE 3.5

The join of two edges is a tetrahedron.

The images under these embeddings of $\mathcal{K}$ and $\mathcal{L}$ are geometric simplicial complexes for which the geometric realizations are disjoint. We can define a new geometric simplicial complex $\mathcal{K} * \mathcal{L}$ by taking all convex hulls conv (σ, τ), where σ is a simplex of $\mathcal{K}$ and τ is a simplex of $\mathcal{L}$. It is a matter of simple linear algebra to show that the open intervals (x, y), where $x \in \text{Im } \varphi$ and $y \in \text{Im } \psi$, never intersect, and so $\mathcal{K} * \mathcal{L}$ satisfies the conditions for the geometric simplicial complex. It is easy to see that the topological spaces $|\mathcal{A} * \mathcal{B}|$ and $|\mathcal{K} * \mathcal{L}|$ are homeomorphic.

An important example is taking the join of $\mathcal{K}$ with a single vertex. When $\mathcal{K}$ is pure of dimension d, $v * \mathcal{K}$ is called a *cone* over $\mathcal{K}$, and v is called the *apex* of the cone. Notice that $v * \mathcal{K}$ is pure of dimension $d + 1$. As an example, for any vertex v of a pure complex $\mathcal{K}$ of dimension d, we have

$$\text{St}(v) = v * \text{Lk}(v).$$

Another example is taking the join of an m-simplex with an n-simplex, which yields an $(m + n + 1)$-simplex.

There is also a purely topological definition of the join of two topological spaces. Here we simply mention that the simplicial and topological joins commute with the geometric realization, that is, for any two abstract simplicial complexes $\mathcal{A}$ and $\mathcal{B}$, the spaces $|\mathcal{A} * \mathcal{B}|$ and $|\mathcal{A}| * |\mathcal{B}|$ are homeomorphic.

3.4 Carrier maps

The concept of a carrier map is especially important for applications of topology in distributed computing.

Definition 3.4.1. Given two abstract simplicial complexes $\mathcal{A}$ and $\mathcal{B}$, a *carrier map* Φ from $\mathcal{A}$ to $\mathcal{B}$ takes each simplex $\sigma \in \mathcal{A}$ to a subcomplex $\Phi(\sigma)$ of $\mathcal{B}$ such that for all $\sigma, \tau \in \mathcal{A}$, such that $\sigma \subseteq \tau$, we have $\Phi(\sigma) \subseteq \Phi(\tau)$.

We usually use uppercase Greek letters for carrier maps ($\Delta, \Phi, \Xi, \ldots$). Since a carrier map takes simplices of $\mathcal{A}$ to subcomplexes of $\mathcal{B}$, we use *powerset notation* to describe its range and domain: $\Phi : \mathcal{A} \rightarrow 2^{\mathcal{B}}$. Definition 3.4.1 can be rephrased as saying that a carrier map Φ is *monotonic*, implying that the inclusion pattern of the subcomplexes $\Phi(\sigma)$ is the same as the inclusion pattern of the simplices of $\mathcal{A}$, implying that:

$$\Phi(\sigma \cap \tau) \subseteq \Phi(\sigma) \cap \Phi(\tau) \tag{3.4.1}$$

for all $\sigma, \tau \in \mathcal{A}$. For a subcomplex $\mathcal{K} \subseteq \mathcal{A}$, we will use the notation $\Phi(\mathcal{K}) := \bigcup_{\sigma \in \mathcal{K}} \Phi(\sigma)$. In particular, $\Phi(\mathcal{A})$ denotes the image of Φ.

Carrier maps are one of the central concepts in our study, and we will sometimes require additional properties. Here are some of them.

Definition 3.4.2. Assume that we are given two abstract simplicial complexes $\mathcal{A}$ and $\mathcal{B}$ and a carrier map $\Phi : \mathcal{A} \rightarrow 2^{\mathcal{B}}$.

(1) The carrier map Φ is called *rigid* if for every simplex $\sigma \in \mathcal{A}$ of dimension d, the subcomplex $\Phi(\sigma)$ is pure of dimension d.

(2) The carrier map Φ is called *strict* if the equality holds in (3.4.1), i.e., we have $\Phi(\sigma \cap \tau) = \Phi(\sigma) \cap \Phi(\tau)$, for all $\sigma, \tau \in \mathcal{A}$.

Note specifically that for a rigid carrier map, the subcomplex $\Phi(\sigma)$ is nonempty if and only if σ is nonempty, since both must have the same dimension.

Given a strict carrier map $\Phi : \mathcal{A} \to 2^{\mathcal{B}}$, for each simplex $\tau \in \Phi(\mathcal{A})$ there is a unique simplex σ in $\mathcal{A}$ of smallest dimension, such that $\tau \in \Phi(\sigma)$. This σ is called the *carrier* of τ, or $\text{Car}(\tau, \Phi(\sigma))$. (Sometimes we omit $\Phi(\sigma)$ when it is clear from the context.)

Definition 3.4.3.   Given two carrier maps $\Phi : \mathcal{A} \to 2^{\mathcal{B}}$ and $\Psi : \mathcal{A} \to 2^{\mathcal{B}}$, where $\mathcal{A}, \mathcal{B}$, are simplicial complexes, and a simplicial map $\varphi : \mathcal{A} \to \mathcal{B}$, we say that

(1) Φ is *carried* by Ψ, and we write $\Phi \subseteq \Psi$ if $\Psi(\sigma) \supseteq \Phi(\sigma)$ for every $\sigma \in \mathcal{A}$; and
(2) φ is *carried* by Φ if $\varphi(\sigma) \in \Phi(\sigma)$ for every $\sigma \in \mathcal{A}$.

Figure 3.6 shows a carrier map that carries a complex consisting of an edge (top) to a complex consisting of three edges (bottom). It carries each vertex of the edge to the two endpoints and carries the edge to all three edges. There is no simplicial map carried by this carrier map, because such a map would have to send vertices connected by an edge to vertices not connected by an edge.

We can compose carrier maps with simplicial maps as well as with each other.

Definition 3.4.4.   Assume we are given three abstract simplicial complexes $\mathcal{A}, \mathcal{B}$, and $\mathcal{C}$ and a carrier map Φ from $\mathcal{A}$ to $\mathcal{B}$.

(1)   If $\varphi : \mathcal{C} \to \mathcal{A}$ is a simplicial map, then we can define a carrier map $\Phi \circ \varphi$ from $\mathcal{C}$ to $\mathcal{B}$ by setting $(\Phi \circ \varphi)(\sigma) := \Phi(\varphi(\sigma))$ for all $\sigma \in \mathcal{C}$.
(2)   If $\varphi : \mathcal{B} \to \mathcal{C}$ is a simplicial map, then we can define a carrier map $\varphi \circ \Phi$ from $\mathcal{A}$ to $\mathcal{C}$ by setting $(\varphi \circ \Phi)(\sigma) := \varphi(\Phi(\sigma))$ for all $\sigma \in \mathcal{A}$, where $\varphi(\Phi(\sigma)) = \cup_{\tau \in \Phi(\sigma)} \varphi(\tau)$.

It is not difficult to see that composing a rigid simplicial map with a rigid carrier map on the left as well as on the right will again produce a rigid carrier map.

Furthermore, we can also compose carrier maps with each other.

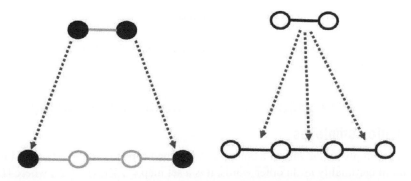

FIGURE 3.6

This carrier map from the top simplex to the bottom carries each vertex to a vertex (left) and the single edge to three edges (right). There is no simplicial map carried by this carrier map.

Definition 3.4.5.   Given two carrier maps $\Phi : \mathcal{A} \to 2^{\mathcal{B}}$ and $\Psi : \mathcal{B} \to 2^{\mathcal{C}}$, where $\mathcal{A}$, $\mathcal{B}$, and $\mathcal{C}$ are simplicial complexes, we define a carrier map $\Psi \circ \Phi : \mathcal{A} \to 2^{\mathcal{C}}$ by setting $(\Psi \circ \Phi)(\sigma) := \cup_{\tau \in \Phi(\sigma)} \Psi(\tau)$, i.e., $(\Psi \circ \Phi)(\sigma) = \Psi(\Phi(\sigma))$ for all $\sigma \in \mathcal{A}$.

Proposition 3.4.6.   *Assume that we are given two carrier maps $\Phi : \mathcal{A} \to 2^{\mathcal{B}}$ and $\Psi : \mathcal{B} \to 2^{\mathcal{C}}$, where $\mathcal{A}$, $\mathcal{B}$, and $\mathcal{C}$ are simplicial complexes.*

(1) If the carrier maps Φ and Ψ are rigid, then so is their composition $\Psi \circ \Phi$.
(2) If the carrier maps Φ and Ψ are strict, then so is their composition $\Psi \circ \Phi$.

Proof.   To show (1), take a d-simplex $\sigma \in \mathcal{A}$. Since Φ is rigid, the subcomplex $\Phi(\sigma)$ is pure of dimension d. Since any carrier map is monotonic, we have $(\Psi \circ \Phi)(\sigma) = \cup_{\tau \in \Phi(\sigma)} \Psi(\tau)$, where the union is taken over all facets of $\Phi(\sigma)$, which is the same as all d-simplices of $\Phi(\sigma)$. For each such d-simplex τ, the subcomplex $\Psi(\tau)$ is a pure d-dimensional complex, since Ψ is rigid. The union of pure d-dimensional complexes is again pure d-dimensional, hence we are done.

Now we show (2). Pick simplices $\sigma, \tau \in \mathcal{A}$. We have

$$
\Psi(\Phi(\sigma)) \cap \Psi(\Phi(\tau)) = \left(\bigcup_{\gamma_1 \in \Phi(\sigma)} \Psi(\gamma_1) \right) \bigcap \left(\bigcup_{\gamma_2 \in \Phi(\tau)} \Psi(\gamma_2) \right) =
$$

$$
= \bigcup_{\gamma_1 \in \Phi(\sigma),\ \gamma_2 \in \Phi(\tau)} \left(\Psi(\gamma_1) \cap \Psi(\gamma_2) \right) = \bigcup_{\gamma_1 \in \Phi(\sigma),\ \gamma_2 \in \Phi(\tau)} \Psi(\gamma_1 \cap \gamma_2) =
$$

$$
= \bigcup_{\gamma \in \Phi(\sigma) \cap \Phi(\tau)} \Psi(\gamma) = \bigcup_{\gamma \in \Phi(\sigma \cap \tau)} \Psi(\gamma) = \Psi(\Phi(\sigma \cap \tau)),
$$

which shows that the composition carrier map is again strict.   □

Finally, if $\mathcal{A}$ and $\mathcal{B}$ are geometric complexes, a continuous map $f : |\mathcal{A}| \to |\mathcal{B}|$ is *carried by* a carrier map $\Phi : \mathcal{A} \to 2^{\mathcal{B}}$ if, for every simplex $\sigma \in \mathcal{A}$, $f(\sigma) \subseteq |\Phi(\sigma)|$.

3.4.1 Chromatic complexes

An *m-labeling*, or simply a *labeling*, of a complex $\mathcal{A}$ is a map carrying each vertex of $\mathcal{A}$ to an element of some domain of cardinality m. In other words, it is a set map $\varphi : V(\mathcal{A}) \to D$, where $|D| = m$.

An *m-coloring*, or simply a *coloring*, of an n-dimensional complex $\mathcal{A}$ is an m-labeling $\chi : V(\mathcal{A}) \to \Pi$ such that χ is injective on the vertices of every simplex of $\mathcal{A}$: for distinct $s_0, s_1 \in \sigma$, $\chi(s_0) \neq \chi(s_1)$. In other words, m-colorings are precisely the rigid simplicial maps into an $(m-1)$-simplex $\chi : \mathcal{A} \to \Delta^{m-1}$. A simplicial complex $\mathcal{A}$ together with a coloring χ is called a *chromatic* complex, written $(\mathcal{A}, \chi)$.

Mathematical Note 3.4.7.   A coloring $\chi : \mathcal{A} \to \Delta^{m-1}$ exists if and only if the 1-skeleton of $\mathcal{A}$, viewed as a graph, is m-colorable in the sense of graph colorings (more precisely, vertex-colorings of graphs).

Definition 3.4.8.   Given two m-chromatic simplicial complexes $(\mathcal{A}, \chi_{\mathcal{A}})$ and $(\mathcal{B}, \chi_{\mathcal{B}})$, a simplicial map $\phi : \mathcal{A} \to \mathcal{B}$ is *color-preserving* if for every vertex $v \in \mathcal{A}$, $\chi_{\mathcal{A}}(v) = \chi_{\mathcal{B}}(\phi(v))$.

Definition 3.4.9.   Assume we are given chromatic simplicial complexes $\mathcal{A}$ and $\mathcal{B}$ and a carrier map $\Phi : \mathcal{A} \to 2^{\mathcal{B}}$. We call Φ *chromatic* if Φ is rigid and for all $\sigma \in \mathcal{A}$ we have $\chi_{\mathcal{A}}(\sigma) = \chi_{\mathcal{B}}(\Phi(\sigma))$, where $\chi_{\mathcal{B}}(\Phi(\sigma)) := \{\chi_{\mathcal{B}}(v) \mid v \in V(\Phi(\sigma))\}$.

When the colors are process names, we often say *name-preserving* instead of *chromatic*.

3.5 Connectivity

We have defined the objects and maps of interest as well as the basic language and constructions to work with them. We are ready to study *topological properties* of these objects, that is, properties that remain invariant under continuous stretching and bending of the object. The first such notion is that of path connectivity and its higher-dimensional analogs.

3.5.1 Path connectivity

Perhaps the most basic topological property of an object is whether it consists of a single connected piece. For simplicial complexes, this topological property can be formalized as follows.

Definition 3.5.1.   Let $\mathcal{K}$ be an arbitrary simplicial complex. An *edge path* (or simply a *path*) between vertices u and v in $\mathcal{K}$ is a sequence of vertices $u = v_0, v_1, \dots, v_\ell = v$ such that each pair $\{v_i, v_{i+1}\}$ is an edge of $\mathcal{K}$ for $0 \le i < \ell$. A path is *simple* if the vertices are distinct.

Definition 3.5.2.   A simplicial complex $\mathcal{K}$ is *path-connected* if there is a path between every two vertices in $\mathcal{K}$. The largest subcomplexes of $\mathcal{K}$ that are path-connected are the *path-connected components* of $\mathcal{K}$.

The path connectivity of $\mathcal{K}$ depends only on the 1-skeleton of $\mathcal{K}$, $\mathrm{skel}^1(\mathcal{K})$, namely, the subcomplex consisting of the set of simplices of $\mathcal{K}$ of dimension 1, at most.

Clearly, the simplicial complex $\mathcal{K}$ is a disjoint union of its path-connected components. Furthermore, any two vertices are connected by a path if and only if they belong to the same path-connected component. A simple but crucial observation is that a simplicial map takes an edge path to an edge path, though the number of edges may decrease. This implies the following proposition.

Proposition 3.5.3.   *An image of a path-connected complex under a simplicial map is again path-connected. In particular, if A and B are simplicial complexes, $\varphi : A \to B$ is a simplicial map, and A is path-connected, then $\varphi(A)$ is contained in one of the connected components of B.*

3.5.2 Simply connected spaces

Before proceeding to higher dimensions, we rephrase notions of connectivity in terms of spheres and disks. A 1-dimensional disk can be thought of as the closed interval $[-1, 1]$, and its boundary, the two points ± 1 on the real line, as a 0-dimensional sphere. A 2-dimensional disk is the set of points in the plane at distance 1 (at most) from the origin, and a 1-dimensional sphere as the points at exactly 1 from the origin. A 2-sphere is an ordinary 2-dimensional sphere in 3-dimensional Euclidean space and is the boundary of an ordinary 3-dimensional ball. An n-sphere, S^n, is a generalization of the surface of an ordinary sphere to arbitrary dimension and is the boundary of an $n + 1$-ball, D^{n+1}.

Given a simplicial complex $\mathcal{K}$, let $|\mathcal{K}|$ denote its polyhedron. We may consider a path in $|\mathcal{K}|$ as a continuous map $f : D^1 \to |\mathcal{K}|$, where $D^1 = [-1, 1]$. We say the path connects the points $f(-1)$ and $f(1)$. See Figure 3.10, in which there is a path connecting $f(a)$ and $f(c)$, where $a = -1$ and $c = 1$. We say that the polyhedron $|\mathcal{K}|$ is *path-connected* if there is a path in $|\mathcal{K}|$ connecting any two points in $|\mathcal{K}|$. The polyhedron $|\mathcal{K}|$ is path-connected if and only if $\mathcal{K}$ is edge-path-connected.

Now, if $|\mathcal{K}|$ is path-connected, then there is a path f between any two points, v_1, v_2. Think of these points as the image, under map $f : S^0 \to |\mathcal{K}|$, of a 0-dimensional sphere, so $f(-1) = v_1$ and $f(1) = v_2$. The existence of the path means that this map from the 0-sphere can be extended to a continuous map of the 1-ball, $f : D^1 \to |\mathcal{K}|$. We say that a path-connected complex is 0-*connected*.[2]

Mathematical Note 3.5.4. This notion generalizes to higher dimensions in a natural way. A *loop* in a complex $\mathcal{K}$ is a path with starting and end vertices that are the same. A loop can be considered a continuous map $f : S^1 \to |\mathcal{K}|$, carrying the 1-sphere S^1 to the polyhedron of $\mathcal{K}$. Usually one also fixes a point x on S^1, fixes a point y in $|\mathcal{K}|$, and considers only the loops that map x to y; this allows loops to be composed. Now, considering all the loops in $|\mathcal{K}|$ based at x up to their continuous deformation and taking the operation of composition, we obtain the so-called *fundamental group*. This group does not depend on the choice of x as long as $|\mathcal{K}|$ is path-connected.

Definition 3.5.5. Let $\mathcal{K}$ be an arbitrary path-connected simplicial complex. The complex $\mathcal{K}$ is 1-*connected* (or *simply connected*) if any continuous map $f : S^1 \to |\mathcal{K}|$ can be extended to the 2-disk $F : D^2 \to |\mathcal{K}|$, where S^1 is the boundary of D^2.

The complex in the right part of Figure 3.10 is 0-connected but not 1-connected.

3.5.3 Higher-dimensional connectivity

We now have the formal framework to extend Definitions 3.5.2 and 3.5.5 to any dimension.

Definition 3.5.6. Let k be any positive integer. The complex $\mathcal{K}$ is k-*connected* if, for all $0 \leq \ell \leq k$, any continuous map $f : S^\ell \to |\mathcal{K}|$ can be extended to $F : D^{\ell+1} \to |\mathcal{K}|$, where the sphere S^ℓ is the boundary of the disk $D^{\ell+1}$.

One way to think about this property is that any map f that cannot be "filled in" represents an n-dimensional "hole" in the complex. Indeed, S^k is ℓ-connected for $\ell < k$, but it is not k-connected.

Notice that Proposition 3.5.3 does not generalize to higher connectivity. An image of a 1-connected complex under a simplicial map is not necessarily 1-connected. For example, a disk D^2 can be mapped to a sphere S^1.

---

[2] We remark that the notion of *path connectivity*, or *0 connectivity*, is different from the notion of *connectivity* for general topological spaces. However, it is the same for the case of simplicial complexes, which is the only case we are considering in this book.

> **Mathematical Note 3.5.7.**   A complex $\mathcal{K}$ is simply connected if and only if its fundamental group $\pi_1(\mathcal{K})$ is trivial, and it is k-connected if and only if its ℓth homotopy group $\pi_\ell(\mathcal{K})$ is trivial, for all $1 \leq \ell \leq k$.

Definition 3.5.8.   A complex $\mathcal{K}$ is *contractible* if there is a continuous map $H : |\mathcal{K}| \times I \rightarrow |\mathcal{K}|$, where I is the unit interval, such that $H(\cdot, 0)$ is the identity map on $|\mathcal{K}|$, and $H(\cdot, 1)$ is a constant map $|\mathcal{K}| \mapsto x$, for some $x \in |\mathcal{K}|$.

Informally, $|\mathcal{K}|$ can be continuously deformed to a single point $x \in |\mathcal{K}|$, where the path of every point under the deformation stays in $|\mathcal{K}|$. An n-connected complex of dimension n is contractible, and every contractible space is n-connected for all n. Examples of contractible spaces include all m-simplices and their subdivisions. Also, all cones over simplicial complexes are contractible.

3.6 Subdivisions

Informally, a *subdivision* of a complex $\mathcal{A}$ is constructed by "dividing" the simplices of $\mathcal{A}$ into smaller simplices to obtain another complex, $\mathcal{B}$. Subdivisions can be defined for both geometric and abstract complexes.

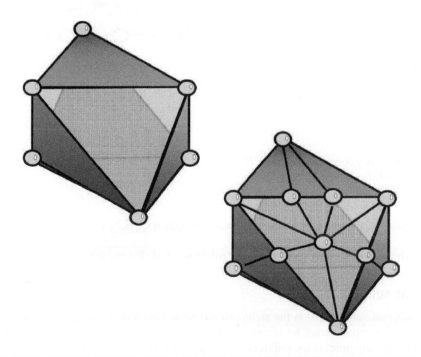

FIGURE 3.7

A geometric complex and its simplicial subdivision.

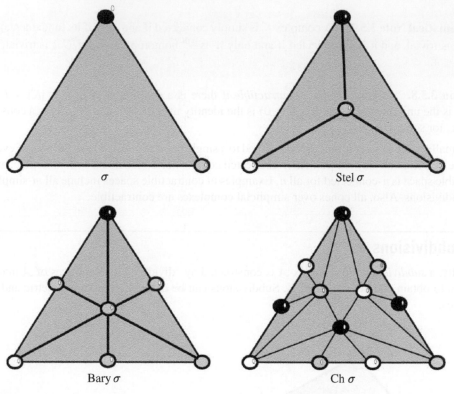

FIGURE 3.8

A simplex σ (upper left), the stellar subdivision stel σ (upper right), the barycentric subdivision Bary σ (lower left), and the standard chromatic subdivision Ch σ (lower right).

Definition 3.6.1. A geometric complex $\mathcal{B}$ is called a subdivision of a geometric complex $\mathcal{A}$ if the following two conditions are satisfied:

(1) $|\mathcal{A}| = |\mathcal{B}|$;
(2) Each simplex of $\mathcal{A}$ is the union of finitely many simplices of $\mathcal{B}$.

Figure 3.7 shows a geometric complex and a subdivision of that complex.

3.6.1 Stellar subdivision

Perhaps the simplest subdivision is the *stellar* subdivision. Given an n-simplex $\sigma = \{s_0, \ldots, s_n\}$, the complex $\text{stel}(\sigma, b)$ is constructed by taking a cone with apex b over the boundary complex $\partial\sigma$. (See Figure 3.8.) Though any point in the interior can be used as the apex, we will typically use the *barycenter* $b = \sum_i s_i/(n + 1)$. Any subdivision can be constructed from repeated stellar subdivisions.

3.6.2 Barycentric subdivision

In classical combinatorial topology, the *barycentric* subdivision is perhaps the most widely used. Given a complex $\mathcal{K}$, the complex Bary $\mathcal{K}$ is constructed inductively over the skeletons of $\mathcal{K}$. We start by taking the vertices of K. At the next step we insert a barycenter in each edge of K and take cones, with apexes at barycenters, over the ends of each edge. In general, to extend the barycentric subdivision from the $(n-1)$-skeleton to the n-skeleton of K, we insert a barycenter b in each simplex σ of K and take a cone with apex at b over Bary $\partial\sigma$, the already subdivided boundary of σ. (See Figure 3.8.)

The barycentric subdivision has an equivalent, purely combinatorial definition.

Definition 3.6.2.   Let $\mathcal{A}$ be an abstract simplicial complex. Its *barycentric* subdivision Bary $\mathcal{A}$ is the abstract simplicial complex whose vertices are the nonempty simplices of $\mathcal{A}$. A $(k+1)$-tuple $(\sigma_0, \ldots, \sigma_k)$ is a simplex of Bary $\mathcal{A}$ if and only if the tuple can be indexed so that $\sigma_0 \subset \cdots \subset \sigma_k$.

Of course, the barycentric subdivision of a geometric realization of an abstract simplicial complex $\mathcal{A}$ is a geometric realization of the barycentric subdivision of $\mathcal{A}$.

3.6.3 Standard chromatic subdivision

For our purposes, however, the barycentric subdivision has a flaw: The barycentric subdivision of a chromatic complex is not itself chromatic. To remedy this shortcoming, we introduce the *standard chromatic subdivision*, the chromatic analog to the barycentric subdivision. (See Figure 3.8.)

Given a chromatic complex $(\mathcal{K}, \chi)$, the complex Ch $\mathcal{K}$ is constructed inductively over the skeletons of $\mathcal{K}$. We start by taking the vertices of K. At the next step, for each edge $\eta = \{s_0, s_1\}$, instead of taking the barycenter we take two interior points slightly displaced from the barycenter:

$$c_0 = \frac{1-\epsilon}{2}s_0 + \frac{1+\epsilon}{2}s_1$$
$$c_1 = \frac{1+\epsilon}{2}s_0 + \frac{1-\epsilon}{2}s_1$$

for some $0 < \epsilon < 1$. Define the *central* edge to be $\{c_0, c_1\}$, and define $\chi(c_i) = \chi(s_i)$. We join each central vertex to the vertex of a complementary color, so that Ch η consists of three edges: $\{s_0, c_1\}$, the central edge $\{c_0, c_1\}$, and $\{c_0, s_1\}$.

In general, to extend the standard chromatic subdivision from the $(n-1)$-skeleton to the n-skeleton of K, for each n-simplex $\sigma = \{s_0, \ldots, s_n\}$, we take $n+1$ interior points displaced from the barycenter:

$$c_0 = \frac{1-\epsilon}{n+1}s_0 + \sum_{j\neq 0} \frac{1+\epsilon/n}{n+1}s_j,$$
$$c_1 = \frac{1-\epsilon}{n+1}s_1 + \sum_{j\neq 1} \frac{1+\epsilon/n}{n+1}s_j,$$
$$\cdots$$
$$c_n = \frac{1-\epsilon}{n+1}s_n + \sum_{j\neq n} \frac{1+\epsilon/n}{n+1}s_j$$

for some $0 < \epsilon < 1$. Define the *central* simplex κ to be $\{c_0, \ldots, c_n\}$, and define $\chi(c_i) = \chi(s_i)$. The complex Ch σ consists of simplices of the form $\alpha \cup \beta$, where α is a face of the central simplex, and β is a simplex of Ch τ, where τ is a proper face of σ whose colors are disjoint from α's: $\chi(\alpha) \cap \chi(\tau) = \emptyset$. Note that Ch $\mathcal{K}$ is a chromatic complex by construction.

Like the barycentric subdivision, the standard chromatic subdivision also has a purely combinatorial definition.

Definition 3.6.3. Let $(\mathcal{A}, \chi)$ be a chromatic abstract simplicial complex. Its *standard chromatic subdivision* Ch $\mathcal{A}$ is the abstract simplicial complex of which the vertices have the form (i, σ_i), where $i \in [n]$, σ_i is a nonempty face of σ, and $i \in \chi(\sigma_i)$. A $(k + 1)$-tuple $(\sigma_0, \ldots, \sigma_k)$ is a simplex of Ch $\mathcal{A}$ if and only if

- The tuple can be indexed so that $\sigma_0 \subseteq \cdots \subseteq \sigma_k$, and
- For $0 \leq i, j \leq n$, if $i \in \chi(\sigma_j)$, then $\sigma_i \subseteq \sigma_j$.

Finally, to make the subdivision chromatic, we define the coloring $\chi : \text{Ch } \mathcal{K}$ to be $\chi(i, \sigma) = i$.

We can now extend the notion of subdivision to abstract simplicial complexes.

Definition 3.6.4. Let $\mathcal{A}$ and $\mathcal{B}$ be abstract simplicial complexes. We say that $\mathcal{B}$ *subdivides* the complex $\mathcal{A}$ if there exists a homeomorphism $h : |\mathcal{A}| \to |\mathcal{B}|$ and a carrier map $\Phi : \mathcal{A} \to 2^{\mathcal{B}}$ such that, for every simplex $\sigma \in \mathcal{A}$, the restriction $h|_{|\sigma|}$ is a homeomorphism between $|\sigma|$ and $|\Phi(\sigma)|$.

The carrier map Φ defining a subdivision must be strict and rigid. Recall that for a strict carrier map, for each simplex τ of $\mathcal{B}$ the unique simplex σ in $\mathcal{A}$ of smallest dimension, such that $\Phi(\sigma)$ contains τ, is called the *carrier* of τ. Thus, we often express subdivisions using operator notation, such as Div $\mathcal{A}$, where Div is the carrier map. For a simplex τ in Div $\mathcal{A}$, the carrier of τ, denoted $\text{Car}(\tau, \mathcal{A})$, is the minimal simplex σ of $\mathcal{A}$ such that $\tau \in \text{Div}(\sigma)$. When $\mathcal{A}$ is clear from context, we write $\text{Car}(\tau)$. Figure 3.9 shows a simplex σ in a subdivision, along with its carrier.

3.6.4 Subdivision operators

The barycentric and standard chromatic subdivisions have a useful property not shared by the stellar subdivision. They can be constructed inductively over skeletons of a simplicial complex using a standard subdivision at each step. We now restate this property more precisely.

Definition 3.6.5. A *boundary-consistent subdivision of simplices* is a sequence of geometric complexes $(\mathcal{S}_i)_{i \geq 1}$ such that

(1) For all $n \geq 1$, the complex $\mathcal{S}_n$ is a geometric subdivision of the standard n-simplex.
(2) Let Δ^n be the standard n-simplex, σ a k-simplex in the boundary complex $\partial \Delta^n$, and $\varphi : \Delta^k \to \sigma$ the characteristic map of σ. Then the induced subdivision $\varphi(\mathcal{S}_k)$ coincides with the restriction of $\mathcal{S}_n$ to σ.

When we have such a string of simplex subdivisions, we can use it to subdivide arbitrary geometric simplices.

Definition 3.6.6. Let $\mathcal{K}$ be a geometric simplicial complex with an ordered set of vertices[3] and let $(\mathcal{S}_i)_{i\geq 1}$ be a boundary-consistent subdivision of simplices. We obtain a subdivision of $\mathcal{K}$, which we call $\mathcal{S}(\mathcal{K})$, by replacing each k-simplex σ of $\mathcal{K}$ with the induced subdivision $\varphi(\mathcal{S}_k)$, where $\varphi : \Delta^k \to \mathcal{K}$ is the characteristic map of σ.

We call $\mathcal{S}(\,\cdot\,)$ the *subdivision operator* associated to the sequence $(\mathcal{S}_i)_{i\geq 1}$.

Let $\mathcal{A}$ be an abstract simplicial complex. Given a boundary-consistent subdivision of simplices $(\mathcal{S}_i)_{i\geq 1}$, we can take a geometric realization $\mathcal{K}$ of $\mathcal{A}$ and then consider the geometric simplicial complex $\mathcal{S}(\mathcal{K})$. Clearly, the underlying abstract simplicial complex of $\mathcal{S}(\mathcal{K})$ does not depend on the choice of the geometric realization of $\mathcal{A}$. We call that abstract simplicial complex $\mathcal{S}(\mathcal{A})$.

3.6.5 Mesh-shrinking subdivision operators

Recall that a geometric n-simplex σ is the convex hull of $n+1$ affinely independent points in a Euclidean space. Its *diameter* diam σ is the length of its longest edge.

Definition 3.6.7. Let $\mathcal{K}$ be a geometric simplicial complex. The *mesh* of $\mathcal{K}$, denoted mesh $\mathcal{K}$, is the maximum diameter of any of its simplices, or, equivalently, the length of its longest edge.

Assume that we are given a boundary-consistent subdivision of simplices $(\mathcal{S}_i)_{i\geq 1}$. Interpreting the subdivision $\mathcal{S}_i$ itself as a geometric simplicial complex, we can iterate the associated subdivision operator, resulting in a subdivision $\mathcal{S}_i^N$ of Δ^i, for every i, $N \geq 1$. We set $c_{i,N} := \text{mesh } \mathcal{S}_i^N$.

Definition 3.6.8. We say that the subdivision operator Div corresponding to a boundary-consistent subdivision of simplices $(\mathcal{S}_i)_{i\geq 1}$ is *mesh-shrinking* if $\lim_{N\to\infty} c_{i,N} = 0$ for all $i \geq 1$.

---

[3] It is enough to have a consistent order on the set of vertices of each simplex, meaning that the restriction of the chosen order of vertices a simplex σ to a boundary simplex τ gives the chosen order on that simplex.

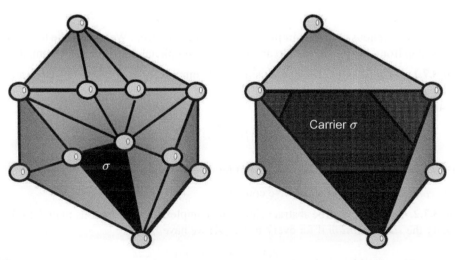

FIGURE 3.9

Simplex in subdivided complex with carrier.

Proposition 3.6.9.   *Assume $\mathcal{K}$ is a finite geometric simplicial complex of dimension n, and* Div *is a mesh-shrinking subdivision operator given by* $(\mathcal{S}_i)_{i \geq 1}$. *Then we have*

$$\lim_{N \to \infty} \text{mesh Div}^N \mathcal{K} = 0. \tag{3.6.1}$$

Proof.   Since $\mathcal{K}$ is finite, it is enough to consider the case when $\mathcal{K}$ is a geometric n-simplex σ. In this case, let $\varphi : \Delta^n \to \sigma$ be the characteristic linear isomorphism. Since φ is a linear map, there is a bound on the factor by which it can increase distances. In other words, there exists a constant c, such that

$$d(\varphi(x), \varphi(y)) \leq c \cdot d(x, y), \quad \text{for all } x, y \in \Delta^n, \tag{3.6.2}$$

where $d(\cdot, \cdot)$ is distance. Since Div is mesh-shrinking, we have $\lim_{N \to \infty} \text{mesh } S_n^N = 0$, which, together with (3.6.2), implies that $\lim_{N \to \infty} \text{mesh Div}^N K = 0$. $\square$

3.7 Simplicial and continuous approximations

In Section 3.2 we saw how to go back and forth between simplicial maps of complexes and continuous maps of their geometric realizations. Assume $\mathcal{A}$ is an abstract simplicial complex. Recall that any point x in $|\mathcal{A}|$ has a unique expression in terms of barycentric coordinates:

$$x = \sum_{i \in I} t_i \cdot s_i,$$

where $I \subseteq [n]$ is an index set, $0 \leq t_i \leq 1$, $\sum_i t_i = 1$, and $\{s_i | i \in I\}$ is a simplex of $\mathcal{A}$. Any simplicial map $\varphi : \mathcal{A} \to \mathcal{B}$ can be turned into a piece-wise linear map $|\varphi| : |\mathcal{A}| \to |\mathcal{B}|$ by extending over barycentric coordinates:

$$|\varphi|(x) = \sum_i t_i \cdot \varphi(s_i).$$

Going from a continuous map to a simplicial map is more involved. We would like to "approximate" a continuous map from one polyhedron to another with a simplicial map on related complexes.

Definition 3.7.1.   Let $\mathcal{A}$ and $\mathcal{B}$ be abstract simplicial complexes, let $f : |\mathcal{A}| \to |\mathcal{B}|$ be a continuous map, and let $\varphi : \mathcal{A} \to \mathcal{B}$ be a simplicial map. The map φ is called a *simplicial approximation* to f if, for every simplex α in $\mathcal{A}$, we have

$$f(\text{Int}|\alpha|) \subseteq \bigcap_{a \in \alpha} \text{St}^\circ(\varphi(a)) = \text{St}^\circ(\varphi(\alpha)), \tag{3.7.1}$$

where St° denotes the open star construction, and $\text{Int}|\alpha|$ denotes the interior of $|\alpha|$ (see Section 3.3.)

The *star condition* is a useful alternative condition.

Definition 3.7.2.   Let $\mathcal{A}$ and $\mathcal{B}$ be abstract simplicial complexes. A continuous map $f : |\mathcal{A}| \to |\mathcal{B}|$ is said to satisfy the *star condition* if for every $v \in V(\mathcal{A})$ we have

$$f(\text{St}^\circ(v)) \subseteq \text{St}^\circ(w) \tag{3.7.2}$$

for some vertex $w \in V(\mathcal{B})$.

Proposition 3.7.3.   *Assume that $\mathcal{A}$ and $\mathcal{B}$ are abstract simplicial complexes. A continuous map $f : |\mathcal{A}| \to |\mathcal{B}|$ satisfies the star condition if and only if it has a simplicial approximation.*

Proof.   Assume first that f has a simplicial approximation $\varphi : \mathcal{A} \to \mathcal{B}$. Given a vertex $v \in V(\mathcal{A})$, we pick a simplex $\alpha \in \text{St}^\circ(v)$. Since φ is a simplicial approximation, we have $f(\text{Int}|\alpha|) \subseteq \text{St}^\circ(\varphi(\alpha)) \subseteq \text{St}^\circ(\varphi(v))$. Varying α, we can conclude that $f(\text{St}^\circ(v)) \subseteq \text{St}^\circ(\varphi(v))$, and hence the star condition is satisfied for $w = \varphi(v)$.

In the other direction, assume that f satisfies the star condition. For every $v \in V(\mathcal{A})$ we let $\varphi(v)$ to denote any vertex w making the inclusion (3.7.2) hold. Let now $\sigma \in \mathcal{A}$, with $\sigma = \{v_0, \dots, v_t\}$. We have $\text{Int}|\sigma| \subseteq \text{St}^\circ(v_i)$, hence $f(\text{Int}|\sigma|) \subseteq f(\text{St}^\circ(v_i)) \subseteq \text{St}^\circ(\varphi(v_i))$, for all $i = 0, \dots, k$. This implies that $f(\text{Int}|\sigma|) \subseteq \cap_{i=1}^{k} \text{St}^\circ(\varphi(v_i))$. By definition of the open star, the latter intersection is nonempty if and only if there exists a simplex containing the vertices $\varphi(v_i)$ for all i, which is the same as to say that $\{\varphi(v_1), \dots, \varphi(v_t)\}$ is a simplex of $\mathcal{B}$. This means that $\varphi : V(\mathcal{A}) \to V(\mathcal{B})$ can be extended to a simplicial map $\varphi : \mathcal{A} \to \mathcal{B}$, and we have just verified that (3.7.1) is satisfied.   □

The following fact will be useful later on.

Proposition 3.7.4.   *Assume $\mathcal{A}$ and $\mathcal{B}$ are abstract simplicial complexes, $f : |\mathcal{A}| \to |\mathcal{B}|$ is a continuous map, and $\varphi : \mathcal{A} \to \mathcal{B}$ is a simplicial approximation of f. For an arbitrary simplex $\alpha \in \mathcal{A}$, let $\mathcal{C}_\alpha$ denote the minimal simplicial subcomplex of $\mathcal{B}$ for which the geometric realization contains $f(|\alpha|)$. Then $\varphi(\alpha)$ is a simplex of $\mathcal{C}_\alpha$.*

Proof.   By definition, if φ is a simplicial approximation of f, and $x \in \text{Int}|\alpha|$, then $f(x) \in \text{St}^\circ(\varphi(\alpha))$, meaning that $f(x)$ is contained in $\text{Int}|\sigma_x|$, where σ_x is a simplex of $\mathcal{B}$ such that $\varphi(\alpha) \subseteq \sigma_x$, and we

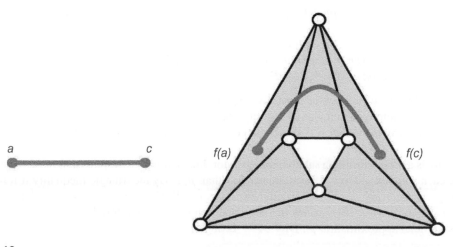

FIGURE 3.10

The continuous map f carries the edge $\{a, b\}$ into an annulus. This map has no simplicial approximation because it is impossible to find two vertices $\varphi(a)$ and $\varphi(b)$ such that the intersections of their stars contain the image of the edge.

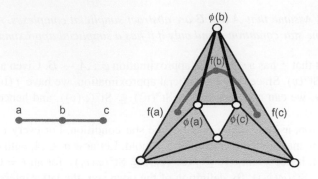

FIGURE 3.11

The continuous map f carries the edge $\{a, b\}$ into an annulus, along with a simplicial approximation φ of f.

choose a minimal such σ_x. Since $f(x) \in |\mathcal{C}_\alpha|$, we must have $\sigma_x \in \mathcal{C}_\alpha$, for all $x \in \mathrm{Int}|\alpha|$, hence $|\mathcal{C}_\alpha| \supseteq \cup_{x \in \mathrm{Int}|\alpha|} |\sigma_x|$, we conclude that $|\mathcal{C}_\alpha| \supseteq |\varphi(\alpha)|$. $\qquad\square$

Not every continuous map $f : |\mathcal{A}| \to |\mathcal{B}|$ has a simplicial approximation. In Figure 3.10, a continuous map f carries an edge $\eta = \{a, b\}$ into an annulus $|\mathcal{A}|$. It is easy to check that there is no simplicial map $\varphi : \eta \to \mathcal{A}$ such that $f(|\eta|) \subseteq \mathrm{St}^\circ\varphi(a) \cap \mathrm{St}^\circ\varphi(b)$. The images $f(a)$ and $f(b)$ are too far apart for a simplicial approximation to exist.

Nevertheless, we can always find a simplicial approximation defined over a sufficiently refined subdivision of $\mathcal{A}$. In Figure 3.11, f carries a *subdivision* of the edge $\eta = \{a, b\}$ into an annulus $|\mathcal{A}|$. It is easy to check that the simplicial map φ shown in the figure is a simplicial approximation to f.

Theorem 3.7.5 (Finite simplicial approximation of continuou maps using mesh-shrinking subdivisions[4].)

Let $\mathcal{A}$ and $\mathcal{B}$ be simplicial complexes. Assume that $\mathcal{A}$ is finite and that Div is a mesh-shrinking subdivision operator. Given a continuous map $f : |\mathcal{A}| \to |\mathcal{B}|$, there is an $N > 0$ such that f has a simplicial approximation $\varphi : \mathrm{Div}^N \mathcal{A} \to \mathcal{B}$.

Proof. Note that $(\mathrm{St}^\circ v)_{v \in V(\mathcal{B})}$ is an open covering of $|\mathcal{B}|$, hence $(f^{-1}(\mathrm{St}^\circ v))_{v \in V(\mathcal{B})}$ is an open covering of $|\mathcal{A}|$. Since the simplicial complex $\mathcal{A}$ is finite, the topological space $|\mathcal{A}|$ is a compact metric space, hence it has a *Lebesgue number* $\rho > 0$ such that every closed set X of diameter less than ρ lies entirely in one of the sets $f^{-1}(\mathrm{St}^\circ v)$.

Since Div is a mesh-shrinking subdivision operator, Inequality 3.6.1 implies that we can pick $N > 0$ such that each simplex in $\mathrm{Div}^N \mathcal{A}$ has diameter less than $\rho/2$. By the triangle inequality it follows that $\mathrm{diam}|\mathrm{St}w| < \rho$ for every $w \in V(\mathcal{A})$. Then there exists $v \in V(\mathcal{B})$ such that $\mathrm{St}^\circ w \subseteq f^{-1}(\mathrm{St}^\circ v)$. Hence the map $f : |\mathrm{Div}^N \mathcal{A}| \to |\mathcal{B}|$ satisfies the star condition (3.7.2); therefore by Proposition 3.7.3 there exists a simplicial approximation $\varphi : \mathrm{Div}^N \mathcal{A} \to \mathcal{B}$ of f. $\qquad\square$

We now proceed with approximations of carrier maps.

Definition 3.7.6. Let $\mathcal{A}$ and $\mathcal{B}$ be simplicial complexes, and $\Phi : \mathcal{A} \to 2^{\mathcal{B}}$ be a carrier map.

---

[4]In this book we restrict our attention to what is called the *finite simplicial approximation*, so we will drop the word *finite*.

(1) We say that a continuous map $f : |\mathcal{A}| \to |\mathcal{B}|$ is a *continuous approximation* of Φ if, for every simplex $\alpha \in \mathcal{A}$, we have $f(|\alpha|) \subseteq |\Phi(\alpha)|$.

(2) We say that Φ *has a simplicial approximation* if there exists a subdivision of $\mathcal{A}$, called Div $\mathcal{A}$, and a simplicial map $\varphi : \text{Div } \mathcal{A} \to \mathcal{B}$ such that $\varphi(\text{Div } \alpha)$ is a subcomplex of $\Phi(\alpha)$ for all $\alpha \in \mathcal{A}$.

Under certain connectivity conditions, both types of approximations must exist, as the next theorem explains.

Theorem 3.7.7 (Continuous and simplicial approximations of carrier maps.)

Assume $\mathcal{A}$ and $\mathcal{B}$ are simplicial complexes such that $\mathcal{A}$ is finite. Assume furthermore that $\Phi : \mathcal{A} \to 2^{\mathcal{B}}$ is a carrier map such that for every simplex $\alpha \in \mathcal{A}$, the subcomplex $\Phi(\alpha)$ is ($\dim(\alpha) - 1$)-connected. Then we can make the following conclusions:

(1) The carrier map Φ has a continuous approximation.

(2) The carrier map Φ has a simplicial approximation.

Proof. We start by proving (1). For $0 \leq d \leq n$, we inductively construct a sequence of continuous maps $f_d : |\text{skel}^d \mathcal{A}| \to |\mathcal{B}|$ on the skeletons of $\mathcal{A}$.

For the base case, let f_0 send any vertex a of $\mathcal{A}$ to any vertex of $\Phi(a)$. This construction is well defined because $\Phi(a)$ is (-1)-connected (nonempty) by hypothesis.

For the induction hypothesis, assume we have constructed

$$f_{d-1} : |\text{skel}^{d-1}(\mathcal{A})| \to |\Phi(\mathcal{A})|.$$

This map sends the boundary of each d-simplex α^d in $\text{skel}^d \mathcal{A}$ to $\Phi(\alpha^d)$. By hypothesis, $\Phi(\alpha^d)$ is ($d-1$)-connected, so this map of the ($d-1$)-sphere $\partial \alpha^d$ can be extended to a continuous map of the d-disk $|\alpha^d|$:

$$f_d : |\alpha^d| \to |\Phi(\alpha^d)|.$$

These extensions agree on the ($d-1$)-skeleton, so together they define a continuous map,

$$f_d : |\text{skel}^d \mathcal{A}| \to |\mathcal{B}|,$$

where for each $\alpha^d \in \text{skel}^d \mathcal{A}$, $f_d(|\alpha^d|) \subseteq |\Phi(\alpha^d)|$.

When $n = \dim \mathcal{A}$, the map f_n is a continuous approximation to Φ.

We now proceed with proving (2). As we just proved, the carrier map Φ has a continuous approximation $f : |\mathcal{A}| \to |\mathcal{B}|$. Let Div be an arbitrary mesh-shrinking subdivision (for example, the barycentric subdivision will do). By Theorem 3.7.5, there exists $N \geq 0$, and a simplicial map $\varphi : \text{Div}^N \mathcal{A} \to \mathcal{B}$ such that φ is a simplicial approximation of f.

To show that φ is also a simplicial approximation for Φ, we need to check that $\varphi(\text{Div}^N \alpha)$ is a subcomplex of $\Phi(\alpha)$ for all simplices $\alpha \in \mathcal{A}$. Pick a simplex $\tau \in \text{Div}^N \alpha$. Since $\varphi : \text{Div}^N \mathcal{A} \to \mathcal{B}$ is a simplicial approximation of f, we know by Proposition 3.7.4 that $\varphi(\tau)$ is a simplex of C_τ, where C_τ is the minimal simplicial subcomplex of $\mathcal{B}$ containing $f(|\tau|)$. In particular, since $f(|\tau|) \subseteq f(|\alpha|) \subseteq |\Phi(\alpha)|$, we see that C_τ is a subcomplex of $\Phi(\alpha)$, hence $\varphi(\tau)$ is a subcomplex of $\Phi(\alpha)$. Since this is true for all $\tau \in \text{Div}^N \alpha$, we conclude that $\varphi(\text{Div}^N \alpha)$ is a subcomplex of $\Phi(\alpha)$ for all $\alpha \in \mathcal{A}$. $\square$

Lemma 3.7.8.   *If $\Phi : \mathcal{A} \to 2^{\mathcal{B}}$ is a carrier map, and $f : |\mathcal{A}| \to |\mathcal{B}|$ is a continuous map carried by Φ, then any simplicial approximation $\phi : \mathrm{Bary}^N \mathcal{A} \to \mathcal{B}$ of f is also carried by Φ.*

Proof.   Let $\mathcal{A} \subset \mathcal{B}$ be complexes. If v is a vertex in $\mathcal{B}$ but not in $\mathcal{A}$, then the open star of v in $|\mathcal{B}|$ does not intersect $|\mathcal{A}|$.

Suppose, by way of contradiction, that σ is a simplex of $\mathcal{A}$, v is a vertex of σ, and $f(v) \in |\Phi(\sigma)|$ but $\phi(v) \notin \Phi(\sigma)$. Because ϕ is a simplicial approximation of f, $f(v) \in \mathrm{St}^{\circ}(\phi(v), \mathcal{B})$, implying that $f(v)$ is not in $\Phi(\sigma)$, contradicting the hypothesis that f is carried by Φ. $\square$

3.8 Chapter notes

A broad, introductory overview to topology is provided by Armstrong [7]. A combinatorial development similar to what we use appears in Henle [77]. A more advanced and modern overview of combinatorial topology can be found in Kozlov [100]. For a standard introduction to algebraic topology, including further information on simplicial approximations, see Munkres [124].

3.9 Exercises

Exercise 3.1.   Let σ be a simplex in a complex $\mathcal{C}$. The *deletion* of $\sigma \in \mathcal{C}$, written $\mathrm{dl}(\sigma, \mathcal{C})$, is the subcomplex of $\mathcal{C}$ consisting of all simplices of $\mathcal{C}$ that do not have common vertices with σ. Prove that

$$\mathrm{Lk}(\sigma, \mathcal{C}) = \mathrm{dl}(\sigma, \mathcal{C}) \cap \mathrm{St}(\sigma, \mathcal{C})$$
$$\mathcal{C} = \mathrm{dl}(\sigma, \mathcal{C}) \cap \mathrm{St}(\sigma, \mathcal{C})$$

Exercise 3.2.

(a) Show that a join of two simplices is again a simplex.

(b) Show that a join of $n+1$ copies of the 0-dimensional sphere is a simplicial complex homeomorphic to an n-dimensional sphere.

(c) Show that a join of an m-dimensional sphere with an n-dimensional sphere is homeomorphic to an $(m + n + 1)$-dimensional sphere for all $m, n \geq 0$.

Exercise 3.3.   Give an example of a rigid carrier map that is not strict.

Exercise 3.4.   Let $\mathcal{A}$ and $\mathcal{B}$ be simplicial complexes and $\Phi : \mathcal{A} \to 2^{\mathcal{B}}$ a rigid carrier map. Assume that $\mathcal{A}$ is pure of dimension d, and Φ is surjective, meaning that every simplex of $\mathcal{B}$ belongs to $\Phi(\sigma)$ for some $\sigma \in \mathcal{A}$. Prove that $\mathcal{B}$ is pure of dimension d.

Exercise 3.5.   Let $\mathcal{A}$ and $\mathcal{B}$ be simplicial complexes and $\Phi : \mathcal{A} \to 2^{\mathcal{B}}$ a surjective carrier map. Assume that $\mathcal{A}$ is connected and $\Phi(\sigma)$ is connected for all $\sigma \in \mathcal{A}$. Prove that $\mathcal{B}$ is also connected.

Exercise 3.6.   Prove that composing a rigid simplicial map with a rigid carrier map on the left as well as on the right will again produce a rigid carrier map.

Exercise 3.7.   Let $\mathcal{A}$ and $\mathcal{B}$ be simplicial complexes and $\Phi : \mathcal{A} \to 2^{\mathcal{B}}$ a surjective carrier map. Prove that if Φ is strict, then for each simplex τ of $\mathcal{B}$ there is a unique simplex σ in $\mathcal{A}$ of smallest dimension such that $\Phi(\sigma)$ contains τ. Thus, if $\mathcal{B}$ is a subdivision of $\mathcal{A}$ with carrier map Φ, the *carrier* of a simplex in $\mathcal{B}$ is well defined.

Exercise 3.8. Consider a task $(\mathcal{I}, \mathcal{O}, \Delta)$. The induced carrier map Δ' is defined as follows: If τ is a simplex of $\mathcal{P}$, let $\sigma \in \mathcal{I}$ be the carrier of τ; then $\Delta'(\tau) = \Delta(\sigma)$. Prove that Δ' is a chromatic carrier map. We say that the diagram commutes (and hence $\mathcal{P}$ via δ solves the task) if the carrier map defined by the composition of Ξ and δ is carried by Δ, or equivalently, if δ is carried the carrier map Δ' induced by Δ. Prove that these two conditions are indeed equivalent.

Exercise 3.9. Prove that the geometric and combinatorial definitions of the barycentric subdivision given in Section 3.6 are indeed equivalent.

Exercise 3.10. Prove that the barycentric subdivision is mesh-shrinking.

This page is intentionally left blank

Colorless Tasks

This page intentionally left blank

Colorless Wait-Free Computation

4

CHAPTER OUTLINE HEAD

We saw in Chapter 2 that we can construct a combinatorial theory of two-process distributed systems using only graph theory. In this chapter, we turn our attention to distributed systems that encompass more than two processes. Here we will need to call on *combinatorial topology*, a higher-dimensional version of graph theory.

Just as in Chapter 2, we outline the basic connection between distributed computing and combinatorial topology in terms of two formal models: a conventional *operational model*, in which systems consist of communicating state machines whose behaviors unfold over time, and the *combinatorial model*, in which all possible behaviors are captured statically using topological notions.

As noted, distributed computing encompasses a broad range of system models and problems to solve. In this chapter, we start with one particular system model (shared memory) and focus on a restricted (but important) class of problems (so-called "colorless" tasks). In later chapters, we will introduce other

Distributed Computing Through Combinatorial Topology. http://dx.doi.org/10.1016/B978-0-12-404578-1.00004-8

models of computation and broader classes of problems, but the concepts and techniques introduced in this chapter will serve as the foundations for our later discussions.

4.1 Operational model

Keep in mind that the operational model, like any such model, is an abstraction. As with the classical study of Turing machines, our aim (for now) is not to try to represent faithfully the way a multicore architecture or a cloud computing service is constructed. Instead, we start with a clean, basic abstraction and later show how it includes specific models of interest.

4.1.1 Overview

A distributed system is a set of communicating state machines called *processes*. It is convenient to model a process as a sequential automaton with a possibly infinite set of states. Remarkably, the set of computable tasks in a given system does not change if the individual processes are modeled as Turing machines or as even more powerful automata with infinite numbers of states, capable of solving "undecidable" problems that Turing machines cannot. The important questions of distributed computing are concerned with communication and dissemination of knowledge and are largely independent of the computational power of individual processes.

For the time being, we will consider a model of computation in which processes communicate by reading and writing a shared memory. In modern shared-memory multiprocessors, often called *multicores*, memory is a sequence of individually addressable *words*. Multicores provide instructions that read or write individual memory words[1] in a single atomic step.

For our purposes, we will use an idealized version of this model, recasting conventional read and write instructions into equivalent forms that have a cleaner combinatorial structure. Superficially, this idealized model may not look like your laptop, but in terms of task solvability, these models are equivalent: Any algorithm in the idealized model can be translated to an algorithm for the more realistic model, and vice versa.

Instead of reading an individual memory word, we assume the ability to read an arbitrarily long sequence of contiguous words in a single atomic step, an operation we call a *snapshot*. We combine writes and snapshots as follows. An *immediate snapshot* takes place in two contiguous steps. In the first step, a process writes its view to a word in memory, possibly concurrently with other processes. In the very next step, it takes a snapshot of some or all of the memory, possibly concurrently with other processes. It is important to understand that in an immediate snapshot, the snapshot step takes place *immediately after* the write step.

Superficially, a model based on immediate snapshots may seem unrealistic. As noted, modern multicores do not provide snapshots directly. At best, they provide the ability to atomically read a small, constant number of contiguous memory words. Moreover, in modern multicores, concurrent read and

---

[1]For now we ignore synchronization instructions such as *test-and-set* and *compare-and-swap*, which are discussed in Chapter 5.

write instructions are typically interleaved in an arbitrary order.[2] Nevertheless, the idealized model includes immediate snapshots for two reasons. First, immediate snapshots simplify lower bounds. It is clear that any task that is impossible using immediate snapshots is also impossible using single-word reads and writes. Moreover, we will see that immediate snapshots yield simpler combinatorial structures than reading and writing individual words. Second, perhaps surprisingly, immediate snapshots do not affect task solvability. It is well known (see Section 4.4, "Chapter Notes") that one can construct a wait-free snapshot from single-word reads and writes, and we will see in Chapter 14 how to construct a wait-free immediate snapshot from snapshots and single-word write instructions. It follows that any task that can be solved using immediate snapshots can be solved using single-word reads and writes, although a direct translation may be impractical.

In Chapter 5, we extend our results for shared-memory models to message-passing models.

As many as n of the $n + 1$ processes may fail. For now, we consider only *crash failures*, that is, failures in which a faulty process simply halts and falls silent. Later, in Chapter 6, we consider *Byzantine* failures, where faulty processes may communicate arbitrary, even malicious, information.

Processes execute *asynchronously*. Each process runs at an arbitrary speed, which may vary over time, independently of the speeds of the other processes. In this model, failures are undetectable: A nonresponsive process may be slow, or it may have crashed, but there is no way for another process to tell. In later chapters, we will consider *synchronous* models, whereby processes take steps at the same time, and *semi-synchronous* models, whereby there are bounds on how far their executions can diverge. In those models, failures are detectable.

Recall from Chapter 1 that a *task* is a distributed problem in which each process starts with a private input value, the processes communicate with one another, and then each process halts with a private output value.

For the next few chapters, we restrict our attention to *colorless tasks*, whereby it does not matter which process is assigned which input or which process chooses which output, only which *sets* of input values were assigned and which *sets* of output values were chosen.

The consensus task studied in Chapter 2 is colorless: All processes agree on a single value that is some process's input, but it is irrelevant which process's input is chosen or how many processes had that input. The colorless tasks encompass many, but not all, of the central problems in distributed computing. Later, we will consider broader classes of tasks.

A *protocol* is a program that solves a task. For now, we are interested in protocols that are *wait-free*: Each process must complete its computation in a bounded number of steps, implying that it cannot wait for any other process. One might be tempted to consider algorithms whereby one process sends some information to another and waits for a response, but the wait-free requirement rules out this technique, along with other familiar techniques such as barriers and mutual exclusion. The austere severity of the wait-free model helps us uncover basic principles more clearly than less demanding models. Later, we will consider protocols that tolerate fewer failures or even irregular failure patterns.

We are primarily interested in lower bounds and computability: which tasks are computable in which models and in the communication complexity of computable tasks. For this reason we assume without loss of generality that processes employ "full-information" protocols, whereby they communicate

---

[2]Some newer multicore architectures support *transactional memory*, a model much closer to immediate snapshots.

to each other everything they "know." For clarity, however, in the specific protocols presented here, processes usually send only the information needed to solve the task at hand.

4.1.2 Processes

There are $n + 1$ *processes*, each with a unique *name* taken from a universe of names Π. We refer to the process with name $P \in \Pi$ as "process P."

In the simplest and most common case, the universe of names Π is just $[n] = \{0, 1, \ldots, n\}$. Often we refer to the process with name i as the ith process (even when $|\Pi|$ is larger than $n + 1$). Some situations, however, become interesting only when there are more possible names than processes.

The ith process is an automaton whose set of states Q_i includes a set of *initial states* Q_i^{in} and a set of *final states* Q_i^{fin}. We do not restrict Q_i to be finite because we allow processes to start with input values taken from a countable domain such as the integers, and we allow them to change state over potentially infinite executions.

Each process "knows" its name, but it does not know *a priori* the names of the participating processes. Instead, each process includes its own name in each communication, so processes learn the names of other participating processes dynamically as the computation unfolds.

Formally, each process state q has an immutable *name* component, with a value taken from Π, denoted $\text{name}(q)$. If the process goes from state q to state q' in an execution, then $\text{name}(q) = \text{name}(q')$.

Each process state q also includes a mutable *view* component, denoted $\text{view}(q)$, which typically changes from state to state over an execution. This component represents what the process "knows" about the current computation, including any local variables the process may use.

A state q is defined by its name and its view, so we may write q as the pair (P, v), where $\text{name}(q) = P$ and $\text{view}(q) = v$.

Remark 4.1.1. There are two equivalent ways of thinking about processes: There could be $n + 1$ processes with distinct names from Π, or there could be $|\Pi| > n$ potential processes but at most $n + 1$ of them participate in an execution.

4.1.3 Configurations and executions

We now turn our attention to computation, expressed in terms of structured state transitions.

A *configuration* C is a set of process states corresponding to the state of the system at a moment in time. Each process appears at most once in a configuration: If s_0, s_1 are distinct states in C_i, then $\text{name}(s_0) \neq \text{name}(s_1)$. An *initial configuration* C_0 is one where every process state is an initial state, and a *final configuration* is one where every process state is a final state. Name components are immutable: Each process retains its name from one configuration to the next. We use $\text{names}(C)$ for the set of names of processes whose states appear in C, and $\text{active}(C)$ for the subset whose states are not final.

Sometimes a configuration also includes an *environment*, usually just the state of a shared memory. In later chapters, the environment will encompass other kinds of of communication channels, such as messages in a network.

An *execution* defines the order in which processes communicate. Formally, an *execution* is an alternating (usually, but not necessarily, finite) sequence of configurations and sets of process names:

$$C_0, S_0, C_1, S_1, \ldots, S_r, C_{r+1},$$

satisfying the following conditions:

- C_0 is the initial configuration, and
- S_i is the set of names of processes whose states change between configuration C_i and its successor C_{i+1}.

We refer to the sequence $S_0, S_1, \ldots, S_r$ as the *schedule* that generates the execution. We may consider a prefix of an execution and say it is a *partial execution*. We refer to each triple C_i, S_i, C_{i+1} as a *concurrent step*. If $P \in S_i$ we say that P *takes a step*. In this chapter, P's step is an immediate snapshot, as discussed next, but in other chapters, we will consider other kinds of steps.

The processes whose states appear in a step are said to *participate* in that step, and similarly for executions. It is essential that only the processes that participate in a step change state. In this way, the model captures the restriction that processes change state only as a result of explicit communication occurring within the schedule.

Crashes are implicit. If an execution's last configuration is not final because it includes processes whose states are not final, then those processes are considered to have crashed. This definition captures an essential property of asynchronous systems: It is ambiguous whether an active process has failed (and will never take a step) or whether it is just slow (and will be scheduled in the execution's extension). As noted earlier, this ambiguity is a key aspect of asynchronous systems.

4.1.4 Colorless tasks

Having described at a high level *how* computation works, we now consider *what* we are computing. We are interested in computing the distributed analogs of sequential functions, called *tasks*. As noted, for now we restrict our attention to a subset of tasks called *colorless tasks*.

First, a colorless task specifies which combinations of input values can be assigned to processes. Each process is assigned a value from a domain of *input values* V^{in}. More precisely, an *input assignment* for a set of processes Π is a set of pairs $\{(P_j, v_j) | P_j \in \Pi, v_j \in V^{in}\}$, where each process $P_j \in \Pi$ appears exactly once, but the input values v_j need not be distinct.

For colorless tasks, it is unimportant which process is assigned which input value. Formally, an input assignment $A = \{(P_j, v_j) | P_j \in \Pi, v_j \in V^{in}\}$ defines a *colorless input assignment* $\sigma = \{v_j | (P_j, v_j) \in A\}$, constructed by discarding the process names from the assignment. An input assignment defines a unique colorless input assignment, but not vice versa. For example, the input assignments $\{(P, 0), (Q, 0), (R, 1)\}$ and $\{(P, 0), (Q, 1), (R, 1)\}$ both produce the colorless input assignment $\{0, 1\}$.

We do not require that every value in a colorless input assignment be assigned to a process; $\{(P, 0), (Q, 0), (R, 0)\}$ also corresponds to the colorless input assignment $\{0, 1\}$. This is consistent with the intuitive notion of a colorless task, where we allow a process to adopt as its own input value any of the other processes' observed input values. In the same way, a colorless task specifies which combinations of output values can be chosen by processes. Each process chooses a value from a domain of *output values* V^{out}. We define *(colorless) output assignments* by analogy with (colorless) input assignments.

Informally, a colorless task is given by a set of colorless input assignments $\mathcal{I}$, a set of colorless output assignments $\mathcal{O}$, and a relation Δ that specifies, for each input assignment, which output assignments can be chosen. Note that a colorless task specification is independent of the number of participating processes or their names.

Definition 4.1.2.   A *colorless task* is a triple $(\mathcal{I}, \mathcal{O}, \Delta)$, where

- $\mathcal{I}$ is a set of colorless input assignments,
- $\mathcal{O}$ is a set of colorless output assignment,
- $\Delta : \mathcal{I} \rightarrow 2^{\mathcal{O}}$ is a map carrying each colorless input assignment to a set of colorless output assignments.

Here is a simple but important example, which we will revisit soon. In the *binary consensus task*, each participating process is assigned a binary input value, either 0 or 1, and all participating processes must agree on one process's input value. An input assignment assigns a binary value to each participating process. There are three possible colorless input assignments, depending on which input values are assigned:

$$\mathcal{I} = \{\{0\}, \{1\}, \{0, 1\}\}.$$

Because the processes must agree, there are only two possible colorless output assignments:

$$\mathcal{O} = \{\{0\}, \{1\}\}$$

The carrier map Δ ensures that the processes agree on some process's input:

$$\Delta(I) = \begin{cases} \{\{0\}\} & \text{if } I = \{0\} \\ \{\{1\}\} & \text{if } I = \{1\} \\ \{\{0\}, \{1\}\} & \text{If } I = \{0, 1\}. \end{cases}$$

4.1.5 Protocols for colorless tasks

We consider protocols where computation is split into two parts: a task-independent *full-information protocol* and a task-dependent *decision*. In the task-independent part, each process repeatedly communicates its view to the others, receives their views in return, and updates its own state to reflect what it has learned. When enough communication layers have occurred, each process chooses an output value by applying a task-dependent decision map to its final view. Recall that the protocol is *colorless* in the sense that each process keeps track of the set of views it received, not which process sent which view.

Specifically, each process executes a *colorless layered immediate snapshot protocol*, the pseudo-code for which is shown in Figure 4.1. (For brevity, we will often say *colorless layered protocol* when there is no danger of ambiguity.) To reflect the layer-by-layer structure of protocols, we structure the memory as a *two-dimensional* array $\mathsf{mem}[\ell][i]$, where row ℓ is shared only by the processes participating in layer ℓ, and column i is written only by P_i. In this way, each layer uses a "clean" region of memory disjoint from the memory used by other layers. Initially, P_i's view is its input value [3]. During layer ℓ, P_i performs an immediate snapshot: It writes its current view to $\mathsf{mem}[\ell][i]$ and in the very next step takes a snapshot of that layer's row, $\mathsf{mem}[\ell][*]$. In our examples, we write this step as:

---

immediate
    mem $[\ell][i]$:= view
    snap:= snapshot (mem $[\ell][*]$)

---

[3] Our pseudo-code uses syntax view: Value to declare a variable view of type Value.

```
// There are N  layers
shared mem: array[0..N−1][0..n] of Value
protocol ColorlessLayered (input: Value): Value
   view: Value := input          // initial  view is input value
   for ℓ := 0 to N − 1 do
     immediate
        mem[ℓ][i] := view
        snap := snapshot(mem[ℓ][*])
     view := set of Values in snap
   return δ(view)                // apply decision map to final view
```

FIGURE 4.1

Colorless layered immediate snapshot protocol: Pseudo-code for P_i.

Discarding process names, P_i takes as its new view the *set* of views it observed in its most recent immediate snapshot. Finally, after completing all layers, P_i chooses a decision value by applying a deterministic decision map δ to its final view. An execution produced by a (colorless) layered immediate snapshot protocol where, in each layer, each process writes and then takes a snapshot is called a *(colorless) layered execution.*

In the task-independent part of the protocol, protocols for colorless tasks are allowed to use process names. For example, in the protocol pseudo-code of Figure 4.1, each process uses its own index to choose where to store its view. In the task-dependent part of the protocol, however, the decision map is not allowed to depend on process names. The decision map keeps track of only the *set of values* in each snapshot, but not *which process* wrote which value, nor even how many times each value was written. This condition might seem restrictive, but for colorless tasks, there is no loss of generality (see Exercise 4.9).

More precisely, a configuration defines a unique *colorless configuration* by discarding process names, taking only the configuration's set of views. Each configuration defines a unique colorless configuration, but not vice versa. The output values chosen by processes in any final configuration must be a function of that final *colorless* configuration.

Consider the single-layer colorless immediate snapshot executions in which processes P, Q, and R, with respective inputs p, q, and r, each perform an immediate snapshot. A partial set of the colorless configurations reachable in such executions appears in Figure 4.2. The initial colorless configuration is $C_0 = \{p, q, r\}$. Colorless configurations are shown as boxes, and process steps are shown as arrows. Arrows are labeled with the names of the participating processes, and black boxes indicate final colorless configurations. For example, if P and Q take simultaneous immediate snapshots, they both observe the view $\{p, q\}$, resulting in the colorless configuration $\{\{p, q\}, r\}$. If R now takes an immediate snapshot, it will observe the view $\{p, q, r\}$, resulting in the colorless configuration $\{\{p, q\}, \{p, q, r\}\}$ (see Figure 4.3).

Figure 4.3 shows three single-layer snapshot executions for processes P, Q, R, with respective inputs p, q, r. Time runs from top to bottom, and the bottom line of each table shows the result of each process's snapshot. In the first execution, the processes are scheduled in distinct steps: first P, then Q, then R. In the second, Q and R's steps are merged. This perturbation changes Q's view but not P and R's views. In the third execution, Q and R's step is merged with P's. This perturbation,

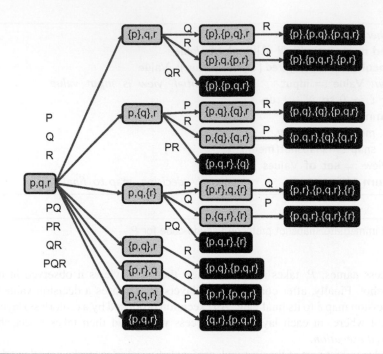

FIGURE 4.2

Colorless configurations for processes P, Q, R with respective inputs p, q, r, with final configurations shown in black.

too, changes only P's view; Q and R's views are the same. The observation that we can "perturb" colorless layered executions to change the view of only one process at a time will turn out to be important. Figure 4.4 shows an example of a snapshot execution that is not immediate, because P's snapshot is delayed until after the other processes have finished. Later we shall see that allowing nonimmediate snapshots does not affect the power of the model (see Exercise 4.14).

A final colorless configuration τ is *reachable* from a colorless initial configuration σ if there exists a colorless layered execution $C_0, S_0, C_1, S_1, \ldots, S_r, C_{r+1}$, where σ corresponds to C_0 and C_{r+1} corresponds to τ. For example, suppose $\sigma = \{p, q, r\}$ and $\tau = \{\{q\}, \{p, q\}, \{p, q, r\}\}$. Figure 4.2 shows that τ is reachable from σ through the sequential execution in which P, Q, R, respectively, start with inputs p, q, r and run in one-at-a-time order.

Given a set of colorless input assignments $\mathcal{I}$ for the pseudo-code of Figure 4.1, we represent its behavior as a *protocol-triple* (usually just *protocol*) $(\mathcal{I}, \mathcal{P}, \Xi)$, where $\mathcal{P}$ is the set of colorless final configurations reachable from the input configurations defined by $\mathcal{I}$. Thus, if σ is an input assignment in $\mathcal{I}$, we take any input configuration where processes start with input values taken from σ, and we add to $\mathcal{P}$ all the reachable colorless final configurations and denote them by $\Xi(\sigma)$. The map Ξ carries each colorless input assignment σ to the set of reachable colorless final configurations from σ.

Protocols and tasks are linked as follows: The processes choose their output values for a protocol $(\mathcal{I}, \mathcal{P}, \Xi)$ using a *decision map* δ that maps each process's final view to an output value in V^{out}. We say

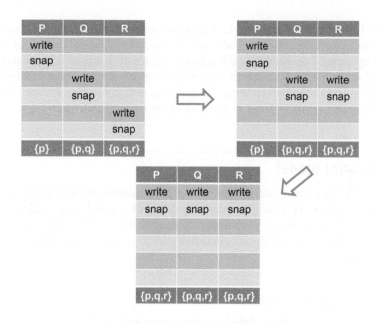

FIGURE 4.3

Three single-layer immediate snapshot executions for processes P, Q, R, with respective inputs p, q, r.

P	Q	R
write		
	write	
	snap	
		write
snap		snap
{p,q,r}	{p,q}	{p,q,r}

FIGURE 4.4

A snapshot execution that is not an immediate snapshot execution.

that a process *chooses* or *decides* the output value u with final view v if $\delta(v) = u$. The map δ extends naturally from final views to final configurations (which are sets of final views).

A colorless protocol $(\mathcal{I}, \mathcal{P}, \Xi)$ with decision map δ *solves* a colorless task $(\mathcal{I}, \mathcal{O}, \Delta)$ if, for every $\sigma \in \mathcal{I}$ and every colorless final configuration $\tau \in \mathcal{P}$ reachable from σ, that is, such that $\tau \in \Xi(\sigma)$, $\delta(\tau)$ is a colorless output assignment O in $\mathcal{O}$ allowed by the task's specification:

$$O \in \Delta(\tau).$$

Colorless initial configurations and colorless input assignments are often both just sets of input values (recall that sometimes a configuration may also specify the state of the environment), so we will abuse notation slightly by using $\mathcal{I}$ to stand for both a protocol's set of colorless initial configurations and a task's set of input assignments. By contrast, a protocol's set of colorless final configurations (usually written $\mathcal{P}$) and a task's set of colorless output assignments (usually written $\mathcal{O}$) are not the same. They are related by the decision map $\delta : \mathcal{P} \rightarrow \mathcal{O}$ and should not be confused.

4.2 Combinatorial model

The operational model may seem natural in the sense that it matches our experience that computations unfold in time. Nevertheless, a key insight underlying this book is that the essential nature of concurrent computing can be understood better by recasting the operational model in static, combinatorial terms, allowing us to transform questions about concurrent and distributed computing into questions about combinatorial topology.

4.2.1 Colorless tasks revisited

Consider a colorless task $(\mathcal{I}, \mathcal{O}, \Delta)$ as defined in the operational model of Section 4.1.4. Each colorless input or output assignment is just a set of values and as such can be viewed as a simplex. The set of all possible colorless input or output assignments forms a simplicial complex because, as discussed shortly, as sets they are closed under containment. We call $\mathcal{I}$ and $\mathcal{O}$ the (colorless) *input* and *output complexes*, respectively. We can reformulate the map Δ to carry each simplex of the input complex $\mathcal{I}$ to a subcomplex of $\mathcal{O}$, making Δ a carrier map, by Property 4.2.1.

Informally, in a colorless task, the processes start on the vertices of a single simplex σ in $\mathcal{I}$, and they halt on the vertices of a single simplex $\tau \in \Delta(\sigma)$. Multiple processes can start on the same input vertex and halt on the same output vertex.

We can now reformulate the operational task definition in combinatorial terms.

Definition 4.2.1.   A *colorless task* is a triple $(\mathcal{I}, \mathcal{O}, \Delta)$, where

- $\mathcal{I}$ is an *input complex*, where each simplex is a subset of V^{in},
- $\mathcal{O}$ is an *output complex*, where each simplex is a subset of V^{out},
- $\Delta : \mathcal{I} \rightarrow 2^{\mathcal{O}}$ is a carrier map.

Δ is a carrier map because it satisfies the monotonicity condition 4.2.1: $\sigma \subseteq \tau$ implies $\Delta(\sigma) \subseteq \Delta(\tau)$.

Reformulating task definitions in terms of simplicial complexes helps capture some important aspects of the model. Recall that a simplicial complex is closed under containment: If $\sigma \in \mathcal{I}$, and $\sigma' \subset \sigma$, then $\sigma' \in \mathcal{I}$. This property captures the notion that it must be possible for nonfaulty processes to choose output values even if some processes crash before taking any steps. (Such processes are said *not to participate*.) The extreme case is a *solo execution* where a single process chooses an output value without ever hearing from any of the others. It follows that the wait-free condition requires that if $\mathcal{I}$ contains a particular colorless configuration (set of values), then it must contain every subset of that configuration. Formally, $\mathcal{I}$ must be *closed under containment*. The motivation is clear: If the processes with inputs from σ participate in an execution, then the remaining processes with inputs in $\sigma \setminus \sigma'$ may fail before taking any steps, and the remaining processes will run as if the initial colorless configuration were σ'. By similar reasoning, $\mathcal{O}$ must also be closed under containment.

Just because process P finishes without hearing from process Q, it does not mean Q crashed, because Q may just be slow to start. The task specification Δ must ensure that any output value chosen by P remains compatible with decisions taken by late-starting processes. Formally, the carrier map Δ is *monotonic*: If $\sigma' \subseteq \sigma$ are colorless input assignments, then $\Delta(\sigma') \subseteq \Delta(\sigma)$. Operationally, the processes with inputs in σ', running by themselves, may choose output values $\tau' \in \Delta(\sigma')$. If the remaining processes with inputs in $\sigma \setminus \sigma'$ then start to run, it must be possible for them to choose an output assignment $\tau \in \Delta(\sigma)$ such that $\tau' \subseteq \tau$. Because $\sigma \cap \sigma'$ is a subset of both σ and σ', $\Delta(\sigma \cap \sigma') \subseteq \Delta(\sigma)$ and $\Delta(\sigma \cap \sigma') \subseteq \Delta(\sigma')$, and therefore

$$\Delta(\sigma \cap \sigma') \subseteq \Delta(\sigma) \cap \Delta(\sigma'). \tag{4.2.1}$$

Although the tasks that concern us here are all monotonic, it is not difficult to find tasks that are not. Here is a simple example. In the *uniqueness* task, the input complex $\mathcal{I}$ is arbitrary. Each process chooses as output the number of distinct input values assigned to processes: $\Delta(\sigma) = \{|\sigma|\}$, for $\sigma \in \mathcal{I}$. It is not hard to see why this task has no wait-free protocol. In a two-process execution, where P has input 0 and Q has input 1, then P must choose the incorrect value 1 in a solo execution where it completes the protocol before Q takes a step. Formally, Δ is not monotonic:

$$\Delta(\{0\}) \not\subseteq \Delta(\{0, 1\}),$$

because

$$\Delta(\{0\}) = \{1\},$$

while

$$\Delta(\{0, 1\}) = \{2\}.$$

4.2.2 Examples of colorless tasks

Here are examples of simple colorless tasks. When we revisit these tasks later, we will see that some have colorless layered protocols, whereas others do not.

Consensus

Perhaps the most important example of a task is *consensus*. As described informally in Chapter 2, each process starts with an input value. All processes must agree on a common output value, which must be some process's input value.

In the *binary consensus task*, each participating process is assigned a binary input value, either 0 or 1, and all participating processes must agree on one process's input value. An input assignment assigns a binary value to each participating process. There are three possible colorless initial assignments, depending on which input values are assigned:

$$\mathcal{I} = \{\{0\}, \{1\}, \{0, 1\}\}$$

Because the processes must agree, there are only two possible colorless output assignments:

$$\mathcal{O} = \{\{0\}, \{1\}\}.$$

The carrier map Δ requires the processes to agree on some process's input:

$$\Delta(I) = \begin{cases} \{\{0\}\} & \text{if } I = \{0\}, \\ \{\{1\}\} & \text{if } I = \{1\}, \\ \{\{0\}, \{1\}\} & \text{if } I = \{0, 1\}. \end{cases}$$

Formally, the input complex $\mathcal{I}$ is an edge with vertices labeled 0 and 1. The output complex $\mathcal{O}$ for binary consensus consists of two disjoint vertices, labeled 0 and 1. If all processes start with input 0, they must all decide 0, so the carrier map Δ carries input vertex 0 to output vertex 0. Similarly, Δ carries input vertex 1 to output vertex 1. If the processes have mixed inputs, then they can choose either output value, but they must agree, meaning they must choose the same output vertex.

It is easy to check that Δ is a carrier map. To see that it satisfies monotonicity, note that if $\sigma \subset \tau$, then the set of values in σ is contained in the set of values of τ.

If there can be $c > 2$ possible input values, we call this task simply *consensus* or *c-consensus*. The input complex consists of a $(c - 1)$-simplex and its faces, and the output complex is a set of c disjoint vertices. In each case, the input complex is connected, whereas the output complex is not, a fact that will be important later.

Set agreement

One way to relax the consensus task is the *k-set agreement* task. Like consensus, each process's output value must be some process's input value. Unlike consensus, which requires that all processes agree, k-set agreement imposes the more relaxed requirement that no more than k distinct output values be chosen. Consensus is a 1-set agreement.

The k-set agreement task has a trivial protocol if k is greater than or equal to the number of processes; a process outputs its input without any communication. We will prove later that this task is not solvable by a colorless layered protocol for any smaller values of k. We will also study under what circumstances set agreement has a solution in other models.

If there are c possible input values, then just as for consensus, the input complex consists of a single $(c - 1)$-simplex σ and its faces, whereas the output complex consists of the $(k - 1)$-skeleton of σ. In general, "k-set agreement" refers to a family of tasks. The input complex $\mathcal{I}$ can be arbitrary, and the output complex is $\text{skel}^{k-1}\mathcal{I}$, the $(k - 1)$-skeleton of the input complex. The task's carrier map carries each input simplex σ to $\text{skel}^{k-1}\sigma$. In Exercise 4.6, we ask you to show that the skeleton operator is indeed a carrier map. We write the k-set agreement task as $(\mathcal{I}, \text{skel}^{k-1}\mathcal{I}, \text{skel}^{k-1})$, where the first skeleton operator denotes a subcomplex and the second a carrier map.

Approximate agreement

Using colorless layered protocols, it is impossible for processes that start with different input values to reach consensus on a single value, but one might ask whether processes can agree on values that are *sufficiently close*. Each process is assigned input 0 or 1. If all processes start with the same value, they must all decide that value; otherwise, they must decide values that lie between 0 and 1, all within ϵ of each other, for a given $\epsilon > 0$. This task can be solved using a colorless layered protocol, but as ϵ gets smaller, more and more layers are needed.

Here is a discrete version of this task. As before, the input complex $\mathcal{I}$ is a single edge with vertices labeled 0 and 1. For the output complex, we subdivide the unit interval into $t + 1$ equal pieces, placing vertices uniformly at a distance of ϵ apart. If we assume for simplicity that $t = \frac{1}{\epsilon}$ is a natural number, then the $(t + 1)$ output vertices are labeled with $\frac{i}{t}$, where $0 \leq i \leq t$. Vertices $\frac{i}{t}$ and $\frac{j}{t}$ form a simplex if and only if $|i - j| \leq 1$.

If all processes start with input 0, they must all decide 0, so the carrier map Δ carries input vertex 0 to output vertex 0. Similarly, Δ carries input vertex 1 to output vertex 1. If the processes have mixed inputs, then they can choose any simplex (vertex or edge) of $\mathcal{O}$.

$$\Delta(\sigma) = \begin{cases} \{\{0\}\} & \text{if } \sigma = \{0\} \\ \{\{1\}\} & \text{if } \sigma = \{1\} \\ \mathcal{O} & \text{if } \sigma = \{0, 1\} . \end{cases}$$

Barycentric agreement

Along with consensus and k-set agreement, one of the most important tasks for analyzing distributed systems is the *barycentric agreement* task. Here processes start on the vertices of a simplex σ in an arbitrary input complex $\mathcal{I}$, and they decide on the vertices of a single simplex in the barycentric subdivision Bary σ.

Formally, the *barycentric agreement* task with input complex $\mathcal{I}$ is the task $(\mathcal{I}, \text{Bary } \mathcal{I}, \text{Bary})$, where the subdivision operator Bary is treated as a carrier map (see Exercise 4.6). We will see later (Theorem 4.2.8) that this task is solved by a single-layer colorless immediate snapshot protocol. This task can be generalized to the *iterated barycentric agreement* protocol $(\mathcal{I}, \text{Bary}^N \mathcal{I}, \text{Bary}^N)$ for any $N > 0$. This task has a straightforward colorless N-layer protocol. Despite the triviality of the solutions, the barycentric task will be essential for later chapters.

Robot convergence tasks

Consider a collection of robots placed on the vertices of a simplicial complex. If they are all placed on the same vertex, they stay there, but if they are placed on distinct vertices, they must all move to the vertices of a single simplex, chosen by task-specific rules. The robots communicate through a colorless layered protocol, and eventually each one chooses a final vertex and halts. Whether a particular convergence task has a solution depends on the rules governing where the robots are allowed to meet. Formally, a robot convergence task for a complex $\mathcal{K}$ is given by $(\mathcal{I}, \mathcal{K}, \Delta)$, where each vertex in $\mathcal{I}$ corresponds to a possible starting vertex in $\mathcal{K}$, each simplex in $\mathcal{I}$ to a set of possible simultaneously starting vertices, and Δ encodes the convergence rules.

The *loop agreement* task (explained in more detail in Chapter 5) is one example of a convergence task. This task is defined by a complex $\mathcal{O}$; three of its vertices, v_0, v_1, and v_2; and disjoint simple paths connecting each pair of vertices in $\mathcal{O}$. If the robots start on the same vertex v_i, they stay there. If some start on v_i and the rest on v_j, they converge to the vertices of an edge on the path connecting v_i and v_j. If some start on each of the three vertices, they converge to the vertices of some simplex of $\mathcal{O}$. The input complex $\mathcal{I}$ for loop agreement is just the 2-simplex $\{0, 1, 2\}$; the carrier map carries each vertex i to $\{v_i\}$, each edge $\{i, j\}$ to the path linking v_i and v_j, and the triangle $\{0, 1, 2\}$ to all of $\mathcal{O}$.

4.2.3 Protocols revisited

Like tasks, protocols can also be recast in terms of simplicial complexes.

Definition 4.2.2.   A (colorless) *protocol* is a triple $(\mathcal{I}, \mathcal{P}, \Xi)$ where

- $\mathcal{I}$ is a simplicial complex, called the *input complex*, where each simplex is a colorless input assignment,
- $\mathcal{P}$ is a simplicial complex, called the *protocol complex*, where each simplex is a colorless final configuration,
- $\Xi : \mathcal{I} \to 2^{\mathcal{P}}$ is a strict carrier map, called the *execution map*, such that $\mathcal{P} = \cup_{\sigma \in \mathcal{I}} \Xi(\sigma)$.

The carrier map Ξ is *strict*:

$$\Xi(\sigma \cap \sigma') = \Xi(\sigma) \cap \Xi(\sigma'). \tag{4.2.2}$$

Here is the intuition behind this equality: $\Xi(\sigma) \cap \Xi(\sigma')$ is the set of colorless final configurations in which no process can "tell" whether the execution started with inputs from σ or from σ'. In any such execution, only the processes with inputs from $\sigma \cap \sigma'$ can participate, because the others "know" which was the starting configuration. But these executions are exactly those with final configurations $\Xi(\sigma \cap \sigma')$, corresponding to executions in which only the processes with inputs from $\sigma \cap \sigma'$ participate.

As reformulated in the language of simplicial complexes, a protocol $(\mathcal{I}, \mathcal{P}, \Xi)$ *solves* a task $(\mathcal{I}, \mathcal{O}, \Delta)$ if there is a simplicial map

$$\delta : \mathcal{P} \to \mathcal{O}$$

such that $\delta \circ \Xi$ is carried by Δ. Here is why we require δ to be simplicial. Each simplex in the protocol complex $\mathcal{P}$ is a colorless final configuration, that is, the set of final states that can be reached in some execution. The tasks' colorless output assignments are the simplices of $\mathcal{O}$. If δ were to carry some final configuration to a set of vertices that did not form a simplex of $\mathcal{O}$, then that configuration is the final state of an execution where the processes choose an illegal output assignment.

4.2.4 Protocol composition

Two protocols for the same set of processes can be *composed* in a natural way. Informally, the processes participate in the first protocol, then they participate in the second, using their final views from the first as their inputs to the second. For example, a colorless layered protocol is just the composition of a sequence of colorless single-layer protocols.

Definition 4.2.3 (Composition of protocols).   Assume we have two protocols $(\mathcal{I}, \mathcal{P}, \Xi)$ and $(\mathcal{I}', \mathcal{P}', \Xi')$, where $\mathcal{P} \subseteq \mathcal{I}'$. Their *composition* is the protocol $(\mathcal{I}, \mathcal{P}'', \Xi'')$, where Ξ'' is the composition of Ξ and Ξ', $(\Xi' \circ \Xi)(\sigma) = \Xi'(\Xi(\sigma))$, for $\sigma \in \mathcal{I}$, and $\mathcal{P}'' = \Xi''(\mathcal{I})$.

The result of the composition is itself a protocol because, by Proposition 3.4.6, strict carrier maps compose.

It is sometimes convenient to speak of composing a protocol with a task, by which we mean composing that protocol with an arbitrary protocol that solves the task. Formally, we exploit the observation that any task whose carrier map is strict can itself be treated as a protocol.

Definition 4.2.4 (Composition of a protocol and a task). Given a protocol $(\mathcal{I}, \mathcal{P}, \Xi)$ and a task $(\mathcal{P}', \mathcal{O}, \Delta)$, where $\mathcal{P} \subseteq \mathcal{P}'$ and Δ is strict, their *composition* is the protocol $(\mathcal{I}, \mathcal{O}', \Delta \circ \Xi)$, where $(\Delta \circ \Xi)(\sigma) = \Delta(\Xi(\sigma))$, for $\sigma \in \mathcal{I}$, and $\mathcal{O}' = (\Delta \circ \Xi)(\mathcal{I})$.

Informally, the processes participate in the first protocol, using their output vertices as inputs to some protocol that solves the task.

Similarly, it is also convenient to speak of composing a task with a protocol.

Definition 4.2.5 (Composition of a task and a protocol). Given a task $(\mathcal{I}, \mathcal{O}, \Delta)$, where Δ is strict, and a protocol $(\mathcal{I}', \mathcal{P}, \Xi)$, where $\mathcal{O} \subseteq \mathcal{I}'$, their *composition* is the protocol $(\mathcal{I}, \mathcal{P}', \Xi \circ \Delta)$, where $(\Xi \circ \Delta)(\sigma) = \Xi(\Delta(\sigma))$, for $\sigma \in \mathcal{I}$, and $\mathcal{P}' = (\Xi \circ \Delta)(\mathcal{I})$.

Informally, the processes participate in some protocol that solves the task, then use their output vertices as inputs to the second protocol.

Redefining tasks and protocols in the language of combinatorial topology makes it easier to prove certain kinds of properties. For example, in analyzing colorless protocols, two kinds of protocols serve as useful building blocks: protocols for barycentric agreement and protocols for k-set agreement. We can reason about such protocols in a *model-independent* way, asking what the implications are if such protocols exist. Separately, for models of interest (like colorless layered protocols), we can use *model-specific* arguments to show that such protocols do or do not exist.

The following *Protocol Complex Lemma* illustrates a useful connection between the discrete and continuous structures of tasks and protocols. This lemma holds for any computational model where there are protocols that solve barycentric agreement.

Let $\Xi : \mathcal{I} \to 2^{\mathcal{P}}$ be a carrier map and $f : |\mathcal{P}| \to |\mathcal{O}|$ a continuous map. We use $(f \circ \Xi) : |\mathcal{I}| \to |\mathcal{O}|$ to denote the continuous map $(f \circ \Xi)(\sigma) = f(|\Xi(\sigma)|)$, for $\sigma \in \mathcal{I}$.

In one direction, the lemma provides a way to find a protocol for a colorless task $(\mathcal{I}, \mathcal{O}, \Delta)$. To show that a protocol-triple $(\mathcal{I}, \mathcal{P}, \Xi)$ solves the task, we must find a simplicial map δ from $\mathcal{P}$ to $\mathcal{O}$ carried by Δ. However, it is sometimes easier to find a continuous map $f : |\mathcal{P}| \to |\mathcal{O}|$ (carried by Δ) and then obtain δ through simplicial approximation, which is possible in any model where barycentric agreement can be solved.

In the other direction, the lemma says that any simplicial map δ from $\mathcal{P}$ to $\mathcal{O}$ carried by Δ approximates a continuous $|\delta|$ (recall Section 3.2.3) carried by Δ. Thus, intuitively, simplicial decision maps and continuous decision maps are interchangeable in such models.

Lemma 4.2.6 (Protocol complex lemma). *Assume that for any input complex $\mathcal{I}$ and any $N > 0$, there is a protocol that solves the barycentric agreement task $(\mathcal{I}, \mathrm{Bary}^N \mathcal{I}, \mathrm{Bary}^N)$. Then a task $(\mathcal{I}, \mathcal{O}, \Delta)$ has a protocol if and only if there exists a protocol $(\mathcal{I}, \mathcal{P}, \Xi)$ with a continuous map:*

$$f : |\mathcal{P}| \to |\mathcal{O}| \tag{4.2.3}$$

such that $(f \circ \Xi)$ is carried by Δ.

Proof. **Protocol implies map:** If $(\mathcal{I}, \mathcal{P}, \Xi)$ solves $(\mathcal{I}, \mathcal{O}, \Delta)$, then the protocol's simplicial decision map

$$\delta : \mathcal{P} \to \mathcal{O},$$

is carried by Δ. The simplicial map δ induces a continuous map

$$|\delta| : |\mathcal{P}| \to |\mathcal{O}|$$

also carried by Δ, as explained in Section 3.2.3.

Map implies protocol: If there is a continuous map

$$f : |\mathcal{P}| \to |\mathcal{O}|,$$

such that $(f \circ \Xi)$ is carried by Δ, then by Theorem 3.7.5, f has a simplicial approximation,

$$\phi : \mathrm{Bary}^N \mathcal{P} \to \mathcal{O},$$

for some $N > 0$. By hypothesis, there is a protocol that solves the barycentric agreement task $(\mathcal{P}, \mathrm{Bary}^N \mathcal{P}, \mathrm{Bary}^N)$. Consider the composition $(\mathcal{I}, \mathrm{Bary}^N \mathcal{P}, (\mathrm{Bary}^N \circ \Xi))$ (Definition 4.2.4). To show that this composite protocol solves $(\mathcal{I}, \mathcal{O}, \Delta)$, we must show that ϕ is a decision map for the task.

By hypothesis, $(f \circ \Xi)$ is carried by Δ:

$$(f \circ \Xi)(\sigma) \subseteq \Delta(\sigma).$$

By Lemma 3.7.8, so is its simplicial approximation:

$$(\phi \circ \mathrm{Bary}^N \circ \Xi)(\sigma) \subseteq \Delta(\sigma)$$
$$\phi(\mathrm{Bary}^N \circ \Xi)(\sigma)) \subseteq \Delta(\sigma).$$

It follows that ϕ is a decision map for the composite protocol. $\square$

It is sometimes convenient to reformulate the Protocol Complex Lemma in the following equivalent discrete form.

Lemma 4.2.7 (Discrete protocol complex lemma).   *Assume that for any input complex $\mathcal{I}$ and any $N > 0$, there is a protocol that solves the barycentric agreement task $(\mathcal{I}, \mathrm{Bary}^N \mathcal{I}, \mathrm{Bary}^N)$. Then a task $(\mathcal{I}, \mathcal{O}, \Delta)$ has a protocol if and only if there exists a protocol $(\mathcal{I}, \mathcal{P}, \Xi)$, a subdivision $\mathrm{Div}\,\mathcal{P}$ of $\mathcal{P}$, and a simplicial map*

$$\phi : \mathrm{Div}\,\mathcal{P} \to \mathcal{O} \tag{4.2.4}$$

carried by Δ.

Proof.   It is enough to show that Conditions 4.2.3 and 4.2.4 are equivalent.

A simplicial map $\varphi : \mathrm{Div}\,\mathcal{P} \to \mathcal{O}$ carried by Δ yields a continuous map $|\varphi| : |\mathrm{Div}\,\mathcal{P}| \to |\mathcal{O}|$ also carried by Δ. Since $|\mathrm{Div}\,\mathcal{P}|$ is homeomorphic to $|\mathcal{P}|$, we see that Condition 4.2.3 implies Condition 4.2.4.

On the other hand, assume we have a continuous map $f : |\mathcal{P}| \to |\mathcal{O}|$ carried by Δ. By Theorem 3.7.5, f has a simplicial approximation: $\varphi : \mathrm{Bary}^N \mathcal{P} \to \mathcal{O}$ also carried by Δ, for some $N > 0$, satisfying Condition 4.2.4. $\square$

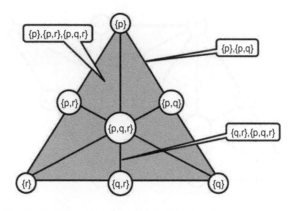

FIGURE 4.5

Single-layer colorless immediate snapshot protocol complex for three or more processes and input values p, q, r.

4.2.5 Single-layer colorless protocol complexes

Although one may list all possible executions of a colorless layered protocol, as in Figure 4.2 (see also Figure 4.4), it may be difficult to perceive an underlying structure. By contrast, an intriguing structure emerges if we display the same information as a simplicial complex. Figure 4.5 shows the protocol complex encompassing the complete set of final configurations for a single-layer protocol with at least three processes. The input complex $\mathcal{I}$ consists of simplex $\{p, q, r\}$ and its faces. To ease comparison, selected simplices are labeled with their corresponding final configurations. The "corner" vertex is labeled with p, the "edge" vertex is labeled with p, q, and the "central" vertex is labeled p, q, r. In this example, it is clear that the protocol complex for a single-input simplex is its *barycentric subdivision*.

In particular, consider the 2-simplex labeled at upper right. It corresponds to a final colorless configuration $\{\{p\} \{p, q\} \{p, q, r\}\}$, which occurs at the end of any "fully sequential" execution, where processes with input p concurrently take immediate snapshots, then processes with input q do the same, followed by processes with input r. In any such execution, there are exactly three final views, corresponding to the three vertices of the 2-simplex labeled at upper right.

Similarly, the simplex corresponding to the fully concurrent execution is the vertex in the center. The view $\{p, q, r\}$ is the single final view in any execution where at least one process has each of the three inputs, and they all take immediate snapshots.

Figure 4.5 shows the output complex for a single-layer protocol whose input complex consists of a single simplex and its faces. The carrier map Ξ defines the subcomplex $\Xi(\sigma)$ consisting of all final configurations in executions starting from that initial configuration σ. Here $\Xi(\sigma)$ consists of all simplices whose vertices are labeled with subsets of vertices of σ. For instance, $\Xi(\{p\})$ is equal to the vertex $\{p\}$. For the input simplex $\{p, q\}$, $\Xi(\{p, q\})$ consists of the edge $\{\{p\} \{p, q\}\}$ and the edge $\{\{q\} \{p, q\}\}$ (and their vertices). For the input simplex $\{p, q, r\}$, $\Xi(\{p, q, r\})$ consists of the entire subdivision $\mathcal{P}$.

What happens if we add one more possible input value, r, with the restriction that if some process has input r, no process can have input s, and vice versa? With this addition, the input complex consists of two 2-simplices (triangles) that share an edge. These triangles represent mutually exclusive initial

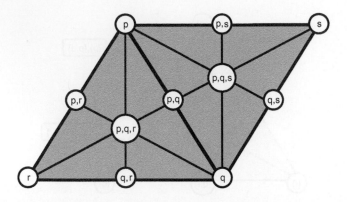

FIGURE 4.6

Single-layer protocol complex for two input simplices $\{p, q, r\}$ and $\{p, q, s\}$ and three or more processes.

configurations. Figure 4.6 shows the resulting single-layer colorless protocol complex, where each vertex is labeled with a view. As one would expect, the protocol complex is a barycentric subdivision of the input complex. The vertices along the boundary dividing the two triangles are views of executions where processes with inputs r or s did not participate, perhaps because they crashed, perhaps because there were none, or perhaps because they started after the others finished.

   We are ready to state the most important property of the colorless single-layer immediate snapshot protocol complex.

Theorem 4.2.8. *For any colorless single-layer $(n + 1)$ -process immediate snapshot protocol $(\mathcal{I}, \mathcal{P}, \Xi)$, the protocol complex $\mathcal{P}$ is the barycentric subdivision of the n-skeleton of $\mathcal{I}$, and the execution map Ξ is the composition of the barycentric subdivision and n-skeleton operators.*

Proof.   Each process P_i takes an immediate snapshot, writing its input to mem[0][i], and then taking a snapshot, retaining the set of non-null values that it observes. Every value written is an input value, which is a vertex of $\mathcal{I}$. All processes start with vertices from some simplex σ in $\mathcal{I}$, so the set of input values read in each snapshot forms a non-empty face of σ. Because snapshots are atomic, if P_i assembles face σ_i and P_j assembles face σ_j, then $\sigma_i \subseteq \sigma_j$ or vice versa. So the sets of views assembled by the processes form a chain of faces

$$\emptyset \subset \sigma_{i_0} \subseteq \cdots \subseteq \sigma_{i_n} \subseteq \sigma.$$

These chains can have lengths of at most $n+1$, and the complex consisting of such simplices is precisely the barycentric subdivision of the n-skeleton of σ (Section 3.6.2). Taking the complex over all possible inputs, we have $\mathcal{P}$ is Bary skel^n $\mathcal{I}$ and $\Xi(\cdot) =$ Bary skel$^n(\cdot)$.   □

4.2.6 Multilayer protocol complexes

In a colorless single-layer protocol, each process takes a pair of steps: It writes and takes a snapshot. There are several ways to generalize this model to allow processes to take additional steps. One approach is to allow processes to write and take a snapshot more than once in a layer. We will consider this extension

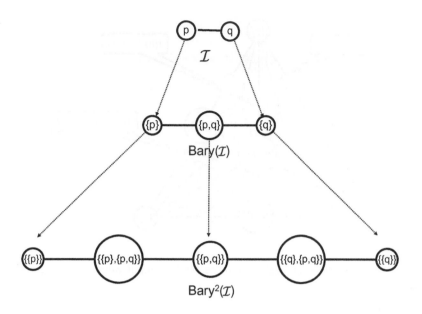

FIGURE 4.7

Input and protocol complex for two input values *p* and *q*: zero, one, and two layers.

in later chapters. For now, however, we construct *colorless layered* protocols by composing colorless single-layer protocols, as shown in Figure 4.1.

For example, Figure 4.7 shows colorless protocol complexes for one and two layers of a protocol where input values can be either *p* or *q*. In the first layer, each process writes its input to memory and takes a snapshot that becomes its input value for the second layer. The two-layer protocol complex is called the *iterated barycentric subdivision* of the input complex.

In Figure 4.8 we see the protocol complex for three or more processes, after two layers, when the inputs are *p*, *q*, and *r*. It is obtained by subdividing the protocol complex of Figure 4.5. Equivalently, it is the single-layer protocol complex when the input is the protocol complex of Figure 4.5.

Theorem 4.2.9. *Any colorless $(n + 1)$-process N-layer immediate snapshot protocol $(\mathcal{I}, \mathcal{P}, \Xi)$ is the composition of N single-layer protocols, where the protocol complex $\mathcal{P}$ is $\mathrm{Bary}^N \mathrm{skel}^n \mathcal{I}$ and $\Xi(\cdot) = \mathrm{Bary}^N \mathrm{skel}^n(\cdot)$.*

Proof. By a simple induction, using Theorem 4.2.8 as the base. □

Corollary 4.2.10. *For any input complex $\mathcal{I}$, $n > 0$, and $N > 0$, there is an $(n + 1)$-process colorless layered protocol that solves the barycentric agreement task $(\mathcal{I}, \mathrm{Bary}^N \mathrm{skel}^n \mathcal{I}, \mathrm{Bary}^N)$.*

One nice property of colorless layered protocols is the following "manifold" property.

Lemma 4.2.11. *Let $(\mathcal{I}, \mathcal{P}, \Xi)$ be a colorless layered protocol where $\mathcal{I}$ is pure and $d = \dim \mathcal{I}$. If σ is a d-simplex of $\mathcal{I}$, and τ is a $(d - 1)$-dimensional simplex of $\Xi(\sigma)$, then τ is contained in either one or two d-dimensional simplices of $\Xi(\sigma)$.*

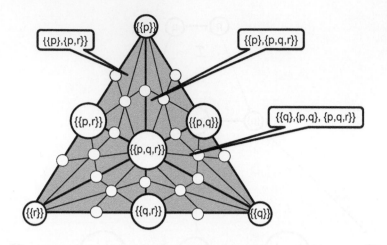

FIGURE 4.8

Protocol complexes for two layers, at least three processes, and input values p, q, and r. Views are shown for selected vertices.

The proof of this important property of colorless layered protocols is a simple consequence of the observation that $\Xi(\,\cdot\,)$ is a subdivision. We will discuss this topic further in Chapter 9.

4.3 The computational power of wait-free colorless immediate snapshots

Recall that in a colorless layered protocol, the processes share a two-dimensional memory array, where the rows correspond to layers and the columns to processes. In layer ℓ, process P_i takes an immediate snapshot: writing to $\mathsf{mem}[\ell][i]$ and immediately taking a snapshot of memory row ℓ. These protocols are communication-closed, in the sense that information flows from earlier layers to later ones but not in the other direction.

4.3.1 Colorless task solvability

We are now ready for the main result concerning the computational power of wait-free protocols in read-write memory.

Theorem 4.3.1. *The colorless task $(\mathcal{I}, \mathcal{O}, \Delta)$ has a wait-free $(n+1)$-process layered protocol if and only if there is a continuous map*

$$f : |\mathrm{skel}^n\, \mathcal{I}| \to |\mathcal{O}| \tag{4.3.1}$$

carried by Δ.

Proof. By Theorem 4.2.9, for any input complex $\mathcal{I}$, $n > 0$, and $N > 0$, there is an $(n+1)$-process colorless layered protocol that solves the barycentric agreement task $(\mathcal{I}, \mathrm{Bary}^N\, \mathcal{I}, \mathrm{Bary}^N)$. It follows that we can apply Lemma 4.2.6: a protocol $(\mathcal{I}, \mathcal{P}, \Xi)$ solves the task $(\mathcal{I}, \mathcal{O}, \Delta)$ if and only if there is a continuous map

$$f : |\text{Bary}^N \text{skel}^n \mathcal{I}| \rightarrow |\mathcal{O}|$$

carried by Δ. Finally, since $|\text{Bary}^N \text{skel}^n \mathcal{I}| = |\text{skel}^n \mathcal{I}|$, we have

$$f : |\text{skel}^n \mathcal{I}| \rightarrow |\mathcal{O}|$$

carried by Δ. $\qquad\qquad\qquad\qquad\qquad\qquad\qquad\qquad\qquad\qquad\qquad\qquad\qquad\quad$ $\square$

Applying the Discrete Protocol Complex Lemma 4.2.7 we get the following result.

Corollary 4.3.2. *For all $n > 0$, the colorless task $(\mathcal{I}, \mathcal{O}, \Delta)$ has a wait-free $(n+1)$-process colorless layered protocol if and only if there is a subdivision $\text{Div}\,\mathcal{I}$ of $\mathcal{I}$ and a simplicial map*

$$\phi : \text{Div}\,\mathcal{I} \rightarrow \mathcal{O}$$

carried by Δ.

4.3.2 Applications

Set Agreement We can use this result to prove that using colorless layered protocols, even the weakest nontrivial form of set agreement is impossible: there is no n-set agreement protocol if processes may be assigned $n+1$ distinct input values. We start with an informal explanation. Consider the executions where each process is assigned a distinct input value in the range $0, \ldots, n$. Because at most only n of these processes can be chosen, the processes must collectively "forget" at least one of them.

This task is $(\sigma, \text{skel}^{n-1}\sigma, \text{skel}^{n-1})$, where the input complex is a single n-simplex σ, the output complex is the $(n-1)$-skeleton of σ, and the carrier map is the $(n-1)$-skeleton operator, carrying each proper face of σ to itself. As illustrated in Figure 4.9, any continuous map $f : |\sigma| \rightarrow |\text{skel}^{n-1}\sigma|$ acts like (is homotopic to) the identity map on the boundary $\text{skel}^{n-1}\sigma$ of σ. Informally, it is impossible to extend f to the interior of σ, because f wraps the boundary of the "solid" simplex σ around the "hole" in the middle of $\text{skel}^{n-1}\sigma$. Of course, this claim is not (yet) a proof, but the intuition is sound.

To prove this claim formally, we use a simple form of a classic result called *Sperner's Lemma*. (Later, in Chapter 9, we will prove and make use of a more general version of this lemma.) Let σ be an n-simplex. A *Sperner coloring* of a subdivision $\text{Div}\,\sigma$ is defined as follows: Each vertex of σ is labeled with a distinct color, and for each face $\tau \subseteq \sigma$, each vertex of $\text{Div}\,\tau$ is labeled with a color from τ. Figure 4.10 shows a Sperner coloring in which each "corner" vertex is given a distinct color (black, white, or gray), each edge vertex is given a color from one of its two corners, and each interior vertex is given one of the three corner colors.

Fact 4.3.3 (Sperner's Lemma for subdivisions). Any Sperner labeling of a subdivision $\text{Div}\,\sigma$ must include an odd number of n-simplices labeled with all $n+1$ colors. (Hence there is at least one.)

Here is another way to formulate Sperner's Lemma: A Sperner labeling of a subdivision $\text{Div}\,\sigma$ is just a simplicial map $\phi : \text{Div}\,\sigma \rightarrow \sigma$ carried by the carrier map $\text{Car}(\cdot, \text{Div}\,\sigma)$ (see Exercise 4.8). It follows that ϕ carries some n-simplex $\tau \in \text{Div}\,\sigma$ to all of σ, and we have the following.

Fact 4.3.4 (Sperner's Lemma for carrier maps). There is no simplicial map ϕ from a subdivision $\text{Div}\,\sigma$ to $\text{skel}^{n-1}\sigma$ carried by the carrier map $\text{Car}(\cdot, \text{Div}\,\sigma)$.

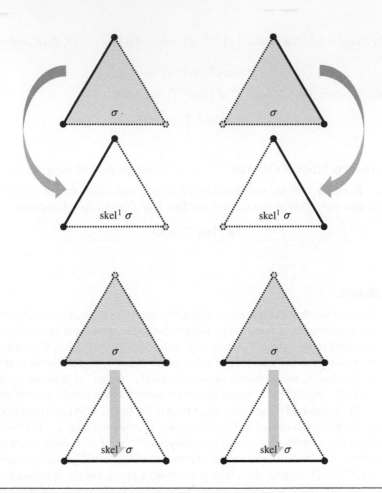

FIGURE 4.9

Why there is no colorless layered protocol for $(n+1)$-process n-set agreement: The map f is well-defined on the boundary of σ, but there is no way to extend it to the interior.

Recall that the k-set agreement task with input complex $\mathcal{I}$ is $(\mathcal{I}, \text{skel}^{k-1}\,\mathcal{I}, \text{skel}^{k-1})$, where the skeleton operator is considered as a strict carrier map (see Exercise 4.8).

Lemma 4.3.5. *There is no continuous map*

$$f : |\text{skel}^k\,\mathcal{I}| \rightarrow |\text{skel}^{k-1}\,\mathcal{I}| \tag{4.3.2}$$

carried by skel^{k-1}.

Proof. Assume by way of contradiction there is such a map f. It has a simplicial approximation $\phi : \text{Div}\,\text{skel}^k\,\mathcal{I} \rightarrow \text{skel}^{k-1}\,\mathcal{I}$ carried by skel^{k-1}, contradicting Fact 4.3.4. $\square$

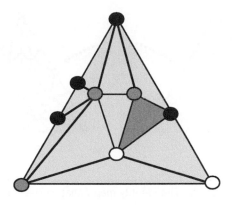

FIGURE 4.10

A subdivided triangle with a Sperner labeling. Sperner's Lemma states that at least one triangle (highlighted) must be labeled with all three colors.

Theorem 4.3.6.   *There is no wait-free $(n+1)$-process, colorless layered immediate snapshot protocol for n-set agreement.*

Proof.   If a protocol exists for $(\sigma, \text{skel}^{n-1}\sigma, \text{skel}^{n-1})$, then by Condition 4.3.2, there is a subdivision Div of σ and a simplicial decision map

$$\phi : \text{Div}\,\sigma \to \text{skel}^{n-1}\sigma$$

carried by skel^{n-1}, contradicting Fact 4.3.4.   □

If we think of Div σ as the protocol complex of all executions starting from input simplex σ, then each $(n+1)$-colored simplex represents an execution where the processes (illegally) choose $n+1$ distinct values.

> **Mathematical Note 4.3.7.**   The continuous version of Sperner's Lemma for carrier maps is essentially the No-Retraction Theorem, which is equivalent to the Brouwer fixed-point theorem, stating that there is no continuous map
>
> $$f : |\sigma| \to |\text{skel}^{n-1}\sigma|$$
>
> such that the restriction of f to $|\text{skel}^{n-1}\sigma|$ is identity. This connection is discussed further in Chapter 8.

Approximate agreement   We next consider a variation of approximate agreement whereby the process starts on the vertices of a simplex σ in $\mathcal{I}$ and must converge to points in σ that lie within ϵ of each other, for a given $\epsilon > 0$.

Here is an informal explanation as to why ϵ-agreement, unlike set agreement, has a colorless layered protocol. Given any $\epsilon > 0$, there is an $N > 0$ such that the diameter of any simplex in $\text{Bary}^N \sigma$ is less

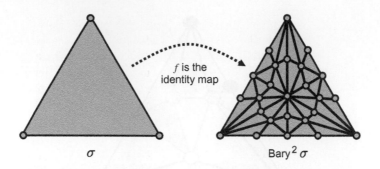

f is the
identity map

σ $\text{Bary}^2\,\sigma$

FIGURE 4.11

Why approximate agreement is possible: The identity map *f* that carries the boundary of σ to $\text{skel}^{n-1}\sigma$ can be extended to the interior.

than ϵ. Consider the task $(\sigma, \text{Bary}^N\sigma, \text{Bary}^N)$, where each process has as input a vertex of a geometric n-simplex σ and chooses as output a vertex in $\text{Bary}^N\sigma$, such that if $\tau \subseteq \sigma$ is the set of input vertices, then the chosen output vertices lie in a simplex of $\text{Bary}^N\tau$. Recast in geometric terms, the processes choose points within an ϵ-ball within the convex hull of the inputs.

As illustrated in Figure 4.11, the identity map from $|\sigma|$ to $|\text{Bary}^N\sigma|$ is carried by the carrier map $\text{Bary}^N : \sigma \to 2^{\text{Bary}^N\sigma}$, so this task does have a colorless layered protocol.

Recall that the protocol complex for a colorless N-layered protocol is the repeated barycentric subdivision $\text{Bary}^N\,\mathcal{I}$. Because barycentric subdivision is mesh-shrinking (Section 3.6.5), we can solve ϵ-agreement simply by running this protocol until the mesh of the subdivision is less than ϵ.

4.4 Chapter notes

Wait-free atomic snapshot algorithms were first proposed by Afek et al. [2] and by Anderson [6]. This algorithm is described and analyzed in Attiya and Welch [17] and in Herlihy and Shavit [92]. The snapshot algorithm in Exercise 4.12 is presented in a recursive form by Gafni and Rajsbaum [68].

The first formal treatment of the consensus task is due to Fischer, Lynch, and Paterson [55], who proved that this task is not solvable in a message-passing system, even if only one process may crash and processes have direct communication channels with each other. The result was later extended to shared memory by Loui and Abu-Amara [110] and by Herlihy [78]. Approximate agreement was shown to be solvable by Dolev, Lynch, Pinter, Stark, and Weihl [47].

Chaudhuri [37] was the first to investigate k-set agreement, where a partial impossibility result was shown. The loop agreement family of tasks was introduced by Herlihy and Rajsbaum [80] to study decidability, later extended [83] to hierarchies of loop agreement tasks. Degenerate loop agreement was considered by Liu, Pu, and Pan [108] and rendezvous tasks and higher dimensional loop agreement by Liu, Xu, and Pan [109]. The family was extended to general colorless tasks by Borowsky, Gafni, Lynch, and Rajsbaum [27] to identify the tasks for which the BG Simulation [23] can be used. Colorless tasks are technically easier to study than general tasks; indeed, many papers explicitly focus on colorless tasks [5,44,64,85,114,86]. Colorless protocols and their behavior in environments where more than one process may run solo are studied by Rajsbaum, Raynal, and Stainer [130].

In 1993, three papers were published together [23,90,134] showing that there is no wait-free protocol for set agreement using shared read-write memory or message passing. Herlihy and Shavit [90] introduced the use of simplicial complexes to model distributed computations. Borowsky and Gafni [23] and Saks and Zaharoughu [134] introduced layered executions. The first paper called them "immediate executions"; the second called them "block executions." Immediate snapshots as a model of computation were considered by Borowsky and Gafni [24].

Attiya and Rajsbaum [16] later used layered immediate snapshot executions in a combinatorial model to show the impossibility of k-set agreement by showing there is a strict carrier map on a protocol complex that is an orientable manifold. A proof that layered executions induce a subdivision of the input complex appears in Kozlov [101]. In these models, processes continually read and write a single shared memory, in contrast to the layered immediate snapshot model, where a clean memory is used for each layer.

In the terminology of Elrad and Francez [51], the layered immediate snapshot model is a *communication-closed layered model*. One of the earliest such models is due to Borowsky and Gafni [26] (see also the survey by Rajsbaum [128]). Instances of this model include the layered immediate snapshot memory model and analogous message-passing models. Other high-level abstract models have been considered by Gafni [61] using failure detector notions, and by Moses and Rajsbaum [120] for situations where at most one process may fail. Various cases of the message-passing model have been investigated by multiple researchers [3,36,103,120,136,138].

Sperner's Lemma implies that there no continuous function from the unit disk to its boundary, which is the identity on the boundary. This is a version of the *No-Retraction Theorem* [124], a form of the Brouwer Fixed-Point Theorem.

Dwork, Lynch, and Stockmeyer [49] have shown that consensus is solvable in semisynchronous environments, where message delivery time has an upper and lower bound. The commit/adopt abstraction of Exercise 4.10 was used by Yang, Neiger, and Gafni [145] for semisynchronous consensus, and a similar technique is used in Lamport's Paxos protocol [106]. Consensus is the basis for the *state machine* approach to building fault-tolerant, distributed systems [104,139].

4.5 Exercises

Exercise 4.1. Explicitly write out the approximate agreement protocol described in Section 4.2.2. Prove it is correct. (*Hint:* Use induction on the number of layers.)

Exercise 4.2. Consider the following protocol intended to solve k-set agreement for $k \leq n$. Each process has an *estimate*, initially its input. For r layers, each process communicates its estimate, receives estimates from others, and replaces its estimate with the smallest value it sees. Describe an execution where this protocol decides $k + 1$ or more distinct values.

Exercise 4.3. In the two-process, two-layer *lonely halting* protocol, each process has input value 1 and writes 0 each time it does not see another's input. Each process communicates its initial state in the first layer, and its state at the end of the layer becomes its input to the second layer, except that a process halts after the first layer if it does not see the other.

Draw a picture of this protocol complex in the style of Figure 4.7.

Exercise 4.4. In the ϵ-*approximate agreement* task, processes are assigned as input points in a high-dimensional Euclidean space $\mathbb{R}^N$ and must decide on points that lie within the convex hull of their inputs, and within ϵ of one another, for some given $\epsilon > 0$. Explain how the iterated barycentric agreement task of Section 4.2.2 can be adapted to solve this task.

Exercise 4.5. Here is another robot convergence task. In the *Earth Agreement* task, robots are placed at fixed positions on (a discrete approximation of) the Earth and must converge to nearby points on the Earth's surface.

The input complex is a 3-simplex $\tau^3 = \{0, 1, 2\}$ (the Earth), and the output complex is $\mathrm{skel}^2\tau^3$ (the Earth's surface). The robots start at any of the four vertices of τ^3. If they all start on one or two vertices, each process halts on one of the starting vertices. If they start on three or more vertices, then they converge to at most three vertices (not necessarily the starting vertices). The task's carrier map is

$$\Delta(\sigma) = \begin{cases} \sigma & \text{if } \dim \sigma \leq 1 \\ \mathrm{skel}^2\tau & \text{if } \dim \sigma > 1 \end{cases}$$

Show that there is a colorless single-layer immediate snapshot protocol for this task. Explain why this task is not equivalent to 3-set agreement with four input values.

Now consider the following variation. Let the output complex be Div $\mathrm{skel}^2\tau$, where Div is an arbitrary subdivision. As before, the robots start at any of the four vertices of τ^3. If they start on a simplex σ of dimension 0 or 1, then they converge to a single simplex of the subdivision Div σ. If they start on three or more vertices, then they converge to any simplex of Div $\mathrm{skel}^2\tau^3$. This carrier map is

$$\Delta(\sigma) = \begin{cases} \text{Div } \sigma & \text{if } \dim \sigma \leq 1 \\ \text{Div } \mathrm{skel}^2\tau & \text{if } \dim \sigma > 1 \end{cases}$$

Show that there is a colorless immediate snapshot protocol for this task. (*Hint:* Use the previous protocol for the first layer.)

Let us change the carrier map slightly to require that if the processes start on the vertices of a 2-simplex σ, then they converge to a simplex of Div σ. The new carrier map is

$$\Delta(\sigma) = \begin{cases} \text{Div } \sigma & \text{if } \dim \sigma \leq 2 \\ \text{Div } \mathrm{skel}^2\tau & \text{if } \dim \sigma > 2 \end{cases}$$

Show that this task has no colorless immediate snapshot protocol.

Exercise 4.6. Prove that skel^k and Bary are strict carrier maps.

Exercise 4.7. Is it true that for any complex $\mathcal{A}$, $\mathrm{skel}^k \mathrm{Bary}^n \mathcal{A} = \mathrm{Bary}^n \mathrm{skel}^k \mathcal{A}$?

Exercise 4.8. Recall that if $\mathcal{K}$ is a simplex, Div $\mathcal{K}$ a subdivision, and τ a simplex of Div $\mathcal{K}$, there is a unique simplex κ in $\mathcal{K}$, called the *carrier*, $\mathrm{Car}(\tau, \mathrm{Div}\,\mathcal{K})$, of minimal dimension such that τ is a simplex of Div κ. Show that the carrier is a carrier map.

```
// There are n + 1 layers
shared mem: array[0..n ][0.. n] of value
protocol ColorlessLayeredScan(input: Value): Value
    for ℓ := 0 to n do
        mem[ℓ][i] := input
        view := set of views in scan(mem[ℓ][*])
        if |view| = n + 1 − ℓ then return view
```

FIGURE 4.12

Colorless Layered Scan Protocol: Code for P_i.

Exercise 4.9. Consider a two-process colorless task $(\mathcal{I}, \mathcal{O}, \Delta)$. Assume for each input vertex v, $\Delta(v)$ is a single output vertex. We have seen in this chapter that its combinatorial representation is in terms of an input graph, an output graph, and a carrier map Δ.

1. In Chapter 2 we described tasks with chromatic graphs, where each vertex is associated to a process. Describe the chromatic task corresponding to the previous colorless task: its chromatic input and output graphs and its carrier map.
2. Prove that the chromatic task is solvable by a layered read-write (chromatic) protocol in the form of Figure 2.6 if and only if the colorless task is solvable by a colorless layered immediate snapshot protocol in the form of Figure 4.1.

Exercise 4.10. The *commit-adopt* task is a variation on consensus where each process is assigned an input value and each chooses as output a pair (D, v), where D is either COMMIT or ADOPT, and v is one of the input values in the execution. Moreover, (i) if a process decides (COMMIT, v), then every decision is $(\cdot, v)$, and (ii) if every process has the same input value v, then (COMMIT, v) is the only possible decision. Define this task formally as a colorless task and show it is solvable by a 2-layer colorless protocol but not by a 1-layer colorless protocol.

Exercise 4.11. Prove Sperner's Lemma for the special case where the protocol complex is the first barycentric subdivision.

Exercise 4.12. Consider the protocol in Figure 4.12, where a non-atomic scan operation reads one by one (in arbitrary order) the memory words $\text{mem}[\ell][i]$ for $0 \leq i \leq n$ (instead of using an atomic snapshot as in Figure 4.1).

Let input_i denote the input value of P_i and v_i the view returned by P_i. Prove that the views returned by the protocol satisfy the following properties. (i) For any two views v_i, v_j, $v_i \subseteq v_j$ or $v_j \subseteq v_i$. (ii) If $\text{input}_i \in v_j$ then $v_i \subseteq v_j$.

Exercise 4.13. Consider the colorless protocol of Exercise 4.12, where a non-atomic scan is used instead of an atomic snapshot. Draw the protocol complex after one round for two and for three processes. Is it a subdivision of the input complex? If not, does it contain one?

Exercise 4.14. Show that if we replace immediate snapshot with write followed by snapshot, the single-layer protocol complex is still a barycentric subdivision of the input complex.

This page is intentionally left blank

Solvability of Colorless Tasks in Different Models

5

CHAPTER OUTLINE HEAD

In Chapter 4 we considered colorless layered immediate snapshot protocols and identified the colorless tasks that such protocols can solve while tolerating crash failures by any number of processes. This chapter explores the circumstances under which colorless tasks can be solved using other computational models.

We consider models with different communication mechanisms and different fault-tolerance requirements. We show that the ideas of the previous chapter can be extended to characterize the colorless tasks that can be solved when up to t out of $n + 1$ processes may crash, when the processes communicate by shared objects that solve k-set agreement, or when the processes communicate by message passing.

Once we have established necessary and sufficient conditions for a task to have a protocol in a particular model, it is natural to ask whether it is *decidable* whether a given task satisfies those conditions. We will see that the answer to that question depends on the model.

Distributed Computing Through Combinatorial Topology. http://dx.doi.org/10.1016/B978-0-12-404578-1.00005-X

5.1 Overview of models

Recall from Chapter 4 that a colorless task is one where only the *sets* of input or output values matter, not which process has which. For such tasks, an initial configuration remains an initial configuration if one participating process exchanges its own input value for another's, and the same holds true for final configurations. Consensus and k-set agreement are examples of colorless tasks.

In a model where the processes communicate by layered snapshots and any number of processes can fail by crashing, a protocol must be wait-free. A process cannot wait for another process to take a step, because it cannot tell whether that process has crashed or is merely slow. We have seen that a colorless task $(\mathcal{I}, \mathcal{O}, \Delta)$ has a wait-free $(n+1)$-process layered immediate snapshot protocol if and only if there is a continuous map

$$f : |\text{skel}^n \mathcal{I}| \rightarrow |\mathcal{O}| \tag{5.1.1}$$

carried by Δ (Theorem 4.3.1). Informally, this characterization says that wait-free layered snapshot protocols transform (sets of at most $n + 1$ different) inputs to outputs in a continuous way.

In this chapter we consider several other models for which the computational power can be measured by a parameter p, $1 \leq p \leq n$. The colorless tasks solvable in a model with parameter p are exactly those for which there is a continuous map

$$f : |\text{skel}^p \mathcal{I}| \rightarrow |\mathcal{O}|$$

carried by Δ. Thus, the wait-free layered snapshot model is the weakest, having $p = n$, whereas a model with $p = 0$ can solve any colorless task.

Sometimes the wait-free condition may be too demanding. Instead of tolerating failures by an arbitrary subset of processes, we may be willing to tolerate fewer failures. A protocol is *t-resilient* if it tolerates halting failures by as many as t, $0 \leq t \leq n$ processes. (A wait-free protocol is n-resilient.) We say that a colorless task *has a t-resilient protocol* in a model if, for all $n \geq t$, there is a t-resilient $(n+1)$-process protocol for that task. In Section 5.2 we will see that a colorless task $(\mathcal{I}, \mathcal{O}, \Delta)$ has a t-resilient layered snapshot protocol if and only if there is a continuous map

$$f : |\text{skel}^t \mathcal{I}| \rightarrow |\mathcal{O}| \tag{5.1.2}$$

carried by Δ. Not surprisingly, the t-resilient Condition 5.1.2 is strictly weaker than its wait-free counterpart, Condition 5.1.1, since the map needs to be defined only over the t-skeleton of the input complex. The lower the dimension t, the easier it is to satisfy this condition and the more tasks that can be solved. In a sense, these two conditions capture the cost of fault tolerance. For colorless tasks, solvability is determined by the number of processes that can fail, whereas the total number of processes is irrelevant.

We also show (Section 5.3) that if we augment layered snapshot protocols by also allowing processes to communicate through k-set agreement objects, then a colorless task $(\mathcal{I}, \mathcal{O}, \Delta)$ has a wait-free layered protocol if and only if there is a continuous map

$$f : |\text{skel}^{k-1} \mathcal{I}| \rightarrow |\mathcal{O}|$$

carried by Δ. Adding k-set agreement objects, $1 \leq k \leq n$, increases the computational power of layered snapshot protocols by lowering the dimension of the skeleton on which a map must exist.

It follows that fault tolerance and communication power are, in a sense, interchangeable for colorless computability. A *t*-resilient layered colorless protocol and a wait-free layered protocol augmented by $(t + 1)$-set agreement objects, are equivalent: They can solve the same colorless tasks. Notice that in the extreme case, where $t = 0$, any colorless task is solvable, either because there are no failures or because the processes can reach consensus (Exercise 5.2). More generally, let p be an integer, $0 \le p \le n$. Then, for any t, k such that $p = \min(k - 1, t)$, there is a *t*-resilient *k*-set agreement layered snapshot protocol for a task $(\mathcal{I}, \mathcal{O}, \Delta)$ if and only if there is a continuous map

$$f : |\text{skel}^p \mathcal{I}| \rightarrow |\mathcal{O}|$$

carried by Δ (see Exercise 5.4).

The previous chapter's techniques extend even to the case where process failures are not independent. In Section 5.4, we show how to exploit knowledge of which potential failures are correlated and which are not. A parameter c captures the power of such a model for solving colorless tasks. This parameter is the size of the smallest *core* in the system, a minimal set of processes that will not all fail in any execution. The result for *t*-resilient solvability readily generalizes to dependent failures. A colorless task $(\mathcal{I}, \mathcal{O}, \Delta)$ has a layered protocol with minimal core size c if and only if there is a continuous map

$$f : |\text{skel}^c \mathcal{I}| \rightarrow |\mathcal{O}|$$

carried by Δ.

Next, in Section 5.5 we consider message-passing protocols. The layered snapshot model might appear to be stronger; once a process writes a value to shared memory, that value is there for all to see, whereas a value sent in a message is visible only to the process that received the message. Perhaps surprisingly, as long as a majority of processes is nonfaulty (that is, $2t < n + 1$), the two models are equivalent: Any task that has a *t*-resilient layered immediate snapshot protocol has a *t*-resilient message-passing protocol, and vice versa.

Once we have established necessary and sufficient conditions for a task to have a protocol in a particular model, it is natural to ask whether it is *decidable* whether a given task satisfies those conditions. We will see in Section 5.6 that the answer depends on the model. Essentially, for any model in which solvable tasks are exactly those for which there is a continuous map

$$f : |\text{skel}^p \mathcal{I}| \rightarrow |\mathcal{O}|$$

carried by Δ, then solvability is decidable if and only if $p \le 1$.

5.2 *t*-Resilient layered snapshot protocols

Recall that wait-free protocols tolerate crash failures by all processes but one (that is, n out of $n + 1$). Sometimes this level of resilience is excessive, especially if there are many processes. Instead, it may be enough to tolerate only t failures among $n + 1$ processes, where $0 \le t \le n$, a property called *t-resilience*.

A *colorless t-resilient layered immediate snapshot protocol* (*t-resilient layered protocol* when clear from context) is structured as shown in Figure 5.1. As in the wait-free case, the processes share a two-dimensional memory array $\text{mem}[\ell][i]$, where row ℓ is shared only by the processes participating in layer ℓ, and column i is written only by P_i. During layer ℓ, P_i writes its current view to $\text{mem}[\ell][i]$,

```
// N  is number of layers,  n + 1  the number of processes
ResilientLayeredSnapshot (vᵢ: Value):  Value
    view: Value := vᵢ              // initial  view is input value
    M: array of value
    for ℓ := 0 to N − 1 do
        do  // collect  values from at  least  n + 1 − t  processes
            immediate
                mem[ℓ][i] := view
                M := snapshot(mem[ℓ][∗])
            until |names(M)| ≥ n + 1 − t
        view := values(M)      // discard  process  names
    return δ(view)             // apply  decision  map to final  view
```

FIGURE 5.1

t-Resilient layered immediate snapshot protocol: Pseudo-code for P_i.

waits for $n + 1 - t$ views (including its own) to be written to that layer's row, and then takes a snapshot of that row. The waiting step introduces no danger of deadlock because at least $n + 1 - t$ nonfaulty processes will eventually reach each level and write their views.

Notice that the wait-free layered snapshot protocol of Figure 4.1, where $t = n$, is a degenerate form of the t-resilient protocol of Figure 5.1. In the wait-free protocol, once P_i has written to $\mathsf{mem}[\ell][i]$, it can proceed immediately because $n + 1 - t = 1$, and one view (its own) has already been written.

Right away we can see that even an $(n - 1)$-resilient protocol can solve colorless tasks that cannot be solved by a wait-free protocol (and in a single layer). The pseudo-code in Figure 5.2 solves $(t + 1)$-set agreement if at most t processes may fail. In contrast, we know from Theorem 4.3.6 that there is no $(t + 1)$-set agreement protocol if $t + 1$ processes can fail when $t = n$. More generally, this impossibility

```
SetAgree(vᵢ: Value):  Value
    view: Value := vᵢ              // initial  view is input value
    M: Array of Value = ∅
    do  // collect  values from at  least  n + 1 − t  processes
        immediate
            mem[0][i] := view
            M := snapshot(mem[0][∗])
    until |names(M)| ≥ n + 1 − t
    view := values(M)     // discard  process  names
    return min(view)      // minimum value from latest snapshot
```

FIGURE 5.2

t-Resilient single-layer snapshot protocol for $(t + 1)$-set agreement.

holds for any value of t (Theorem 5.2.9), so each additional level of resilience allows us to solve a harder instance of set agreement.

Lemma 5.2.1.   *There exists a t-resilient layered snapshot protocol for $(t + 1)$-set agreement.*

Proof.   As shown in Figure 5.2, each process writes its input, waits until $n + 1 - t$ inputs have been written, and then chooses the least value read. Because there are at least $n + 1 - t$ nonfaulty processes, the waiting step has no danger of deadlock. Because each process can "miss" values from at most t processes, each value chosen will be among the $t + 1$ least input values, so at most $t + 1$ distinct values can be chosen.   □

In Exercise 5.22, we ask you to show that this protocol does not actually require immediate snapshots.

The following lemma will be useful for characterizing the colorless tasks that can be solved, tolerating t failures, by a layered colorless protocol. It is similar to Theorem 4.2.8 for wait-free single-layer colorless immediate snapshot protocol complex, and indeed the proof is similar as well.

By Definition 4.2.2 we can consider the triple $(\mathcal{I}, \mathcal{P}, \Xi)$ for the protocol of Figure 5.1, where $\mathcal{I}$ is the input complex of a task, $\mathcal{P}$ is the protocol complex where each simplex is a colorless final configuration, and $\Xi : \mathcal{I} \to 2^{\mathcal{P}}$ is the strict execution carrier map.

Lemma 5.2.2.   *For any colorless single-layer $(n + 1)$-process t-resilient snapshot protocol $(\mathcal{I}, \mathcal{P}, \Xi)$, we have $\mathrm{Bary\,skel}^t \mathcal{I} \subseteq \mathcal{P}$, and the restriction of the execution map Ξ to this skeleton is the composition of the t-skeleton and barycentric subdivision operators.*

Proof.   Consider all executions of the t-resilient protocol of Figure 5.1 on the input subcomplex $\mathrm{skel}^t \mathcal{I}$. Assume all processes start with vertices from a simplex σ in $\mathrm{skel}^t \mathcal{I}$. The sets of views assembled by the processes form a chain of faces

$$\emptyset \subset \sigma_{i_0} \subseteq \cdots \subseteq \sigma_{i_n} \subseteq \sigma.$$

The inclusion follows because these views are snapshots, and snapshots are atomic: If P_i assembles face σ_i and P_j assembles face σ_j, then $\sigma_i \subseteq \sigma_j$, or vice versa.

These chains can have length at most $t + 1$, because $\sigma \in \mathrm{skel}^t \mathcal{I}$, so indeed the complex consisting of such simplices is contained in the t-skeleton of the barycentric subdivision $\mathrm{Bary}\,\sigma$.

Moreover, any simplex in $\mathrm{Bary}(\sigma)$ can be produced by such a chain. Consider an execution where $n + 1 - t$ processes start with input vertices from σ_{i_0} and at least one starts with each of the other vertices of σ. (There are enough processes because the chain has length at most $t + 1$.) Suppose all the processes with inputs from σ_{i_0} concurrently write to the array and immediately take a snapshot, ending up with views equal to σ_{i_0}. Similarly, all processes with input from $\sigma_{i_1} \setminus \sigma_{i_0}$ write and immediately take a snapshot, and so on.

The complex consisting of such simplices is precisely the barycentric subdivision of the t-skeleton of σ. Taking the complex over all possible inputs, we have $\mathcal{P}$ contains $\mathrm{Bary}(\mathrm{skel}^t(\mathcal{I}))$, and $\Xi(\cdot)$ is the restriction of $\mathrm{Bary}(\mathrm{skel}^t(\cdot))$.   □

A simple induction, with Lemma 5.2.2 as the base, yields the following.

Lemma 5.2.3.   *Any colorless N-layer $(n + 1)$-process t-resilient snapshot protocol $(\mathcal{I}, \mathcal{P}, \Xi)$ is the composition of N single-layer t-resilient protocols, where $\mathrm{Bary}^N \mathrm{skel}^t \mathcal{I} \subseteq \mathcal{P}$, and the restriction of the execution map Ξ to this skeleton is the composition of the barycentric subdivision and t-skeleton operators.*

If a protocol solves a colorless task $(\mathcal{I}, \mathcal{O}, \Delta)$, then we are free to add a preprocessing step to the protocol, where first the processes agree on at most k of their inputs, where $k = t + 1$, using the protocol of Figure 5.2. The following lemma states this formally using the protocol composition Definition 4.2.5.

Lemma 5.2.4 (Skeleton Lemma). *Assume that for any input complex $\mathcal{I}$ there is an $(n + 1)$-process protocol, $n > 0$, that solves the k-set agreement task $(\mathcal{I}, \text{skel}^{k-1}\mathcal{I}, \text{skel}^{k-1})$ for some fixed k.*

Assume furthermore that the protocol $(\mathcal{I}, \mathcal{P}, \Xi)$ solves the colorless task $(\mathcal{I}, \mathcal{O}, \Delta)$ with decision map δ. Then the composition of the k-set agreement task with the protocol $(\mathcal{I}, \mathcal{P}, \Xi)$ also solves $(\mathcal{I}, \mathcal{O}, \Delta)$ using the same decision map δ.

Proof. Recall that by Definition 4.2.5 the task $(\mathcal{I}, \text{skel}^{k-1}\mathcal{I}, \text{skel}^{k-1})$ can be composed with the protocol $(\mathcal{I}, \mathcal{P}, \Xi)$, since $\text{skel}^{k-1}\mathcal{I} \subseteq \mathcal{I}$. The result of the composition is a new protocol $(\mathcal{I}, \mathcal{P}', \Xi \circ \text{skel}^{k-1})$, where $\mathcal{P}' = (\Xi \circ \text{skel}^{k-1})(\mathcal{I}) = \Xi(\text{skel}^{k-1}\mathcal{I})$.

We check that δ is a correct decision map for the task. Pick an arbitrary $\sigma \in \mathcal{I}$. We have

$$\delta((\Xi \circ \text{skel}^{k-1})(\sigma)) = \delta(\Xi(\text{skel}^{k-1}\sigma)) \subseteq \delta(\Xi(\sigma)) \subseteq \Delta(\sigma),$$

where the last inclusion is a corollary of the fact that the protocol $(\mathcal{I}, \mathcal{P}, \Xi)$ solves the task $(\mathcal{I}, \mathcal{O}, \Delta)$. It follows that δ is a decision map for the composite protocol. $\qquad\square$

We may now combine the previous results to show that, for t-resilient colorless task solvability, we may assume without loss of generality that a protocol complex is a barycentric subdivision of the t-skeleton of the input complex.

Lemma 5.2.5. *If there is a t-resilient layered protocol that solves the colorless task $(\mathcal{I}, \mathcal{O}, \Delta)$, then there is a t-resilient layered protocol $(\mathcal{I}, \mathcal{P}, \Xi)$ solving that task whose protocol complex $\mathcal{P}$ is $\text{Bary}^N(\text{skel}^t\mathcal{I})$, and*

$$\Xi(\,\cdot\,) = \text{Bary}^N \circ \text{skel}^t(\,\cdot\,).$$

Proof. By Lemma 5.2.1, there exists a t-resilient layered snapshot protocol for k-set agreement. By the Skeleton Lemma (5.2.4), we can assume without loss of generality that any t-resilient colorless protocol's input complex is $\text{skel}^t\mathcal{I}$. Starting on a simplex σ in $\text{skel}^t\mathcal{I}$, after the first layer each process's view is a vertex of σ, and all their views form a simplex of $\text{Bary}(\sigma)$. After N layers, their views form a simplex of $\text{Bary}^N\sigma$. It follows that $\mathcal{P} \subseteq \text{Bary}^N(\text{skel}^t\mathcal{I})$.

The other direction follows from Lemma 5.2.3. It follows that $\text{Bary}^N\text{skel}^t\mathcal{I} \subseteq \mathcal{P}$. $\qquad\square$

Corollary 5.2.6. *For any input complex $\mathcal{I}$, $n > 0$, and $N > 0$, there is an $(n + 1)$-process t-resilient layered protocol that solves the barycentric agreement task $(\mathcal{I}, \text{Bary}^N\text{skel}^t\mathcal{I}, \text{Bary}^N)$.*

Theorem 5.2.7. *The colorless task $(\mathcal{I}, \mathcal{O}, \Delta)$ has a t-resilient layered snapshot protocol if and only if there is a continuous map*

$$f : |\text{skel}^t\mathcal{I}| \to |\mathcal{O}| \tag{5.2.1}$$

carried by Δ.

Proof. By Lemma 5.2.5, for any t-resilient layered snapshot protocol $(\mathcal{I}, \mathcal{P}, \Xi)$ we may assume the protocol complex $\mathcal{P}$ is $\text{Bary}^N\text{skel}^t\mathcal{I}$. Because layered snapshot protocols solve any barycentric agreement task, we can apply the Protocol Complex Lemma (4.2.6), which states that the protocol solves the task if and only if there is a continuous map

$$f : |\text{Bary}^N \text{skel}^t \mathcal{I}| \to |\mathcal{O}|$$

carried by Δ. The claim follows because $|\text{Bary}^N \text{skel}^t \mathcal{I}| = |\text{skel}^t \mathcal{I}|$.   □

Applying the Discrete Protocol Complex Lemma (4.2.7),

Corollary 5.2.8. *The colorless task* $(\mathcal{I}, \mathcal{O}, \Delta)$ *has a t-resilient layered snapshot protocol if and only if there is a subdivision* Div *of* skel$^t \mathcal{I}$ *and a simplicial map*

$$\phi : \text{Div skel}^t \mathcal{I} \to \mathcal{O}$$

carried by Δ.

Without loss of generality, we can assume that any t-resilient layered protocol consists of one $(t + 1)$-set agreement layer followed by any number of immediate snapshot layers. Moreover, only the first $(t + 1)$-set agreements layer requires waiting; the remaining layers can be wait-free.

Theorem 5.2.9. *There is no t-resilient layered snapshot protocol for t-set agreement.*

Proof.   See Exercise 5.3.   □

An important special case of the previous theorem occurs when $t = 1$, implying that consensus is not solvable by a layered protocol even if only a single process can fail.

5.3 Layered snapshots with *k*-set agreement

Practically all modern multiprocessor architectures provide synchronization primitives more powerful than simple read or write instructions. For example, the *test-and-set* instruction atomically swaps the value *true* for the contents of a memory location. If we augment layered snapshots with *test-and-set*, for example, it is possible to solve wait-free k-set agreements for $k = \lceil \frac{n+1}{2} \rceil$ (see Exercise 5.5). In this section, we consider protocols constructed by *composing* layered snapshot protocols with k-set agreement protocols.

In more detail, we consider protocols in the form of Figure 5.3. The protocol is similar to the colorless wait-free snapshot protocol of Figure 4.1 except that in addition to sharing memory, the objects share an array of k-set agreement objects (Line 3). In each layer ℓ, the processes first join in a k-set agreement protocol with the other processes in that layer (Line 8), and then they run an N_ℓ-layer immediate snapshot protocol (Line 11) for some $N_\ell \geq 0$.

Recall that the k-set agreement protocol with input complex $\mathcal{I}$ is $(\mathcal{I}, \text{skel}^{k-1}\mathcal{I}, \text{skel}^{k-1})$, where the skeleton operator is considered as a strict carrier map (see Exercise 4.8).

Recall also that if $(\mathcal{I}, \mathcal{P}, \Xi)$ and $(\mathcal{P}, \mathcal{P}', \Xi')$ are protocols where the protocol complex for the first is contained in the input complex for the second, then their *composition* is the protocol $(\mathcal{I}, \mathcal{P}', \Xi' \circ \Xi)$, where $(\Xi' \circ \Xi)(\sigma) = \Xi'(\Xi(\sigma))$ (Definition 4.2.3).

Definition 5.3.1. A *k-set layered snapshot protocol* is one composed from layered snapshot and k-set agreement protocols.

Lemma 5.3.2. *Without loss of generality, we can assume the that the first protocol in any such composition is a k-set agreement protocol. (That is, $N_0 > 0$.)*

```
1   // There are N  layers
2   shared mem: array[0..N−1][0..n] of Value
3   shared SA: array [0.. N−1][0..n] of SetAgree //k−set agreement objects
4   protocol ColorlessLayeredSA(input: Value): Value
5     view: Value := input          // initial  view is input value
6     for ℓ := 0 to N − 1 do
7       // k−set agreement with others in  layer
8       sa: View := SA[ℓ].decide(view)
9       // Nₗ−layer immediate snapshot protocol
10      for j := 0 to Nₗ do
11        view := ColorlessLayered (view))
12    return δ(view)              // apply decision map to final view
```

FIGURE 5.3

Colorless layered set agreement protocol: Pseudo-code for P_i.

Proof.   This claim follows directly from the Skeleton Lemma (5.2.4).   □

Lemma 5.3.3.   *If $(\mathcal{I}, \mathcal{P}, \Xi)$ is a k-set layered snapshot protocol, then $\mathcal{P}$ is equal to $\mathrm{Bary}^N \mathrm{skel}^{k-1}\mathcal{I}$ for some $N \geq 0$.*

Proof.   We argue by induction on ℓ, the number of k-set and layered snapshot protocols composed to construct $(\mathcal{I}, \mathcal{P}, \Xi)$. For the base case, when $\ell = 1$, the protocol is just a k-set agreement protocol by Lemma 5.3.2, so the protocol complex $\mathcal{P}$ is just $\mathrm{skel}^{k-1}\mathcal{I}$.

For the induction step, assume that $(\mathcal{I}, \mathcal{P}, \Xi)$ is the composition of $(\mathcal{I}, \mathcal{P}_0, \Xi_0)$ and $(\mathcal{P}_1, \mathcal{P}, \Xi_1)$, where the first protocol is the result of composing $\ell - 1$ k-set or layered snapshot protocols, and $\mathcal{P}_0 \subseteq \mathcal{P}_1$. By the induction hypothesis, $\mathcal{P}_0$ is $\mathrm{Bary}^N \mathrm{skel}^{k-1}\mathcal{I}$ for some $N \geq 0$.

There are two cases. First, if $(\mathcal{P}_1, \mathcal{P}, \Xi_1)$ is a k-set protocol, then

$$\Xi_1(\mathcal{P}_0) = \mathrm{Bary}^N \mathrm{skel}^{k-1}\mathrm{skel}^{k-1}\mathcal{I} = \mathrm{Bary}^N \mathrm{skel}^{k-1}\mathcal{I}.$$

Second, if it is an M-layer snapshot protocol, then

$$\Xi_1(\mathcal{P}_0) = \mathrm{Bary}^M(\mathrm{Bary}^N(\mathrm{skel}^{k-1}\mathcal{I})) = \mathrm{Bary}^{M+N}\mathrm{skel}^{k-1}\mathcal{I}.$$ □

Theorem 5.3.4.   *The colorless task $(\mathcal{I}, \mathcal{O}, \Delta)$ has a k-set layered snapshot protocol if and only if there is a continuous map*

$$f : |\mathrm{skel}^{k-1}\mathcal{I}| \to |\mathcal{O}| \tag{5.3.1}$$

carried by Δ.

Proof.   By Lemma 5.2.5, any k-set layered snapshot protocol $(\mathcal{I}, \mathcal{P}, \Xi)$ has $\mathcal{P} = \mathrm{Bary}^N \mathrm{skel}^{k-1}\mathcal{I}$. By the Protocol Complex Lemma (4.2.6), the protocol solves the task if and only if there is a continuous map

$$f : |\mathrm{Bary}^N \mathrm{skel}^{k-1}\mathcal{I}| \to |\mathcal{O}|$$

carried by Δ. The claim follows because $|\mathrm{Bary}^N \mathrm{skel}^{k-1}\mathcal{I}| = |\mathrm{skel}^{k-1}\mathcal{I}|$.   □

Applying the Discrete Protocol Complex Lemma (4.2.7):

Corollary 5.3.5. *The colorless task $(\mathcal{I}, \mathcal{O}, \Delta)$ has a k-set layered snapshot protocol if and only if there is a subdivision Div of $\text{skel}^{k-1}\mathcal{I}$ and a simplicial map*

$$\phi : \text{Div skel}^{k-1}\mathcal{I} \to \mathcal{O}$$

carried by Δ.

Theorem 5.3.6. *There is no k-set layered snapshot protocol for $(k-1)$-set agreement.*

Proof. See Exercise 5.7. □

The next corollary follows because Theorem 5.3.4 is independent of the order in which k-set agreement layers are composed with immediate snapshot layers.

Corollary 5.3.7. *We can assume without loss of generality that any set agreement protocol consists of a single k-set agreement layer followed by some number of layered immediate snapshot protocols.*

5.4 Adversaries

A *t*-resilient protocol is designed under the assumption that failures are uniform: *Any t* out of $n + 1$ processes can fail. Often, however, failures are correlated. In a distributed system, processes running on the same node, in the same network partition, or managed by the same provider may be more likely to fail together. In a multiprocessor, processes running on the same core, on the same processor, or on the same card may be likely to fail together. It is often possible to design more effective fault-tolerant algorithms if we can exploit knowledge of which potential failures are correlated and which are not.

One way to think about such failure models is to assume that failures are controlled by an *adversary* who can cause certain subsets of processes to fail, but not others. There are several ways to characterize adversaries. The most straightforward is to enumerate the *faulty sets*: all sets of processes that fail in some execution. We will assume that faulty sets are closed under inclusion; if F is a maximal set of processes that fail in some execution, then for any $F' \subset F$ there is an execution in which F' is the actual set of processes that fail. There is a common-sense justification for this assumption: We want to respect the principle that fault-tolerant algorithms should continue to be correct if run in systems that display *fewer failures* than in the worst-case scenario. A model that permits algorithms that are correct *only if* certain failures occur is unlikely to be useful in practice.

Faulty sets can be described as a simplicial complex $\mathcal{F}$, called the *faulty set complex*, the vertices of which are process names and the simplices of which are sets of process names such that exactly those processes fail in some execution.

Faulty sets can be cumbersome, so we use a more succinct and flexible way to characterize adversaries. A *core* is a minimal set of processes that will not all fail in any execution. A core is a simplex that is *not* itself in the faulty set complex, but all of its proper faces are in $\mathcal{F}$. The following dual notion is also useful. A *survivor set* is a minimal set of processes that intersects every core (such a set is sometimes called a *hitting set*). In every execution, the set of nonfaulty processes includes a survivor set.

Here are some examples of cores and survivor sets.

The Wait-Free Adversary. The entire set of processes is the only core, and the singleton sets are the survivor sets.

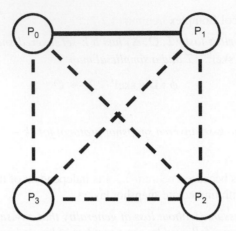

FIGURE 5.4

An irregular adversary: P_0, P_1, P_2, and P_3 can each fail individually, or P_0 and P_1 may both fail. The faulty set complex consists of an edge linking P_0 and P_1, shown as a solid line, and two isolated vertices, P_2 and P_3. There are five cores, shown as dotted lines.

The t-Faulty Adversary. The cores are the sets of cardinality $t + 1$, and the survivor sets are the sets of cardinality $n + 1 - t$.

An Irregular Adversary. Consider a system of four processes, P_0, P_1, P_2, and P_3, where any individual process may fail, or P_0 and P_1 may both fail. Here $\{P_0, P_2\}$ is a core, since they cannot both fail, yet there is an execution in which each one fails. In all, there are five cores:

$$\{\{P_i, P_j\} | 0 \leq i < j \leq 3, \quad (i, j) \neq (0, 1)\}$$

and three survivor sets:

$$\{P_2, P_3\}, \{P_0, P_1, P_3\}, \{P_0, P_1, P_2\}.$$

The set $\{P_2, P_3\}$ is a survivor set, since there is an execution where only these processes are nonfaulty. This adversary is illustrated in Figure 5.4.

Here is how to use cores and survivor sets in designing a protocol. Given a fixed core C, it is safe for a process to wait until it hears from *some* member of C, because they cannot all fail. It is also safe for a process to wait until it hears from *all* members of *some* survivor set, because the set of nonfaulty processes always contains a survivor set. See Exercise 5.14.

Let A be an adversary with minimum core size c. We say that a protocol is A-*resilient* if it tolerates any failure permitted by A. As illustrated in Figure 5.5, an A-resilient layered snapshot protocol differs from a t-resilient protocol as follows. At each layer, after writing its own value, each process waits until all the processes in a survivor set (possibly including itself) have written their views to that layer's memory. As noted, there is no danger of deadlock waiting until a survivor set has written.

Notice that the t-resilient layered snapshot protocol of Figure 5.1 is a degenerate form of the A-resilient protocol of Figure 5.5 because, for the t-resilient protocol, any set of $n + 1 - t$ processes is a survivor set.

```
// N  is  number  of  layers,  n + 1  the  number  of  processes
AdversaryLayeredSnapshot(v_i: value ): value
    view: value := v_i              // initial  view  is  input  value
    M: array of value
    for ℓ := 0 to N − 1 do
      do  // collect  values  from  a  survivor  set
        immediate
            mem[ℓ][i] := view
            M := snapshot(mem[ℓ][*])
        until names(M) contains a survivor set
      view := values(M) // discard  process  names
    return δ(view)        // apply  decision  map  to  final  view
```

FIGURE 5.5

A-resilient layered snapshot protocol: Pseudo-code for P_i.

```
SetAgree(v_i): value
    if  P_i ∈ C  then // write  if  member  of  core
        mem[0][i] := v_i
        return v_i
    else  // wait  for  core  process  to  write
        do
            M := snapshot(mem[0][*])
        until  M[j] ≠ ⊥  for  some  j
    return M[j]
```

FIGURE 5.6

A-resilient layered snapshot protocol for $(c + 1)$-set agreement.

Lemma 5.4.1.   *Let* A *be an adversary with minimum core size* $c + 1$. *There is an* A-*resilient layered snapshot protocol for c-set agreement.*

Proof.   It is a little easier to explain this protocol using writes and snapshots instead of immediate snapshots (see Exercise 5.23). Pick a core C of A of minimal size $c + 1$. Figure 5.6 shows a single-layer protocol. Each process P_i in C writes its input to mem[0][i], while each process not in C repeatedly takes snapshots until it sees a value written (by a process in C). It then replaces its own input value with the value it found. At most $c + 1$ distinct values can be chosen. This protocol must terminate because C is a core, and the adversary cannot fail every process in C.   □

Lemma 5.4.2.   *Without loss of generality, for any N-layer* A-*resilient colorless protocol* $(\mathcal{I}, \mathcal{P}, \Xi)$,

$$\mathcal{P} = \mathrm{Bary}^N \mathrm{skel}^c \mathcal{I} \text{ and } \Xi(\cdot) = \mathrm{Bary}^N \circ \mathrm{skel}^c(\cdot).$$

Proof. By Lemma 5.4.1, there exists an A-resilient layered snapshot protocol for $(c+1)$-set agreement. By the Skeleton Lemma (5.2.4), we can assume without loss of generality that any A-resilient colorless protocol's input complex is $\text{skel}^c \mathcal{I}$. From that point on the rest of the proof is virtually identical to the proof of Lemma 5.2.5. □

Theorem 5.4.3. *The colorless task* $(\mathcal{I}, \mathcal{O}, \Delta)$ *has an* A-*resilient layered snapshot protocol if and only if there is a continuous map*

$$f : |\text{skel}^c \mathcal{I}| \to |\mathcal{O}| \qquad (5.4.1)$$

carried by Δ.

Proof. By Lemma 5.4.2, any t-resilient layered snapshot protocol $(\mathcal{I}, \mathcal{P}, \Xi)$ has $\mathcal{P} = \text{Bary}^N \text{skel}^c \mathcal{I}$. The Protocol Complex Lemma (4.2.6) states that the protocol solves the task if and only if there is a continuous

$$f : |\text{Bary}^N \text{skel}^c \mathcal{I}| \to |\mathcal{O}|$$

carried by Δ. The claim follows because $|\text{Bary}^N \text{skel}^c \mathcal{I}| = |\text{skel}^c \mathcal{I}|$. □

Applying the Discrete Protocol Complex Lemma (4.2.7):

Corollary 5.4.4. *The colorless task* $(\mathcal{I}, \mathcal{O}, \Delta)$ *has an* A-*resilient layered snapshot protocol if and only if there is a subdivision* Div *of* $\text{skel}^c \mathcal{I}$ *and a simplicial map*

$$\phi : \text{Div } \text{skel}^c \mathcal{I} \to \mathcal{O}$$

carried by Δ.

Theorem 5.4.5. *There is no* A-*resilient c-set agreement layered snapshot protocol.*

Proof. See Exercise 5.15. □

5.5 Message-passing protocols

So far we have focused on models in which processes communicate through shared memory. We now turn our attention to another common model of distributed computing, where processes communicate by *message passing*.

There are $n + 1$ asynchronous processes that communicate by sending and receiving messages via a communication network. The network is fully connected; any process can send a message to any other. Message delivery is *reliable*; every message sent is delivered exactly once to its target process after a finite but potentially unbounded delay. Message delivery is *first-in, first-out* (FIFO); messages are delivered in the order in which they were sent.

The operational model is essentially unchanged from the layered snapshot model. The principal difference is that communication is now one-to-one rather than one-to-many. In Exercise 5.11, we ask you to show that barycentric agreement is impossible in a message-passing model if a majority of the process can fail. For this reason, we restrict our attention to t-resilient protocols where t, the number of processes that can fail, is less than half: $2t < n + 1$.

We will see that as long as a majority of processes are nonfaulty, there is a t-resilient message-passing protocol if and only if there is a t-resilient layered snapshot protocol. We will see, however, that message-passing protocols look quite different from their shared-memory counterparts.

For shared-memory protocols, we focused on layered protocols because it is convenient to have a "clean" shared memory for each layer. For message-passing protocols, where there is no shared memory, we will not need to use layered protocols. Later, in Chapter 13, it will be convenient impose a layered structure on asynchronous message-passing executions.

In our examples we use the following notation. A process P sends a message containing values $v_0, \ldots, v_\ell$ to Q as follows:

> **send**$(P, v_0, \ldots, v_\ell)$ **to** Q

We say that a process *broadcasts* a message if it sends that message to all processes, including itself:

> **send**$(P, v_0, \ldots, v_\ell)$ **to** all

Here is how Q receives a message from P:

> **upon receive** $(P, v_0, \ldots, v_\ell)$ **do**
>   ... *// do something with the values received*

Some message-passing protocols require that each time a process receives a message from another, the receiver forwards that message to all processes. Each process must continue to forward messages even after it has chosen its output value. Without such a guarantee, a nonfaulty process that chooses an output and falls silent is indistinguishable from a crashed process, implying that tasks requiring a majority of processes to be nonfaulty become impossible. We think of this continual forwarding as a kind of operating system service running in the background, interleaved with steps of the protocol itself. In our examples, such loops are marked with the **background** keyword:

> **background** *// forward messages forever*
>   **upon receive** (P_j, v) **do**
>   **send** (P_i, v) **to** all

We start with two useful protocols, one for $(t+1)$-set agreement and one for barycentric agreement.

5.5.1 Set agreement

As a first step, each process assembles values from as many other processes as possible. The getQuorum() method shown in Figure 5.7 collects values until it has received messages from all but t processes. It is safe to wait for that many messages because there are at least $n + 1 - t$ nonfaulty processes. It is not safe to wait for more, because the remaining t processes may have crashed.

Figure 5.8 shows a simple protocol for $(t+1)$-set agreement. Each process broadcasts its input value, waits to receive values from a quorum of $n + 1 - t$ messages, and chooses the least value among them. A proof of this protocol's correctness is left as Exercise 5.9. Note that this protocol works for any value of t.

5.5.2 Barycentric agreement

Recall that in the *barycentric agreement* task, each process P_i is assigned as input a vertex v_i of a simplex σ, and after exchanging messages with the others, chooses a face $\sigma_i \subseteq \sigma$ containing v_i, such that for any two participating processes P_i and P_j the faces they choose are ordered by inclusion: $\sigma_i \subseteq \sigma_j$, or

```
getQuorum(): Set of Value
    V: Set of Value := ∅
    q: int := 0   // count how many messages received
    do
        upon receive(Q, v) do
            V := V ∪ {v}   // add value to set
            q := q + 1         // increment count
    until q = n + 1 − t       // stop when enough received
    return V
```

FIGURE 5.7

Return values from at least $n + 1 - t$ processes.

```
SetAgree(vᵢ): value
    send(Pᵢ, vᵢ) to all
    V: Set of Value := getQuorum()
    return min(V)
```

FIGURE 5.8

t-resilient message-passing protocol for $(t + 1)$-set agreement.

vice versa. This task is essentially equivalent to an immediate snapshot, which it is convenient (but not necessary) to assume as a shared-memory primitive operation. In message-passing models, however, we assume send and receive as primitives, and we must build barycentric agreement from them.

Figure 5.9 shows a message-passing protocol for barycentric agreement. Each P_i maintains a set V_i of messages it has received, initially only P_i's input value (Line 2). P_i repeatedly broadcasts V_i, and waits to receive sets from other processes. If it receives V' such that $V' = V_i$ (Line 7), then it increments its count of the number of times it has received V_i. If it receives V' such that $V' \setminus V \neq \emptyset$ (Line 9). It sets V_i to $V_i \cup V'$ and starts over. When P_i has received $n + 1 - t$ identical copies of V_i from distinct processes, the protocol terminates, and P_i decides V_i. As usual, after the protocol terminates, P_i must continue to forward messages to the others (Lines 15–17).

Lemma 5.5.1. *The protocol in Figure 5.9 terminates.*

Proof.   Suppose, by way of contradiction, that P_i runs this protocol forever. Because P_i changes V_i at most n times, there is some time at which P_i's V_i assumes its final value V. For every set V' that P_i received earlier, $V' \subset V$, and for every V' received later, $V' \subseteq V$.

When P_i updates V_i to V, it broadcasts V to the others. Suppose a nonfaulty P_j receives V from P, where $V_j = V'$. P_j must have sent V' to P_i when it first set V_j to V'. Since P_i henceforth does not change V_i, either $V' \subset V$, or $V' = V$. If $V' \subset V$, then P_j will send V back to P_i, increasing its count. If $V' = V$, then P_j already sent V to P_i. Either way, P_i receives a copy of V from at least $n + 1 - t$ nonfaulty processes and terminates the protocol.   □

```
1   BaryAgree(v_i: Vertex): set of Vertex
2     V_i: set of Vertex := {v_i}
3     count := 0
4     while count < n + 1 − t do
5       send(P_i, V_i) to all
6       on receive(P_j, V_j) do
7         if V_i = V_j then        // confirmation
8           count := count + 1
9         else if V_j \ V_i ≠ ∅ then
10          V_i := V_i ∪ V_j      // start over
11          count := 0
12    return V_i
13    // run background code after protocol returns
14    background // forward input value messages forever
15      upon receive(P_j, V_j) do
16        V_i := V_i ∪ V_j
17        send(P_i, V_i) to all
```

FIGURE 5.9

Barycentric agreement message-passing protocol.

Lemma 5.5.2. *In the protocol in Figure 5.9, if P_i decides V_i and P_j decides V_j, then either $V_i \subseteq V_j$, or vice versa.*

Proof. Note that the sequence of sets $V^{(0)}, \ldots, V^{(0)}$ broadcast by any process is strictly increasing: $V^{(i)} \subset V^{(i+1)}$. To decide, P_i received V_i from a set X of at least $n + 1 − t$ processes, and P_i received V_i from a set Y at least $n + 1 − t$ processes. Because t cannot exceed $\frac{n+1}{2}$, X and Y must both contain a process P_k that sent both V_i and V_j, implying they are ordered, a contradiction. □

5.5.3 Solvability condition

We can now characterize which tasks have protocols in the t-resilient message-passing model.

Theorem 5.5.3. *For $2t < n + 1$, $(\mathcal{I}, \mathcal{O}, \Delta)$ has a t-resilient message-passing protocol if and only if there is a continuous map*

$$f : |\mathrm{skel}^t \mathcal{I}| \to |\mathcal{O}|$$

carried by Δ,

Proof. Protocol Implies Map. If a task has an $(n + 1)$-process t-resilient message-passing protocol, then it has an $(n + 1)$-process t-resilient layered snapshot protocol (see Exercise 5.10). The claim then follows from Theorem 5.2.7.

Map Implies Protocol. The map

$$f : |\mathrm{skel}^t \mathcal{I}| \to |\mathcal{O}|,$$

has a simplicial approximation,

$$\phi : \mathrm{Bary}^N \, \mathrm{skel}^t \mathcal{I} \to \mathcal{O},$$

also carried by Δ. We construct a two-step protocol. In the first step, the processes use the $(t + 1)$-set agreement protocol of Figure 5.8 to converge to a simplex σ in $\mathrm{skel}^t \mathcal{I}$, In the second step, they repeat the barycentric agreement protocol of Figure 5.9 to converge to a simplex in $\mathrm{Bary}^N \, \mathrm{skel}^t \mathcal{I}$. Composing these protocols and using ϕ as a decision map yields the desired protocol. $\square$

Theorem 5.5.4.   *For $2t < n + 1$, $(\mathcal{I}, \mathcal{O}, \Delta)$ has a t-resilient message-passing protocol if and only if there is a subdivision Div of $\mathrm{skel}^t \mathcal{I}$ and a simplicial map*

$$\phi : \mathrm{Div} \, \mathrm{skel}^t \mathcal{I} \to \mathcal{O}$$

carried by Δ.

Proof.   See Exercise 5.16. $\square$

Theorem 5.5.5.   *There is no t-resilient message-passing protocol for t-set agreement.*

Proof.   See Exercise 5.17. $\square$

5.6 Decidability

This section uses more advanced mathematical techniques than the earlier sections.

Now that we have necessary and sufficient conditions for a task to have a protocol in various models, it is natural to ask whether we can *automate* the process of deciding whether a given task has a protocol in a particular model. Can we write a program (that is, a Turing machine) that takes a task description as input and returns a Boolean value indicating whether a protocol exists?

Not surprisingly, the answer depends on the model of computation. For wait-free layered snapshot protocols or wait-free k-set layered snapshot protocols for $k \geq 3$, the answer is *no*: There exists a family of tasks for which it is *undecidable* whether a protocol exists. We will construct one such family: the loop agreement tasks, discussed in Chapter 15. On the other hand, for wait-free k-set layered snapshot protocols for $k = 1$ or 2, the answer is *yes*: For every task, it is *decidable* whether a protocol exists. For any model where the solvability question depends only on the 1-skeleton of the input complex, solvability is decidable (see Exercise 5.19).

5.6.1 Paths and loops

Let $\mathcal{K}$ be a finite 2-dimensional complex. Recall from Chapter 3 that an *edge path* between vertices u and v in $\mathcal{K}$ is a sequence of vertices $u = v_0, v_1, \ldots, v_\ell = v$ such that each pair $\{v_i, v_{i+1}\}$ is an edge of $\mathcal{K}$ for $0 \leq i < \ell$. A path is *simple* if the vertices are distinct.

Definition 5.6.1.   An edge path is an *edge loop* if its first and last vertices are the same. An edge loop is *simple* if all the other vertices are distinct. An edge loop's first vertex is called its *base point*.

All edge loops considered here are assumed to be simple.

Informally, we would like to distinguish between edge loops that circumscribe "solid regions" and edge loops that circumscribe holes. To make this notion precise, we must introduce some continuous concepts.

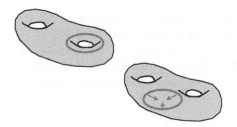

FIGURE 5.10

Noncontractible (left) and contractible (right) continuous loops.

Definition 5.6.2.   Fix a point s on the unit circle S^1. A *continuous loop* in $|\mathcal{K}|$ *with base point x* is a continuous map $\rho : S^1 \to |\mathcal{K}|$ such that $\rho(s) = x$. A continuous loop ρ is *simple* if it has no self-intersections: $\rho(s_0) = \rho(s_1)$ only if $s_0 = s_1$.

All continuous loops considered here are assumed to be simple.

As illustrated in Figure 5.10, a continuous loop in $|\mathcal{K}|$ is *contractible* if it can be continuously deformed to its base point in finite "time," leaving the base point fixed. Formally, we capture this notion as follows.

Definition 5.6.3.   A continuous loop $\rho : S^1 \to |\mathcal{K}|$ in $\mathcal{K}$ is *contractible* if it can be extended to a continuous map $\hat{\rho} : D^2 \to X$, where D^2 denotes the 2-disk for which the boundary is the circle S^1, the input domain for ρ.

A simple continuous loop λ is a *representative* of a simple edge loop ℓ if their geometric images are the same: $|\lambda(S^1)| = |\ell|$.

Definition 5.6.4.   A simple edge loop p is *contractible* if it has a contractible representative.

Although any particular simple edge loop has an infinite number of representatives, it does not matter which one we pick.

Fact 5.6.5.   Either all of an edge loop's representatives are contractible, or none are.

In Exercise 5.18, we ask you to construct an explicit representative of an edge path.

Fact 5.6.6.   The question whether an arbitrary simple edge loop in an arbitrary finite simplicial complex is contractible is undecidable.

Remarkably, the question remains undecidable even for complexes of dimension two (see Section 5.7, "Chapter notes").

Mathematical Note 5.6.7.   The notion of contractibility is a special case of a more general notion called *loop homotopy*. Given two continuous loops with the same base point, we would like to treat them as equivalent if one loop can be continuously deformed to the other in finite "time," leaving their common base point fixed. Formally, two loops $\rho, \rho' : S^1 \to |\mathcal{K}|$ with common base point x are *homotopic* if there is a continuous map $h : S^1 \times [0, 1] \to |\mathcal{K}|$, such that $h(s, 0) = \rho$, $h(s, 1) = \rho'$, $h(0, t) = h(1, t) = x$, for all $s, t \in [0, 1]$. If we think of the second coordinate in $S^1 \times [0, 1]$ as time, then $h(s, 0)$ is ρ, $h(s, 1)$ is ρ', and $h(s, t)$ is the intermediate loop at time t, for $0 < t < 1$. Note that the base point does not move during the deformation.

The *trivial loop* never leaves its base point. It is given by $\tau : S^1 \to |\mathcal{K}|$, where $\tau(s) = x$ for all $s \in S^1$. It is a standard fact that a loop is contractible if and only if it is homotopic to the trivial loop at its base point.

The homotopy classes of loops for a topological space X are used to define that space's *fundamental group*, usually denoted $\pi_1(X)$. These groups are extensively studied in algebraic topology.

5.6.2 Loop agreement

Let Δ^2 denote the 2-simplex for which the vertices are labeled 0, 1, and 2, and let $\mathcal{K}$ denote an arbitrary 2-dimensional complex. We are given three distinct vertices v_0, v_1, and v_2 in $\mathcal{K}$, along with three edge paths p_{01}, p_{12}, and p_{20}, such that each path p_{ij} goes from v_i to v_j. We let p_{ij} denote the corresponding 1-dimensional simplicial subcomplex as well, in which case we let $p_{ij} = p_{ji}$. We assume that the paths are chosen to be nonself-intersecting and that they intersect each other only at corresponding end vertices.

Definition 5.6.8. These edge paths p_{01}, p_{12}, and p_{20} form a simple edge loop ℓ with base point v_0, which we call a *triangle loop*, denoted by the 6-tuple $\ell = (v_0, v_1, v_2, p_{01}, p_{12}, p_{20})$.

In the *loop agreement task*, the processes start on vertices of Δ^2 and converge on a simplex in $\mathcal{K}$, subject to the following conditions. If all processes start on a single vertex i, they converge on the corresponding vertex v_i. If they start on two distinct input vertices, i and j, they converge on some simplex (vertex or edge) along the path p_{ij} linking v_i and v_j. Finally, if the processes start on all three input vertices $\{0, 1, 2\}$, they converge to some simplex (vertex, edge, or triangle) of $\mathcal{K}$. See Figure 5.11 for an illustration. More precisely:

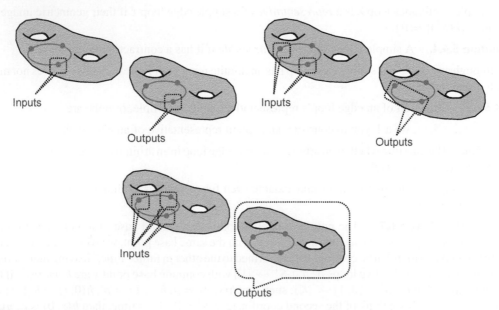

FIGURE 5.11

Loop agreement.

Definition 5.6.9. The *loop agreement* task associated with a triangle loop ℓ in a simplicial complex $\mathcal{K}$ is a triple $(\Delta^2, \mathcal{K}, \Lambda)$, where the carrier map Λ is given by

$$\Lambda(\tau) = \begin{cases} v_i & \text{if } \tau = \{i\}, \\ p_{ij} & \text{if } \tau = \{i, j\}, 0 \le i < j \le 2, \text{ and} \\ \mathcal{K} & \text{if } \tau = \Delta^2. \end{cases}$$

Since the loop agreement task is completely determined by the complex $\mathcal{K}$ and the triangle loop ℓ, we also denote it by $\text{loop}(\mathcal{K}, \ell)$.

5.6.3 Examples of loop agreement tasks

Here are some examples of interesting loop agreement tasks:

- A 2-set agreement task can be formulated as the loop agreement task $\text{Loop}(\text{skel}^1(\Delta^2), \ell)$, where $\ell = (0, 1, 2, ((0, 1)), ((1, 2)), ((2, 0)))$.
- Let $\text{Div}\Delta^2$ be an arbitrary subdivision of Δ^2. In the 2-dimensional *simplex agreement* task, each process starts with a vertex in Δ^2. If $\tau \in \Delta^2$ is the face composed of the starting vertices, then the processes converge on a simplex in $\text{Div}\tau$. This task is the loop agreement task $\text{Loop}(\text{Div}\Delta^2, \ell)$, where $\ell = (0, 1, 2, p_{01}, p_{12}, p_{20})$, with p_{ij} denoting the unique simple edge path from i to j in the subdivision of the edge $\{i, j\}$.
- The 2-dimensional *N-th barycentric simplex agreement* task is simplex agreement for $\text{Bary}^N \Delta^2$, the N-th iterated barycentric subdivision of Δ^2. Notice that 0-barycentric agreement is just the trivial loop agreement task $\text{Loop}(\Delta^2, \ell)$, where $\ell = (0, 1, 2, ((0, 1)), ((1, 2)), ((2, 0)))$, since a process with input i can directly decide s_i.
- In the 2-dimensional ϵ-*agreement* task, input values are vertices of a face τ of σ, and output values are points of $|\tau|$ that lie within $\epsilon > 0$ of one another in the convex hull of the input values. This task can be solved by a protocol for N-barycentric simplex agreement for suitably large N.
- In the 1-dimensional *approximate agreement* task input values are taken from the set $\{0, 1\}$, and output values are real numbers that lie within $\epsilon > 0$ of one another in the convex hull of the input values. This task can be solved by a 2-dimensional ϵ-agreement protocol.

  Of course, not all tasks can be cast as loop agreement tasks.

5.6.4 Decidability for layered snapshot protocols

We now show that a loop agreement task $\text{Loop}(\mathcal{K}, \ell)$ has layered snapshot protocol for $t \ge 2$ if and only if the triangle loop ℓ is contractible in $\mathcal{K}$. Loop contractibility, however, is undecidable, and therefore so is the question whether an arbitrary loop agreement task has a protocol in this model.

We will need the following standard fact.

Fact 5.6.10. There is a homeomorphism from the 2-disk D^2 to $|\Delta^2|$,

$$g : D^2 \to |\Delta^2|,$$

that carries boundary to boundary: $g(S^1) = \text{skel}^1 \Delta^2$.

Theorem 5.6.11.   *For $t \geq 2$, the loop agreement task* $\mathrm{Loop}(\mathcal{K}, \ell)$ *has a t-resilient layered snapshot protocol if and only if the triangle loop ℓ is contractible.*

Proof.   Note that because $\mathcal{K}$ has dimension 2, $\mathrm{skel}^t \mathcal{K} = \mathcal{K}$ for $t \geq 2$.

Protocol Implies Contractible. By Theorem 4.3.1, if the task $(\Delta^2, \mathcal{K}, \ell)$ has a wait-free layered snapshot protocol, then there exists a continuous map $f : |\Delta^2| \rightarrow |\mathcal{K}|$ carried by Λ. Because f is carried by Λ, f satisfies $f(i) = v_i$, for $i = 0, 1, 2$, and $f(\{i, j\}) \subseteq p_{ij}$, for $0 \leq i, j \leq 2$. Composing with the homeomorphism g of Fact 5.6.10, we see that the map $g \circ f : D^2 \rightarrow |\mathcal{K}|$, restricted to the 1-sphere S^1, is a simple continuous loop λ. Moreover, this continuous loop is a representative of ℓ. Since the map λ can be extended to all of D^2, it is contractible, and so is the triangle loop ℓ.

Contractible Implies Protocol. Let $g : D^2 \rightarrow |\Delta^2|$ be the homeomorphism of Fact 5.6.10.

The edge map ℓ induces a continuous map

$$|\ell| : |\mathrm{skel}^1 \Delta^2| \rightarrow |\mathcal{K}|$$

carried by Λ: $|\ell|(i) = v_i$ for $i = 0, 1, 2$, and $|\ell|(\{i, j\}) \subseteq p_{ij}$ for $0 \leq i, j \leq 2$. The composition of g followed by $|\ell|$ is a simple loop:

$$\lambda : S^1 \rightarrow |\mathcal{K}|,$$

also carried by Λ. Because ℓ is contractible, Fact 5.6.5 implies that λ can be extended to

$$f : D^2 \rightarrow |\mathcal{K}|,$$

also carried by Λ. It is easy to check that the composition

$$f \circ g^{-1} : |\Delta^2| \rightarrow |\mathcal{K}|,$$

is also carried by Λ. Theorem 5.2.7 implies that there is a t-resilient layered snapshot protocol for this loop agreement task. $\square$

Corollary 5.6.12.   *It is undecidable whether a loop agreement task has a t-resilient layered snapshot protocol for $t \geq 2$.*

5.6.5 Decidability with *k*-set agreement

Essentially the same argument shows that the existence of a wait-free loop agreement protocol is also undecidable for k-set layered snapshot protocols for $k > 2$.

Corollary 5.6.13.   *A loop agreement task* $\mathrm{Loop}(\mathcal{K}, \ell)$ *has a wait-free k-set layered snapshot protocol for $k > 2$ if and only if the triangle loop ℓ is contractible.*

It follows from Fact 5.6.6 that it is undecidable whether a loop agreement task has a protocol for three processes in this model.

The situation is different in models capable of solving 1-set or 2-set agreement, such as 1-resilient layered snapshot or message-passing protocols, or wait-free k-set layered snapshot protocols for $k = 1$ or 2.

Theorem 5.6.14.   *In any model capable of solving k-set agreement for $k \leq 2$, it is decidable whether a task has a protocol.*

Proof.   In each of these models, a task $(\mathcal{I}, \mathcal{O}, \Delta)$ has a protocol if and only if there exists a continuous map $f : |\text{skel}^{k-1}\mathcal{I}| \to |\mathcal{O}|$ carried by Δ.

When $k = 1$, this map exists if and only if $\Delta(v)$ is nonempty for each $v \in \mathcal{I}$, which is certainly decidable. When $k = 2$, this map exists if and only if, in addition to the nonemptiness condition, for every pair of vertices v_0, v_1 in $\mathcal{I}$ there is a path from a vertex of $\Delta(v_0)$ to a vertex of $\Delta(v_1)$ contained in $\Delta(\{v_0, v_1\})$. This graph-theoretic question is decidable.   □

5.7 Chapter notes

The layered approach used in this chapter was employed by Herlihy, Rajsbaum, and Tuttle [88,89] for message-passing systems. It was used to prove that connectivity is conserved across layers, something we will do later on. In this chapter we used the more direct approach of showing that subdivisions are created in each layer. Earlier work by Herlihy and Rajsbaum [79] and Herlihy and Shavit [91] was based on the "critical state" approach, a style of argument by contradiction pioneered by Fischer, Lynch, and Paterson [55]. This last paper proved that consensus is not solvable in a message-passing system, even if only one process may fail by crashing, a special case of Theorem 5.5.5. Our message-passing impossibility result is simplified by using layering.

In shared-memory systems the wait-free layered approach used in this chapter was introduced as an "iterated model" of computation by Borowsky and Gafni [26]; see the survey by Rajsbaum [128] for additional references. Algorithms in this model can be presented in a recursive form as described by Gafni and Rajsbaum [68] and in the tutorial by Herlihy, Rajsbaum, and Raynal [87]. Fault-tolerant versions of the model were studied by Rajsbaum, Raynal, and Travers [132]. In Chapter 14 we study the relationship of this model with a more standard model in which processes can write and read the same shared array any number of times.

The *BG-simulation* [27] provides a way to transform colorless tasks wait-free impossibilities bounds to t-resilient impossibilities. As we shall see in Chapter 7, the t-resilient impossibility theorems proved directly in this chapter can be obtained by reduction to the wait-free case using this simulation. The BG simulation and layered models are discussed by Rajsbaum and Raynal [129]. Lubitch and Moran [111] provide a direct model-independent t-resilient impossibility proof of consensus.

Early applications of Sperner's lemma to set agreement are due to Chaudhuri [38] and to Chaudhuri, Herlihy, Lynch, and Tuttle [40]. Herlihy and Rajsbaum [79] present critical state arguments to prove results about the solvability of set agreement using set agreement objects. We explore in Chapter 9 why renaming is weaker than n-set agreement, as shown by Gafni, Rajsbaum, and Herlihy [69].

Junqueira and Marzullo [99,98] introduced the core/survivor-set formalism for characterizing general adversaries used here, and they derived the first lower bounds for synchronous consensus against such an adversary. Delporte-Gallet et al. [46] investigate the computational power of more general adversaries in asynchronous shared memory using simulation. By contrast, the analogous impossibility results proved here use direct combinatorial arguments. The colorless task solvability characterization theorem for adversaries was proved by Herlihy and Rajsbaum [84] (and extended in [86], as discussed in Chapter 13).

Biran, Moran, and Zaks [19] showed that task solvability is decidable in a message-passing system where at most one process can fail by crashing, providing a characterization of solvable tasks in terms of graph connectivity, extending earlier work by Moran and Wolfstahl [118]. They further present a setting where the decision problem is NP-hard [20]. Gafni and Koutsoupias [63] were the first to note that three-process tasks are undecidable for wait-free layered snapshot protocols. This observation was generalized to other models by Herlihy and Rajsbaum [80].

The message-passing barycentric agreement protocol of Figure 5.9 is adapted from the *stable vectors* algorithm of Attiya et al. [9]. Attiya et al. [8] showed that it is possible to simulate shared memory using message-passing when a majority of processes are nonfaulty. One could use this simulation to show that our message-passing characterization follows from the shared-memory characterization.

The hierarchy of loop agreement tasks defined by Herlihy and Rajsbaum [83] will be presented in Chapter 15. Several variants and extensions have been studied. Degenerate loop agreement was defined in terms of two vertices of the output complex instead of three, by Liu, Pu, and Pan [108]. More general rendezvous task were studied by Liu, Xu, and Pan [109]. Similar techniques were used by Fraigniaud, Rajsbaum, and Travers [59] to derive hierarchies of tasks motivated by checkability issues.

Contractibility is undecidable because it reduces to the *word problem* for finitely presented groups: whether an expression reduces to the unit element. This problem was shown to be undecidable by S. P. Novikov [126] in 1955, and the *isomorphism problem* (whether two such groups are isomorphic) was shown to be undecidable by M. O. Rabin [127] in 1958. (For a more complete discussion of these problems, see Stillwell [142] or Sergeraert [140].)

Biran, Moran, and Zaks [21] study the round complexity of tasks in a message-passing system where at most one process can fail by crashing. Hoest and Shavit [94] consider nonuniform layered snapshot subdivisions to study the number of layers needed to solve a task in the wait-free case (see Exercise 5.21 about the complexity of solving colorless tasks).

5.8 Exercises

Exercise 5.1. Show that the colorless complex corresponding to independently assigning values from a set V^{in} to a set of $n + 1$ processes is the n-skeleton of a $|V^{in}|$-dimensional simplex. Thus, it is homeomorphic to the n-skeleton of a $|V^{in}|$-disk.

Exercise 5.2. Show that any colorless task $(\mathcal{I}, \mathcal{O}, \Delta)$ such that $\Delta(v)$ is nonempty for every input vertex v is solvable by a 0-resilient layered snapshot colorless protocol and by a wait-free layered snapshot colorless protocol augmented with consensus objects.

Exercise 5.3. Prove Theorem 5.2.9: There is no t-resilient layered snapshot protocol for t-set agreement.

Exercise 5.4. Use the techniques of this chapter to show that there is a t-resilient k-set agreement layered snapshot protocol for a task $(\mathcal{I}, \mathcal{O}, \Delta)$ if and only if there is a continuous map

$$f : |\text{skel}^{\min(k-1,t)}\mathcal{I}| \to |\mathcal{O}|$$

carried by Δ.

Exercise 5.5. Recall that the *test-and-set* atomically swaps 1 into a memory location and returns that location's prior value. Give an $(n + 1)$-process protocol for solving $\lceil \frac{n+1}{2} \rceil$-set agreement using layered snapshots and *test-and-set* instructions.

Exercise 5.6.   Suppose we are given a "black box" object that solves k-set agreement for $m + 1$ processes. Give a wait-free $(n + 1)$-process layered snapshot protocol for K-set agreement, where

$$K = \left\lfloor \frac{n + 1}{m + 1} \right\rfloor + \min(n + 1 \bmod m + 1, k).$$

Exercise 5.7.   Prove Theorem 5.3.6: There is no k-set layered snapshot protocol for $(k - 1)$-set agreement.

Exercise 5.8.   Consider a model where message delivery is reliable, but the same message can be delivered more than once, and messages may be delivered out of order. Explain why that model is or is not equivalent to the one we use.

Exercise 5.9.   Prove that the set agreement protocol of Figure 5.8 is correct.

Exercise 5.10.   Show how to transform any t-resilient message-passing protocol into a t-resilient layered snapshot protocol, even when $t > (n + 1)/2$.

Exercise 5.11.   Show that barycentric agreement is impossible if a majority of the processes can fail: $2t \geq n + 1$. (*Hint:* A *partition* occurs when two disjoint sets of nonfaulty processes both complete their protocols without communicating.)

Exercise 5.12.   Show that a barycentric agreement protocol is impossible if a process stops forwarding messages when it chooses an output value.

Exercise 5.13.   Prove Theorem 5.5.5: There is no wait-free message-passing protocol for $(k - 1)$-set agreement. (*Hint:* Use Sperner's Lemma.)

Exercise 5.14.   Explain how to transform the set of cores of an adversary into the set of survivor sets, and vice versa. (*Hint:* Use disjunctive and conjunctive normal forms of Boolean logic.)

Exercise 5.15.   Prove Theorem 5.4.5: There is no A-resilient c-set agreement layered snapshot protocol.

Exercise 5.16.   Prove Theorem 5.5.4: For $2t < n + 1$, $(\mathcal{I}, \mathcal{O}, \Delta)$ has a t-resilient message-passing protocol if and only if there is a subdivision Div of $\operatorname{skel}^t \mathcal{I}$ and a simplicial map

$$\phi : \operatorname{Div} \operatorname{skel}^t \mathcal{I} \to \mathcal{O}$$

carried by Δ.

Exercise 5.17.   Prove Theorem 5.5.5: There is no t-resilient message-passing protocol for t-set agreement.

Exercise 5.18.   Construct a loop $\rho : S^1 \to |\mathcal{K}|$ that corresponds to the edge loop given by $e_0 = \{v_0, v_1\}$, $e_1 = \{v_1, v_2\}, \ldots, e_\ell = \{v_\ell, v_{\ell+1}\}$, where $v_0 = v_{\ell+1}$. (*Hint:* Start by dividing the circle into $\ell + 1$ equal parts.)

Exercise 5.19.   Consider a model of computation where a colorless task $(\mathcal{I}, \mathcal{O}, \Delta)$ has a protocol $(\mathcal{I}, \mathcal{P}, \Xi)$ if and only if there is a continuous map

$$f : |\operatorname{skel}^1 \mathcal{I}| \to |\mathcal{O}| \tag{5.8.1}$$

carried by Δ. Prove that it is decidable whether a protocol exists for a colorless task in this model.

Exercise 5.20. Consider a model of computation where a colorless task $(\mathcal{I}, \mathcal{O}, \Delta)$ has a protocol $(\mathcal{I}, \mathcal{P}, \Xi)$ if and only if there is a continuous map

$$f : |\mathrm{skel}^1 \mathcal{I}| \to |\mathcal{O}| \tag{5.8.2}$$

carried by Δ. Prove that every loop agreement task is solvable in this model.

Exercise 5.21. Show that for any n, m, and $t \geq 1$, there is a loop agreement task such that any $(n + 1)$-process t-resilient snapshot protocol that solves it, requires more than m layers. In more detail, suppose the number of edges in each path p_{ij} of the triangle loop $\ell = (v_0, v_1, v_2, p_{01}, p_{12}, p_{20})$ of the task is 2^m, $m \geq 0$. Then any t-resilient snapshot protocol that solves it requires at least m layers. (*Hint:* Use Lemma 5.2.3.)

Exercise 5.22. Show that the t-resilient single-layer snapshot protocol for $(t + 1)$-set agreement protocol of Figure 5.2 still works if we replace the immediate snapshot with a nonatomic scan, reading the layer's memory one word at a time.

Exercise 5.23. Rewrite the protocol of Figure 5.6 to use immediate snapshots.

Exercise 5.24. As noted, because message-passing protocols do not use shared memory, there is less motivation to use layered protocol. Figure 5.12 shows a layered message-passing barycentric agreement protocol. Is it correct?

```
BaryAgree(vᵢ: Vertex): set of Vertex
    Vᵢ: set of Vertex := {vᵢ}
    news: Boolean
    for ℓ := 0 to n + 1 do // execute n + 1  layers
      do
          send(Pᵢ, ℓ, Vᵢ) to all  // send view and layer  number
          count := 0  // collect  messages from at  least  n + 1 − t  processes
          while count < n + 1 − t do
              on receive(Q, ℓ, V) do //accepting  only  messages from layer  ℓ
              count := count + 1
              if Vᵢ ≠ V then
                  news := true
                  Vᵢ := Vᵢ ∪ V
          if not news and no decision then //  at  least  n + 1 − t   collected
              decision := Vᵢ // decide but continue  to  next  layer
    return decision
```

FIGURE 5.12

Layered barycentric agreement message-passing protocol.

Exercise 5.25.   In the adversarial model, suppose we drop the requirement that faulty sets be closed under inclusion. Show that without this requirement, that if all and only sets of n out of $n + 1$ processes are faulty sets, then it is possible to solve consensus.

Exercise 5.26.   Let $\mathcal{F}$ be a faulty set complex with vertices V. Show that a set of process names S is a survivor set of $\mathcal{F}$ if and only if $V \backslash S$ is a facet of $\mathcal{F}$.

This page is intentionally left blank

Byzantine-Resilient Colorless Computation

We now turn our attention from the *crash failure* model, in which a faulty process simply halts, to the *Byzantine failure* model, in which a faulty process can display arbitrary, even malicious, behavior. We will see that the colorless task computability conditions in the Byzantine model are similar to those in the crash failure model except that t, the number of failures that can be tolerated, is substantially lower. Indeed, no process can "trust" any individual input value it receives from another, because that other process may be "lying." A process can be sure an input value is genuine only if it receives that value from at least $t + 1$ processes (possibly including itself), because then at least one of those processes is nonfaulty.

6.1 Byzantine failures

In a *Byzantine failure* model, a faulty process can display arbitrary, even malicious, behavior. A Byzantine process can lie about its input value, it can lie about the messages it has received from other processes, it can send inconsistent messages to nonfaulty processes, and it can collude with other faulty processes. A Byzantine failure-tolerant algorithm is characterized by its resilience t, the number of faulty processes with which it can cope.

The Byzantine failure model was originally motivated by hardware systems such as automatic pilots for airplanes or spacecraft, whereby sensors could malfunction in complex and unpredictable ways. Rather then making risky assumptions about the specific ways in which components might fail, the Byzantine failure model simply assumes that faulty components might fail in the worst way possible.

Distributed Computing Through Combinatorial Topology. http://dx.doi.org/10.1016/B978-0-12-404578-1.00006-1

As before, a faulty process may fall silent, but it may also lie about its input or lie about the information it has received from other processes.

As in the colorless model, a task is defined by a triple $(\mathcal{I}, \mathcal{O}, \Delta)$, where the input and output complexes $\mathcal{I}$ and $\mathcal{O}$ define the possible input and output values, and the carrier map $\Delta : \mathcal{I} \to 2^{\mathcal{O}}$ specifies which output value assignments are legal for which input value assignments. It is important to understand that Δ constrains the inputs and outputs of nonfaulty processes only, since a Byzantine process can ignore its input and choose any output it likes.

The principal difference between the Byzantine and crash failure models is that no process can "trust" any individual input value it receives from another, because that other process may be faulty. A process can be sure an input value is genuine only if it receives that value from at least $t + 1$ processes (possibly including itself), because then at least one of those processes is nonfaulty.[1]

In this chapter, we restrict our attention to tasks $(\mathcal{I}, \mathcal{O}, \Delta)$ whose carrier maps are *strict*. For all input simplices $\sigma_0, \sigma_1 \in \mathcal{I}$,

$$\Delta(\sigma_0 \cap \sigma_1) = \Delta(\sigma_0) \cap \Delta(\sigma_1).$$

We will see (in Theorem 6.5.4) that without this restriction, it may be possible to solve the task without any processes "learning" any other process's input.

We will see that the computability conditions for strict tasks in the Byzantine model are similar to those in the crash failure models except that t, the number of failures that can be tolerated, is substantially lower. Namely, for $n + 1 > (\dim(\mathcal{I}) + 2)t$, a strict colorless task $(\mathcal{I}, \mathcal{O}, \Delta)$ has a t-resilient protocol in the asynchronous Byzantine message-passing model if and only if there is a continuous map

$$f : |\text{skel}^t \mathcal{I}| \to |\mathcal{O}|$$

carried by Δ. The analogous condition for the crash failure models, given in Chapter 5 (Theorem 5.2.7), is the same except that it requires that $t < n + 1$ for read-write memory and that $2t < n + 1$ for message passing. Note also that the crash failure model, unlike the Byzantine failure model, places no constraints on $\dim(\mathcal{I})$, the size (minus 1) of the largest simplex in the input complex.

A necessary condition for a strict task $(\mathcal{I}, \mathcal{O}, \Delta)$ to have a protocol is that $n + 1 > (\dim(\mathcal{I}) + 2)t$. Informally, it is easy to see why this additional constraint is required. As noted, a process cannot "trust" any input value proposed by t or fewer processes, because all the processes that proposed that value may be lying. The requirement that $n + 1 > (\dim(\mathcal{I}) + 2)t$ ensures that at least one input value will be proposed by at least $t + 1$ processes, ensuring that each nonfaulty process will observe at least one "trustworthy" input value.

When analyzing Byzantine failure models, it is natural to start with message-passing systems, where faulty processes are naturally isolated from nonfaulty processes. Later we discuss ways to extend Byzantine failures to shared-memory systems, where we will see that the characterization of solvability for strict colorless tasks remains essentially the same.

---

[1] A nonfaulty process can be sure its own input value is authentic, but it cannot, by itself, convince any other process to accept that value.

6.2 Byzantine communication abstractions

The first step in understanding the asynchronous Byzantine communication model is to build higher-level communication abstractions. These abstractions will allow us to reuse, with some modifications, the protocols developed for the crash failure model.

Communication is organized in asynchronous layers, where a layer may involve several message exchanges. Messages have the form (P, tag, v), where P is the sending process, tag is the message type, and v is a sequence of one or more values. A faulty process can provide arbitrary values for tag and v, but it cannot forge another process's name in the first field.

Reliable broadcast is a communication abstraction constructed from simple message passing that forces Byzantine processes to communicate consistently with nonfaulty processes. A process sends a message to all the others by calling *reliable send*, RBSend(P, tag, v), where P is the name of the sending process, tag is a tag, and v a value. A process receives a message by calling *reliable receive*, RBReceive(P, tag, v), which sets P to the name of the sending process, tag to the message's tag, and v to its value. If fewer than a third of the processes are faulty, that is, if $n + 1 > 3t$, then reliable broadcast provides the following guarantees.

Nonfaulty integrity. If a nonfaulty P never reliably broadcasts (P, tag, v) (by calling RBSend(P, tag, v)), then no nonfaulty process ever reliably receives (P, tag, v) (by calling RBReceive(P, tag, v)).

Nonfaulty liveness. If a nonfaulty P does reliably broadcast (P, tag, v), then all nonfaulty processes will reliably receive (P, tag, v).

Global uniqueness. If nonfaulty processes Q and R reliably receive, respectively, (P, tag, v) and (P, tag', v'), then the messages are equal ($tag = tag'$ and $v = v'$) even if the sender P is faulty.

Global liveness. For nonfaulty processes Q and R, if Q reliably receives (P, tag, v), then R will reliably receive (P, tag, v), even if the sender P is faulty.

Figure 6.1 shows the protocol for reliable broadcast. In figures and proofs, we use $*$ as a wildcard symbol to indicate an arbitrary process name.

1. Each process P broadcasts its message v, labeled with the SEND tag (Line 3).
2. The first time a process receives a SEND message from P (Line 12), it broadcasts v with an ECHO tag.
3. The first time a process receives $n - t + 1$ ECHO messages for v from Q (Line 16), it broadcasts Q and v with a READY tag,
4. The first time a process receives $t + 1$ READY messages for v from Q (Line 20), it broadcasts Q and v with a READY tag.
5. The first time a process receives $n - t + 1$ READY messages for v from Q (Line 6), (Q, v) is reliably delivered to that process.

Lemma 6.2.1. *The reliable broadcast protocol satisfies nonfaulty integrity.*

Proof. Suppose nonfaulty P reliably receives (Q, INPUT, v) from a nonfaulty Q. P must have received at least $n + 1 - t$ $(*, \text{READY}, Q, v)$ messages, so at least $n - 2t + 1$ processes sent $(*, \text{ECHO}, Q, v)$ messages. Let R be the first nonfaulty process to send (R, ECHO, Q, v). Any ECHO messages R received

```
1   protocol ReliableBroadcast
2     RBSend(P, v)
3       send(P, SEND, v) to all
4
5     RBReceive()
6       upon receive (*, READY, Q, v) from at least n + 1 − t processes do
7         return (Q, v)
8
9       background   // run forever
10        upon receive (Q, SEND, v) from Q do
11
12          if P never sent a message of the form (P, ECHO, Q, *)
13            send (P, ECHO, Q, v) to all
14
15          upon receive (*, ECHO, Q, v) from n + 1 − t processes do
16            if P never sent a message of the form (P, READY, Q, *)
17              send (P, READY, Q, v) to all
18
19          upon receive (*, READY, Q, v) from at least t + 1 processes do
20            if P never sent a message of the form (P, READY, Q, *)
21              send (P, READY, Q, v) to all
```

FIGURE 6.1

Reliable broadcast.

from processes other than Q came from faulty processes, so it cannot have received more than t, and therefore it did not send its message at Line 10. Instead, it must have received (Q, INPUT, v) directly from Q, implying that Q sent the message.   □

Lemma 6.2.2.   *The reliable broadcast protocol satisfies nonfaulty liveness.*

Proof.   If P broadcasts (P, INPUT, v), that message will eventually be received by $n + 1 − t$ nonfaulty processes. Each one will send $(*, \text{ECHO}, P, v)$ to all processes, and each will eventually receive $n + 1 − t$ such messages and send $(*, \text{READY}, Q, v)$ to all processes. Each nonfaulty process will eventually receive $n + 1 − t$ of these messages and reliably receive (P, INPUT, v).   □

Lemma 6.2.3.   *The reliable broadcast protocol satisfies global uniqueness.*

Proof.   The uniqueness tests at Lines 16 and 20 ensure that any process that broadcasts $(*, \text{ECHO}, P, v)$ or $(*, \text{READY}, P, v)$ will not broadcast $(*, \text{ECHO}, P, v')$ or $(*, \text{READY}, P, v)$ where $v \neq v$.   □

Lemma 6.2.4.   *The reliable broadcast protocol satisfies global liveness.*

```
getQuorum(tag : Tag): Set of Message
    M := ∅
    while |M| < n + 1 - t or Trusted(M) = ∅ do
        upon RBReceive(Q, tag, v) do
            M := M ∪ {(Q, tag, v)}
    return M
```

FIGURE 6.2

Assemble a Byzantine quorum of messages.

Proof.   Suppose nonfaulty Q reliably receives (P, INPUT, v) from P, which may be faulty, and let R be another nonfaulty process. Q must have received at least $n + 1 - t$ $(*, \text{READY}, P, v)$ messages, and at least $n - 2t + 1 \geq t + 1$ of these came from nonfaulty processes. If at least $t + 1$ nonfaulty processes send $(*, \text{READY}, P, v)$ messages, then every nonfaulty process will eventually receive them and will rebroadcast them at Line 21, ensuring that every nonfaulty process will eventually receive at least $n + 1 - t$ $(*, \text{READY}, P, v)$ messages, causing that message to be reliably received by every nonfaulty process.   □

As in the crash failure model, our first step is to assemble a quorum of messages. As noted earlier, a process can recognize an input as genuine only if it receives that input from $t + 1$ distinct processes. Let M be a set of messages reliably received during a protocol execution. We use Good(M) to denote the set of input values that appear in messages of M that were broadcast by nonfaulty processes, and we use Trusted(M) to denote the set of values that appear in $t + 1$ distinct messages. The getQuorum() method shown in Figure 6.2 waits until (1) it has received messages from at least $n + 1 - t$ processes and (2) it recognizes at least one trusted value. It is safe to wait for the first condition to hold because the process will eventually receive messages from a least $n + 1 - t$ nonfaulty processes. It is safe to wait for the second condition to hold because the requirement that $n + 1 > (d + 1)t$ ensures that some value is represented at least $t + 1$ times among the nonfaulty processes' inputs. The process must wait for both conditions because it may receive $n + 1 - t$ messages without any individual value appearing $t + 1$ times.

Lemma 6.2.5.   *Each call to* getQuorum() *eventually returns, and, for any nonfaulty process P_i that receives message set M_i,*

$$|M_i| \geq n + 1 - t \text{ and Trusted}(M_i) \neq ∅.$$

Proof.   Since $n + 1 > 3t$, the processes can perform reliable broadcast. Notice that the $n + 1 - t$ messages sent by the nonfaulty processes can be grouped by their values:

$$n - t + 1 = \sum_{v \in \text{Good}(M)} |(P, v) : \{(P, v) \in M, P \text{ is non-faulty}\}|.$$

By way of contradiction, assume that every value v in Good(M) was reliably broadcast by at most t nonfaulty processes. It follows that $n + 1 - t \leq |\text{Good}(M)| \cdot t$, which contradicts the hypothesis. Hence, at least one value in Good(M) was reliably broadcast by more than $t + 1$ nonfaulty processes. By the nonfaulty liveness of the reliable broadcast, such a value will eventually be reliably received by all nonfaulty processes.   □

```
SetAgree(v): value
    RBSend(P_i, INPUT, v)
    M: Set of Message := getQuorum(INPUT)
    return min Trusted(M)
```

FIGURE 6.3

Byzantine k-set agreement protocol: Code for P_i.

Lemma 6.2.6.   *After executing* getQuorum(), *for any nonfaulty processes P_i and P_j, $|M_i \setminus M_j| \leq t$.*

Proof.   If $|M_i \setminus M_j| > t$, then M_j missed more than t messages in M, the messages reliably broadcast in layer r. However, this contradicts the fact that $|M_j| \geq n + 1 - t$, where M_j was assembled by the reliable broadcast and receive protocols.   □

6.3 Byzantine set agreement

Theorem 6.3.1.   *The* SetAgree() *protocol shown in Figure 6.3 solves k-set agreement for input simplex σ when $\dim(\sigma) = d > 0$, $k > t$, and $n + 1 > (d + 2)t$.*

Proof.   At most $d + 1$ distinct values are reliably broadcast by nonfaulty processes, so $|\text{Good}(M)| \leq d + 1$. As no more than t messages are missed by any nonfaulty P_i, the value chosen is among the $(t + 1)$ least-ranked input values. Because $k > t$, the value chosen is among the k least-ranked inputs.

6.4 Byzantine barycentric agreement

In the Byzantine barycentric agreement protocol shown in Figure 6.4, each process broadcasts an INPUT message with its input value (Line 6). In the background, it collects the input vertices from the messages it receives (Line 15) and forwards them to all processes in a REPORT message (Line 17). Each P_i keeps track of a set B_i of *buddies*—processes that have reported the same set of vertices (Line 11). The protocol terminates when B_i contains at least $n + 1 - t$ processes (Line 8).

Lemma 6.4.1.   *The sequence of M_i message sets reliably broadcast by P_i in REPORT messages is monotonically increasing, and all processes reliably receive those simplices in that order.*

Proof.   Each P_i's simplex σ_i is monotonically increasing by construction, and so is the sequence of reports it reliably broadcasts. Because channels are FIFO, any other nonfaulty process reliably receives those reports in the same order.   □

Lemma 6.4.2.   *Protocol* BaryAgree() *guarantees that nonfaulty processes P_i and P_j have (i) $|M_i \cap M_j| \geq n + 1 - t$, (ii) $\text{Trusted}(M_i \cap M_j) \neq \emptyset$, and (iii) $M_i \subseteq M_j$ or $M_j \subseteq M_i$.*

Proof.   Call Q_i the set of processes whose reports are stored in R_i at some layer ℓ. Since all reports are transmitted via reliable broadcast, and every nonfaulty process collects $n + 1 - t$ reports, $|Q_i \setminus Q_j| \leq t$ with $|Q_i| \geq n + 1 - t$, which implies that $|Q_i \cap Q_j| \geq n + 1 - 2t$. In other words, any nonfaulty

```
1   BaryAgree(v: vertex ): simplex
2     R_i: array 0..n of Set of Message := (∅,...,∅)
3     M_i: Set of Message := ∅
4     B_i: Set of Process := ∅
5
6     RBSend(P_i, INPUT, v)
7     M_i := getQuorum(INPUT)
8     while |B_i| < n + 1 − t do
9       upon RBReceive(P_j, REPORT, M_j) do
10        R_i[j] := M_j
11        B_i := {P_ℓ : R_i[ℓ] = M_i, 0 ≤ ℓ ≤ n}
12    return Trusted(M_i)
13
14    background // run forever
15      upon RBReceive(P_j, INPUT, u) do
16        M_i := M_i ∪ {u}
17        RBSend(P_i, REPORT, M_i)
```

FIGURE 6.4

Byzantine barycentric agreement protocol for P_j.

processes have $n + 1 − 2t > t + 1$ buddies in common, including a nonfaulty P_k. Therefore, $M_i = R'_k$ and $M_j = R''_k$, where R'_k and R''_k are reports sent by P_k possibly at different occasions.

Since the set M_k is monotonically increasing, either $R'_k \subseteq R''_k$ or $R''_k \subseteq R'_k$, guaranteeing property (iii). Both R'_k and R''_k contain R_k, the first report sent by P_k, by Lemma 6.4.1. Lemma 6.2.5 guarantees that $|R_k| \geq n + 1 − t$ and Trusted(R_k) $\neq \emptyset$, implying properties (i) and (ii). □

Theorem 6.4.3. *Protocol* BaryAgree() *solves barycentric agreement when* $n + 1 > (\dim(\mathcal{I}) + 2)t$.

Proof. By Lemma 6.4.2, nonfaulty processes P_i and P_j, we have that $M_i \subseteq M_j$ or $M_j \subseteq M_i$ and also that Trusted($M_i \cap M_j$) $\neq \emptyset$. It follows that Trusted(M_i) $\subset$ Trusted(M_j), or vice versa, so the sets of values decided, which are faces of σ, are ordered by containment. □

6.5 Byzantine task solvability

Here is the main theorem for Byzantine colorless tasks.

Theorem 6.5.1. *For* $n + 1 > (\dim(\mathcal{I}) + 2)t$, *a strict colorless task* $(\mathcal{I}, \mathcal{O}, \Delta)$ *has a t-resilient protocol in the asynchronous Byzantine message-passing model if and only if there is a continuous map*

$$f : |\mathrm{skel}^t \mathcal{I}| \to |\mathcal{O}|$$

carried by Δ.

Proof. Map Implies Protocol.   Given such a map f, by Theorem 3.7.5, f has a *simplicial approximation*

$$\phi : \text{Bary}^N \text{skel}^t \mathcal{I} \rightarrow \mathcal{O}$$

for some $N > 0$, also carried by Δ. Here is the protocol.

1. Call the Byzantine k-set agreement protocol, for $k = t + 1$, choosing vertices on a simplex in $\text{skel}^t \mathcal{I}$.
2. Call the Byzantine barycentric agreement protocol N times to choose vertices in $\text{Bary}^N \text{skel}^t \mathcal{I}$.
3. Use $\phi : \text{Bary}^N \text{skel}^t \mathcal{I} \rightarrow \mathcal{O}$ as the decision map.

Because ϕ and f are carried by Δ, nonfaulty processes starting on vertices of $\sigma \in \mathcal{I}$ finish on vertices of $\tau \in \Delta(\sigma)$. Also, since $\dim(\sigma) \leq \dim(\mathcal{I})$, the preconditions are satisfied for calling the protocols in each step.

   Protocol Implies Map. Given a protocol, we argue by reduction to the crash failure case. By Theorem 5.2.7, if there is a t-resilient protocol in the crash failure model, then there is a continuous map $f : |\text{skel}^t(\mathcal{I})| \rightarrow |\mathcal{O}|$ carried by Δ. But any t-resilient Byzantine protocol is also a t-resilient crash failure protocol, so such a map exists even in the more demanding Byzantine model.   □

Remark 6.5.2.   Because there is no t-resilient message-passing t-set agreement protocol in the crash failure model, there is no such protocol in the Byzantine failure model.

   Any task for which $(\dim(\mathcal{I}) + 2)t \geq n + 1$ can make only weak guarantees. Consider the following *k-weak agreement* task. Starting from input simplex σ, each P_i chooses a set of vertices V_i with the following properties.

- Each V_i includes at least one valid input value: $|\sigma \cap V_i| > 0$, and
- At most $2t + 1$ vertices are chosen: $|\cup_i V_i| \leq 2t + 1$.

This task has a simple one-round protocol: Each process reliably broadcasts its input value, reliably receives values from $n + 1 - t$ processes, and chooses the least $t + 1$ values among the values it receives. It is easy to check that this task is not strict, and there are executions in which no process ever learns another's input value (each process knows only that its set contains a valid value).

   We now show that any strict task that has a protocol when $n + 1 < (\dim(\mathcal{I}) + 2)t$ is trivial in the following sense.

Definition 6.5.3.   A strict colorless task $(\mathcal{I}, \mathcal{O}, \Delta)$ is *trivial* if there is a simplicial map $\delta : \mathcal{I} \rightarrow \mathcal{O}$ carried by Δ.

   In particular, a trivial task can be solved without communication.

Theorem 6.5.4.   *If a strict colorless task $(\mathcal{I}, \mathcal{O}, \Delta)$ has a protocol for $n + 1 \leq (\dim(\mathcal{I}) + 2)t$, then that task is trivial.*

Proof.   Let $\{v_0, \ldots, v_d\}$ be a simplex of $\mathcal{I}$. Consider an execution where each process P_i has input $v_{i \bmod d}$, all faulty processes behave correctly, and each process in $S = \{P_0, \ldots, P_{n-t}\}$ finishes the protocol with output value u_i without receiving any messages from $T = \{P_{n+1-t}, \ldots, P_n\}$. Let

$$S_j = \left\{ P \in S \,|\, P \text{ has input } v_j \right\}.$$

Note that because $n + 1 - t < t(d + 1)$, each $|S_j| < t + 1$.

Note that if $u_i \in \Delta(\sigma_i)$, and $u_i \in \Delta(\sigma_i')$, then $u_i \in \Delta(\sigma_i \cap \sigma_i')$, so σ has a unique minimal face σ_i such that $u_i \in \Delta(\sigma_i)$. If $\sigma_i = \{v_i\}$ for all i, then the task is trivial, so for some i, there is $v_j \in \sigma_i$, for $i \neq j$.

Now consider the same execution except that the processes in S_j and T all start with input v_i, but the processes in S_j are faulty and pretend to have input v_j. To P_i, this modified execution is indistinguishable from the original, so P_i still chooses u_i, implying that $u_i \in \Delta(\sigma_i \setminus \{v_j\})$, contradicting the hypothesis that σ' has minimal dimension.   □

6.6 Byzantine shared memory

Because the study of Byzantine faults originated in systems where controllers communicate with unreliable devices, most of the literature has focused on message-passing systems. Before we can consider how Byzantine failures might affect shared-memory protocols, we need to define a reasonable model.

We will assume that the shared memory is partitioned among the processes so that each process can write only to its own memory locations, although it can read from any memory location. Without this restriction, a faulty process could overwrite all of memory, and any kind of nontrivial task would be impossible. In particular, a faulty process can write anything to its own memory but cannot write to the memory belonging to a nonfaulty process. As in the crash failure case, nonfaulty processes can take immediate snapshots, writing a value to memory and in the very next step taking an atomic snapshot of an arbitrary region of memory.

A natural way to proceed is to try to adapt the shared memory k-set agreement (Figure 5.2) and barycentric agreement protocols from the crash failure model. It turns out, however, there are obstacles to such a direct attack. As usual in Byzantine models, a process can "trust" an input value only if it is written by at least $t + 1$ distinct processes. It is straightforward to write a getQuorum() protocol that mimics the message-passing protocol of Figure 6.2 and a k-set agreement protocol that mimics the one of Figure 6.3 (see Exercise 6.7).

The difficulty arises in trying to solve barycentric agreement. Suppose there are four processes P, Q, R, and S, where S is faulty. P has input value u and Q has input value v. Suppose P and Q each write their values to shared memory, S writes u, and P takes a snapshot. P sees two copies of u and one of v, so it accepts u and rejects v. Now S, which is faulty, overwrites its earlier value of u with v. Q then takes a snapshot, sees two copies of v and one of u, so it accepts v and rejects u. Although P and Q have each accepted sets of valid inputs, their sets are not ordered by containment, even though they were assembled by atomic snapshots!

Instead, the simplest approach to barycentric agreement is to simulate the message-passing model in the read-write model. Each process has an array whose i^{th} location holds the ith message it sent, and $\perp$ if that message has not yet been sent. When P wants to check for a message from Q, it reads through Q's array from the last location it read, "receiving" each message it finds, until it reaches an empty location. We omit the details, which are straightforward.

Theorem 6.6.1. *For $n + 1 > (\dim(\mathcal{I}) + 2)t$, a strict colorless task $(\mathcal{I}, \mathcal{O}, \Delta)$ has a t-resilient protocol in the asynchronous Byzantine read-write model if and only if there is a continuous map*

$$f : |\mathrm{skel}^t \mathcal{I}| \to |\mathcal{O}|$$

carried by Δ.

Proof. Map Implies Protocol.   Given such a map f, by Theorem 6.5.1, the task has a t-resilient message-passing protocol. This protocol can be simulated in read-write memory as described previously.

Protocol Implies Map. If the task has a t-resilient read-write protocol in the Byzantine model, then it has such a protocol in the crash failure model, and the map exists by Theorem 5.2.7.   □

6.7 Chapter notes

Much of the material in this chapter is adapted from Mendes, Tasson, and Herlihy [115]. Barycentric agreement is related to *lattice agreement* [14,53] and to multidimensional ϵ-approximate agreement as studied by Mendes and Herlihy [114] as well as to vector consensus as studied by Vaidya and Garg [143], both in the case of Byzantine message-passing systems. In the 1-dimensional case, Byzantine approximate agreement protocols were considered first by Dolev *et al.* [47] and by Abraham *et al.* [1].

The k-weak consensus task mentioned in Section 6.5 was called to the authors' attention in a private communication from Zohir Bouzid and Petr Kuznetsov.

The Byzantine failure model was first introduced by Lamport, Shostak, and Pease [107] in the form of the *Byzantine Generals* problem, a problem related to consensus. Most of the literature in this area has focused on the synchronous model (see the survey by Fischer [56]), not the (more demanding) asynchronous model considered here.

Our reliable broadcast protocol is adapted from Bracha [28] and from Srikanth and Toueg [141]. The stable vectors protocol is adapted from Attiya *et al.* [9].

Malkhi *et al.* [112] propose several computational models whereby processes that communicate via shared objects (instead of messages) can display Byzantine failures. Their proposals include "persistent" objects that cannot be overwritten and access control lists. De Prisco *et al.* [43] consider the k-set agreement task in a variety of asynchronous settings. Their notion of k-set agreement, however, uses weaker notions of validity than the one used here.

6.8 Exercises

Exercise 6.1.   Consider two possible Byzantine failure models. In the first, up to t faulty processes are chosen in the initial configuration; in the second, all processes start off nonfaulty, but up to t of them are dynamically designated as faulty in the course of the execution. Prove that these two models are equivalent.

Exercise 6.2.   Consider a Byzantine model in which message delivery is reliable, but the same message can be delivered more than once and messages may be delivered out of order. Explain why that model is or is not equivalent to the one we use.

Exercise 6.3.   Prove that the protocol of Figure 6.3 is correct.

Exercise 6.4.   In the crash failure model, show how to transform any t-resilient message-passing protocol into a t-resilient read-write protocol.

Exercise 6.5.   In the asynchronous message-passing model with crash failures, show that a barycentric agreement protocol is impossible if a majority of the processes can crash ($2t \geq n + 1$).

Exercise 6.6.   In the asynchronous message-passing model with crash failures, show that a barycentric agreement protocol is impossible if a process stops forwarding messages when it chooses an output value.

Exercise 6.7.   Write explicit protocols in the Byzantine read-write model for getQuorum() and k-set agreement based on the protocols of Figures 5.2 and 5.9. Explain why your protocols are correct.

Exercise 6.8.   Suppose the reliable broadcast protocol were shortened to deliver a message as soon as it receives $t + 1$ ECHO messages from other processes. Describe a scenario in which this shortened protocol fails to satisfy the reliable broadcast properties.

Exercise 6.9.   Let $(\mathcal{I}, \mathcal{P}, \Xi)$ be a layered Byzantine protocol in which processes communicate by reliable broadcast. Show that:

- Ξ is not monotonic: if $\sigma \subset \tau$, then
$$\Xi(\sigma) \nsubseteq \Xi(\tau).$$

- For any σ_0, σ_1 in $\mathcal{I}$,
$$\Xi(\sigma_0) \cap \Xi(\sigma_1) \subseteq \Xi(\sigma_0 \cap \sigma_1).$$

Exercise 6.10.   Which of the decidability results of Section 5.6 apply to strict tasks in the Byzantine message-passing model?

Exercise 6.11.   Suppose we replace the send and receive statements in the protocols shown in Figures 5.1 and 5.8 with reliable send and receive statements. Explain why the result is not a correct Byzantine $(t + 1)$-set agreement protocol.

This page is intentionally left blank

Simulations and Reductions

7

CHAPTER OUTLINE HEAD

We present here a general combinatorial framework to translate impossibility results from one model of computation to another. Once one has proved an impossibility result in one model, one can avoid reproving that result in related models by relying on reductions. The combinatorial framework explains how the topology of the protocol complexes in the two models have to be related to be able to obtain a reduction. We also describe an operational framework consisting of an explicit distributed simulation protocol that implements reductions. Although this protocol provides algorithmic intuition behind the combinatorial simulation framework and may even be of practical interest, a key insight behind this chapter is that there is often no need to construct such explicit simulations. Instead, we can treat simulation as a task like any other and apply the computability conditions of Chapter 5 to show when a simulation protocol *exists*. These existence conditions are given in terms of the topological properties of the models' protocol complexes instead of devising pair-wise simulations.

7.1 Motivation

Modern distributed systems are highly complex yet reliable and efficient, thanks to heavy use of abstraction layers in their construction. At the hardware level processes may communicate through low-level shared-register operations, but a programmer uses complex shared objects to manage concurrent threads. Also from the theoretical perspective, researchers have devised algorithms to implement

Distributed Computing Through Combinatorial Topology. http://dx.doi.org/10.1016/B978-0-12-404578-1.00007-3

higher-level-of-abstraction shared objects from lower-level-of-abstraction objects. We have already encountered this technique to build larger set agreement boxes from smaller ones (Exercise 5.6) or to implement snapshots from single-writer/multireader registers (Exercise 4.12). We say snapshots can be *simulated* in a wait-free system where processes communicate using single-writer/single-reader registers. Simulations are useful also to deduce the relative power of abstractions; in this case, snapshots are as powerful as single-writer/single-reader registers, but not more powerful. In contrast, a consensus shared black box cannot be simulated in a wait-free system where processes communicate using only read-write registers, as we have already seen.

Software systems are built in a modular fashion using this simulation technique, assuming a black box for a problem has been constructed and using it to further extend the system. However, this technique is also useful to prove impossibility results. In complexity theory, it is common to prove results by *reduction* from one problem to another. For example, to prove that there is not likely to exist a polynomial algorithm for a problem, one may try to show that the problem is NP-complete. Textbooks typically prove from first principles that satisfiability (SAT) is NP-complete. To show that another problem is also NP-complete, it is enough to show that SAT (or some other problem known to be NP-complete) reduces to the problem in question. Reductions are appealing because they are often technically simpler than proving NP completeness directly.

Reductions can also be applied in distributed computing for impossibility results. For example, suppose we know that a colorless task has no wait-free layered immediate snapshot protocol, and we want to know whether it has a t-resilient protocol for some $t < n$. One way to answer this question is to assume that an $(n + 1)$-process, t-resilient protocol exists and devise a wait-free protocol where $t + 1$ processes "simulate" the t-resilient $(n + 1)$-process protocol execution in the following sense: The $(t + 1)$-processes use the code for the protocol to simulate an execution of the $(n + 1)$-processes. They assemble mutually consistent final views of an $(n + 1)$-process protocol execution during which at most t processes may fail. Each process halts after choosing the output value that would have been chosen by one of the simulated processes. Because the task is colorless, any process can choose any simulated process's output, so this simulation yields a wait-free $(t + 1)$-process layered protocol, contradicting the hypothesis that no such protocol exists. Instead of proving directly that no t-resilient protocol exists, we *reduce* the t-resilient problem to the previously solved wait-free problem.

In general, we can use simulations and reductions to translate impossibility results from one model of computation to another. As in complexity theory, once one has proved an impossibility result in one model, one can avoid reproving that result in related models by relying on reductions. One possible problem with this approach is that known simulation techniques, such as the *BG-simulation* protocol presented in Section 7.4, are model-specific, and a new, specialized simulation protocol must be crafted for each pair of models. Moreover, given two models, how do we know if there *is* a simulation before we start to try to design one?

The key insight behind this chapter is that there is often no need to construct explicit simulations. Instead, we can treat simulation as a task like any other and apply the computability conditions of Chapter 5 to show when a simulation protocol *exists*. These existence conditions are given in terms of the topological properties of the models' protocol complexes and are likely to be easier to determine in general than devising pair-wise simulations. Once it is known that a simulation exists, one may then concentrate on finding an efficient one that might be of practical interest.

7.2 Combinatorial setting

So far we have considered several models of computation. Each one is given by a set of process names, Π; a communication medium, such as shared memory or message-passing; a timing model, such as synchronous or asynchronous; and a failure model, given by an adversary, $\mathbb{A}$. For each model of computation, once we fix a colorless input complex $\mathcal{I}$, we may consider the set of final views of a protocol. We have the combinatorial definition of a protocol (Definition 4.2.2), as a triple $(\mathcal{I}, \mathcal{P}, \Xi)$ where $\mathcal{I}$ is an input complex, $\mathcal{P}$ is a protocol complex (of final views), and $\Xi : \mathcal{I} \to 2^{\mathcal{P}}$ is an execution map. For each $\mathcal{I}$, a model of computation may be represented by all the protocols on $\mathcal{I}$.

Definition 7.2.1. A *model of computation* $\mathbb{M}$ on an input complex $\mathcal{I}$ is a (countably infinite) family of protocols $(\mathcal{I}, \mathcal{P}_i, \Xi_i)$, $i \geq 0$.

Consider for instance the $(n+1)$-process, colorless layered immediate snapshot protocol of Chapter 4. If we take the wait-free adversary and any input complex $\mathcal{I}$, the model $\mathbb{M}^n_{WF}(\mathcal{I})$ obtained consists of all protocols $(\mathcal{I}, \mathcal{P}_r, \Xi_r)$, $r \geq 0$, corresponding to having the layered immediate snapshot protocol execute r layers, where $\mathcal{P}_r$ is the complex of final configurations and Ξ_r the corresponding carrier map. Similarly, taking the t-resilient layered immediate snapshot protocol of Figure 5.1 for $n + 1$ processes and input complex $\mathcal{I}$, $\mathbb{M}^n_t(\mathcal{I})$ consists of all protocols $(\mathcal{I}, \mathcal{P}_r, \Xi_r)$, $r \geq 0$, corresponding to executing the protocol for r layers.

Definition 7.2.2. A model of computation $\mathbb{M}$ *solves* a colorless task $(\mathcal{I}, \mathcal{O}, \Delta)$ if there is a protocol in $\mathbb{M}$ that solves that task.

Recall that a protocol $(\mathcal{I}, \mathcal{P}, \Xi)$ *solves* a colorless task $(\mathcal{I}, \mathcal{O}, \Delta)$ if there is a simplicial map $\delta : \mathcal{P} \to \mathcal{O}$ carried by Δ. Operationally, in each execution, processes end up with final views that are vertices of the same simplex τ of $\mathcal{P}$. Moreover, if the input simplex of the execution is σ, then $\tau \in \Xi(\sigma)$. Each process finishes the protocol in a local state that is a vertex of τ and then applies δ to choose an output value. These output values form a simplex in $\Delta(\sigma)$.

For example, the model $\mathbb{M}_{WF}$ solves the iterated barycentric agreement task $(\mathcal{I}, \text{Bary}^N \mathcal{I}, \text{Bary}^N)$ for any $N > 0$. To see this, we must verify that there is some r_N such that the protocol $(\mathcal{I}, \mathcal{P}_{r_N}, \Xi_{r_N}) \in \mathbb{M}_{WF}$ solves $(\mathcal{I}, \text{Bary}^N \mathcal{I}, \text{Bary}^N)$.

A reduction is defined in terms of two models of computation: a model $\mathbb{R}$ (called the *real model*) and a model $\mathbb{V}$ (called the *virtual model*). They have the same input complex $\mathcal{I}$, but their process names, protocol complexes, and adversaries may differ. The real model reduces to the virtual model if the existence of a protocol in the virtual model implies the existence of a protocol in the real model.

For example, the t-resilient layered immediate snapshot model $\mathbb{M}^n_t(\mathcal{I})$ for $n + 1$ processes trivially reduces to the wait-free model $\mathbb{M}^n_{WF}(\mathcal{I})$. Operationally it is clear why. If a wait-free $n + 1$-process protocol solves a task $(\mathcal{I}, \mathcal{O}, \Delta)$ it tolerates failures by n processes. The same protocol solves the task if only t out of the $n + 1$ may crash. Combinatorially, the definition of reduction is as follows.

Definition 7.2.3. Let $\mathcal{I}$ be an input complex and $\mathbb{R}$, $\mathbb{V}$ be two models on $\mathcal{I}$. The (real) model $\mathbb{R}$ *reduces to* the (virtual) model $\mathbb{V}$ if, for any colorless task $\mathcal{T}$ with input complex $\mathcal{I}$, a protocol for $\mathcal{T}$ in $\mathbb{V}$ implies that there is a protocol for $\mathcal{T}$ in $\mathbb{R}$.

We typically demonstrate reduction using simulation.

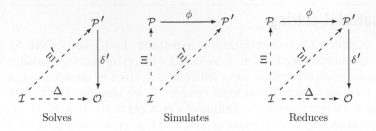

Solves   Simulates   Reduces

FIGURE 7.1

Carrier maps are shown as dashed arrows, simplicial maps as solid arrows. On the left, $\mathcal{P}'$ via δ' solves the colorless task $(\mathcal{I}, \mathcal{O}, \Delta)$. In the middle, $\mathcal{P}$ simulates $\mathcal{P}'$ via ϕ. On the right, $\mathcal{P}$ via the composition of ϕ and δ' solves $(\mathcal{I}, \mathcal{O}, \Delta)$.

Definition 7.2.4. Let $(\mathcal{I}, \mathcal{P}, \Xi)$ be a protocol in $\mathbb{R}$ and $(\mathcal{I}, \mathcal{P}', \Xi')$ a protocol in $\mathbb{V}$. A *simulation* is a simplicial map

$$\phi : \mathcal{P} \to \mathcal{P}'$$

such that for each simplex σ in $\mathcal{I}$, ϕ maps $\Xi(\sigma)$ to $\Xi'(\sigma)$.

The operational intuition is that each process executing the real protocol chooses a simulated execution in the virtual protocol, where each virtual process has the same input as some real process. However, from a combinatorial perspective, it is sufficient to show that there exists a simplicial map $\phi : \mathcal{P} \to \mathcal{P}'$ as above. Note that ϕ may be collapsing: Real processes with distinct views may choose the same view of the simulated execution.

The left-hand diagram of Figure 7.1 illustrates how a protocol solves a task. Along the horizontal arrow, Δ carries each input simplex σ of $\mathcal{I}$ to a subcomplex of $\mathcal{O}$. Along the diagonal arrow, a protocol execution, here denoted Ξ', carries each σ to a subcomplex of its protocol complex, denoted by $\mathcal{P}'$, which is mapped to a subcomplex of $\mathcal{O}$ along the vertical arrow by the simplicial map δ'. The diagram *semi-commutes*: The subcomplex of $\mathcal{O}$ reached through the diagonal and vertical arrows is contained in the subcomplex reached through the horizontal arrow.

Simulation is illustrated in the middle diagram of Figure 7.1. Along the diagonal arrow, Ξ' carries each input simplex σ of $\mathcal{I}$ to a subcomplex of its protocol complex $\mathcal{P}'$. Along the vertical arrow, Ξ carries each input simplex σ of $\mathcal{I}$ to a subcomplex of its own protocol complex $\mathcal{P}$, which is carried to a subcomplex of $\mathcal{P}'$ by the simplicial map ϕ. The diagram semi-commutes: The subcomplex of $\mathcal{P}'$ reached through the vertical and horizontal arrows is contained in the subcomplex reached through the diagonal arrow. Thus, we may view simulation as solving a task. If we consider $(\mathcal{I}, \mathcal{P}', \Xi')$ as a task, where $\mathcal{I}$ is input complex and $\mathcal{P}'$ is output complex, then $\mathcal{P}$ solves this task with decision map ϕ carried by Ξ'.

Theorem 7.2.5. *If every protocol in $\mathbb{V}$ can be simulated by a protocol in $\mathbb{R}$, then $\mathbb{R}$ reduces to $\mathbb{V}$.*

Proof. Recall that if $\mathbb{V}$ has a protocol $(\mathcal{I}, \mathcal{P}', \Xi')$ for a colorless task $(\mathcal{I}, \mathcal{O}, \Delta)$, then there is a simplicial map $\delta' : \mathcal{P}' \to \mathcal{O}$ carried by Δ, that is, $\delta'(\Xi'(\sigma)) \subseteq \Delta(\sigma)$ for each $\sigma \in \mathcal{I}$. If model $\mathbb{R}$ simulates model $\mathbb{V}$, then for any protocol $\mathcal{P}' \in \mathbb{V}$, $\mathbb{R}$ has a protocol $(\mathcal{I}, \mathcal{P}, \Xi)$ in $\mathbb{R}$ and a simplicial map $\phi : \mathcal{P} \to \mathcal{P}'$ such that for each simplex σ in $\mathcal{I}$, $\phi(\Xi(\sigma)) \subseteq \Xi'(\sigma)$.

Let δ be the composition of ϕ and δ'. To prove that $(\mathcal{I}, \mathcal{P}, \Xi)$ solves $(\mathcal{I}, \mathcal{O}, \Delta)$ with δ, we need to show that $\delta(\Xi(\sigma)) \subseteq \Delta(\sigma)$. By construction,

$$\delta'(\phi(\Xi(\sigma))) \subseteq \delta'(\Xi'(\sigma)) \subseteq \Delta(\sigma),$$

so $\mathbb{R}$ also solves $(\mathcal{I}, \mathcal{O}, \Delta)$. □

Theorem 7.2.5 depends only on the *existence* of a simplicial map. Our focus in the first part of this chapter is to establish conditions under which such maps exist. In the second part, we will construct one operationally.

7.3 Applications

In Chapters 5 and 6, we gave necessary and sufficient conditions for solving colorless tasks in a variety of computational models. Table 7.1 lists these models, parameterized by an integer $t \geq 0$. We proved that the colorless tasks that can be solved by these models are the same and those colorless tasks $(\mathcal{I}, \mathcal{O}, \Delta)$ for which there is a continuous map

$$f : |\mathrm{skel}^t \mathcal{I}| \to |\mathcal{O}|$$

carried by Δ. Another way of proving this result is showing that these protocols are equivalent in the simulation sense of Definition 7.2.4.

Lemma 7.3.1. *Consider any input complex $\mathcal{I}$ and any two models $\mathbb{R}$ and $\mathbb{V}$ with $t \geq 0$. For any protocol $(\mathcal{I}, \mathcal{P}', \Xi')$ in $\mathbb{V}$ there is a protocol $(\mathcal{I}, \mathcal{P}, \Xi)$ in $\mathbb{R}$ and a simulation map*

$$\phi : \mathcal{P} \to \mathcal{P}'$$

carried by Ξ'.

Here are some of the implications of this lemma, together with Theorem 7.2.5:

- A $(t+1)$-process wait-free model can simulate an $(n+1)$-process wait-free model, and vice versa. We will give an explicit algorithm for this simulation in the next section.
- If $2t > n+1$, an $(n+1)$-process t-resilient message-passing model can simulate an $(n+1)$-process t-resilient layered immediate snapshot model, and vice versa.
- Any adversary model can simulate any other adversary model for which the minimum core size is the same or larger. In particular, all adversaries with the same minimum core size are equivalent.

Table 7.1 Models that solve the same colorless tasks for each $t \geq 0$.

Processes	Fault Tolerance	Model
$t+1$	Wait-free	Layered immediate snapshot
$n+1$	t-resilient	Layered immediate snapshot
$n+1$	Wait-free	$(t+1)$-set layered immediate snapshot
$n+1$	t-resilient for $2t < n+1$	Message passing
$n+1$	A-resilient, min. core size $t+1$	Layered immediate snapshot with adversary
$n+1$	t-resilient for $n+1 > (\dim \mathcal{I} + 2)t$	Byzantine

- An adversarial model with minimum core size k can simulate a wait-free k-set layered immediate snapshot model.
- A t-resilient Byzantine model can simulate a t-resilient layered immediate snapshot model if t is sufficiently small: $n + 1 > (\dim(\mathcal{I}) + 2)t$.

7.4 BG simulation

In this section, we construct an explicit shared-memory protocol by which $n + 1$ processes running against adversary A can simulate $m + 1$ processes running against adversary A', where A and A' have the same minimum core size. We call this protocol *BG simulation* after its inventors, Elizabeth Borowsky and Eli Gafni. As noted, the results of the previous section imply that this simulation exists, but the simulation itself is an interesting example of a concurrent protocol.

7.4.1 Safe agreement

The heart of the BG simulation is the notion of *safe agreement*. Safe agreement is similar to consensus except it is not wait-free (nor is it a colorless task; see Chapter 11). Instead, there is an *unsafe* region during which a halting process will block agreement. This unsafe region encompasses a constant number of steps. Formally, safe agreement satisfies these conditions:

- *Validity*. All processes that decide will decide some process's input.
- *Agreement*. All processes that decide will decide the same value.

To make it easy for processes to participate in multiple such protocols simultaneously, the safe agreement illustrated in Figure 7.2 is split into two methods: propose(v) and resolve(). When a process joins the protocol with input v, it calls propose(v) once. When a process wants to discover the protocol's result, it calls resolve(), which returns either a value or $\perp$ if the protocol has not yet decided. A process may call resolve() multiple times.

The processes share two arrays: announce[] holds each process's input, and level[] holds each process's *level*, which is 0, 1, or 2. Each P_i starts by storing its input in announce[i], making that input visible to the other processes (Line 9). Next, P_i raises its level from 0 to 1 (Line 10), entering the unsafe region. It then takes a snapshot of the level[] array (Line 11). If any other process is at level 2 (Line 12), it leaves the unsafe region by resetting its level to 0 (Line 13). Otherwise, it leaves the unsafe region by advancing its level to 2 (Line 15). This algorithm uses only simple snapshots because there is no need to use immediate snapshots.

To discover whether the protocol has chosen a value and what that value is, P_i calls resolve(). It takes a snapshot of the level[] array (Line 18). If there is a process still at level 1, then the protocol is unresolved and the method returns $\perp$. Otherwise, P_i decides the value announced by the processes at level 2 whose index is least (Line 22).

Lemma 7.4.1. *At Line 18, once P_i observes that* level[j] $\neq$ 1 *for all j, then no process subsequently advances to level 2.*

Proof. Let k be the least index such that level[k] = 2. Suppose for the sake of contradiction that P_ℓ later sets level[ℓ] to 2. Since level[ℓ] = 1 when the level is advanced, P_ℓ must have set level[ℓ] to 1 after P_i's snapshot, implying that P_ℓ's snapshot would have seen that level[k] is 2, and it would have reset its level to 0, a contradiction. $\qquad\square$

```
1   protocol SafeAgree
2     shared level : array [0.. n] of int  :=  {0,...,0}
3     shared announce: array [0.. n] of value := {⊥,  ...,  ⊥}
4
5     // my snapshot of the  level  array
6     local snap: array [0.. n] of int
7
8     method propose(input: value)
9       announce[i] := input                // make input public
10      level [ i] := 1                      // enter unsafe zone
11      snap:= snapshot(level )
12      if (∃ j | level [j] = 2) then
13        level [ i] := 0                    // leave unsafe zone
14      else
15        level [ i] := 2                    // leave unsafe zone
16
17    method resolve(): value
18      snap := snapshot(level )
19      if (∃ j | level [j] = 1) then
20        return ⊥
21      else
22        return announce[j] for minimal j such that  level [j] = 2
```

FIGURE 7.2

Safe Agreement protocol: Code for P_i.

Lemma 7.4.2.   *If* resolve() *returns a value v distinct from* ⊥, *then all such values are valid and they agree.*

Proof.   Every value written to announce[] is some process's input, so validity is immediate. Agreement follows from Lemma 7.4.1.   □

If a process fails in its unsafe region, it may block another process from eventually returning a value different from ⊥, but only if it fails in this region.

Lemma 7.4.3.   *If all processes are nonfaulty, then all calls to* resolve() *eventually return a value distinct from* ⊥.

Proof.   When each process finishes propose(), its level is either 0 or 2, so eventually no process has level 1. By Lemma 7.4.1, eventually no processes sees another at level 1.   □

7.4.2 The simulation

For BG simulation, the real model $\mathbb{R}$ is an A-resilient snapshot protocol with $n+1$ processes, $P_0, \ldots, P_n$. (It is not layered; see "Chapter notes" section.) The virtual model $\mathbb{V}$ is the A′-resilient layered snapshot protocol with $m + 1$ processes, $Q_0, \ldots, Q_m$. They have the same colorless input complex $\mathcal{I}$, and both adversaries have the same minimum core size $t + 1$. For any given R-layered protocol $(\mathcal{I}, \mathcal{P}', \Xi')$ in $\mathbb{V}$, we need to find a protocol $(\mathcal{I}, \mathcal{P}, \Xi)$ in $\mathbb{R}$ and a simplicial map

$$\phi : \mathcal{P} \to \mathcal{P}'$$

such that, for each simplex σ in $\mathcal{I}$, ϕ maps $\Xi(\sigma)$ to $\Xi'(\sigma)$. We take the code for protocol $(\mathcal{I}, \mathcal{P}', \Xi')$ (as in Figure 5.5) and construct $(\mathcal{I}, \mathcal{P}, \Xi)$ explicitly, with a shared-memory protocol by which the $n + 1$ processes can simulate $(\mathcal{I}, \mathcal{P}', \Xi')$. Operationally, in the BG simulation, an **A**-resilient, $(n + 1)$-process protocol produces output values corresponding to final views an R-layered, **A'**-resilient, $(m + 1)$-process protocol. The processes P_i start with input values, which form some simplex $\sigma \in \mathcal{I}$. They run against adversary **A** and end up with final views in $\mathcal{P}$. If P_i has final view v, then P_i produces as output a view $\phi(v)$, which could have been the final view of a process Q_j in an R-layer execution of the virtual model under adversary **A'**, with input values taken from σ.

The BG-simulation code is shown in Figure 7.3. In the simulated computation, $m + 1$ processes $Q_0, \ldots, Q_m$ share a two-dimensional memory mem$[0..R][0..m]$. At layer 0, the state of each Q_i is its input. At layer r, for $0 \le r \le R$, Q_i writes its current state to mem$[r][i]$, then waits until the set of processes that have written to mem$[r][\cdot]$ constitutes a survivor set for $\mathcal{A}$. Q_i then takes a snapshot of mem$[r][\cdot]$, which becomes its new state. After completing R steps, Q_i halts.

```
1   protocol BGSimulation
2      shared mem: array[0..R][0.. m] of value
3      shared agree: array [0.. R ][0.. m] of SafeAgree
4
5      local pc: array [0.. m] of int := {0,...,0}
6
7      method run(input: value): state        // return simulated view
8         for j := 0 to m do                   // set simulated input to mine
9            agree [0][ j ]. propose(input)
10        do forever
11           for j := 0 to m do                // simulate Q_j
12              r := pc[j]                      // next simulated layer
13              v := agree[r][j]. resolve ()    // resolve last layer's snapshot
14              if v ≠ ⊥ then                   // resolved?
15                 mem[r][j] := v               // write snapshot
16                 if pc[j] = R then            // if simulated state is final ...
17                    return v                  // ... return it, otherwise
18                 if arrived (r, j) then       // survivor set present?
19                    snap := snapshot(mem[r])  // take snapshot
20                    view := values(snap)      // discard process names
21                    agree[r+1][j]. propose(view) // propose it
22                    pc[j] := pc[j] + 1        // step complete
```

FIGURE 7.3

BG Simulation Protocol: Code for P_i.

```
1   method arrived(layer: int, i: int): Boolean
2     shared bitmap: array [0.. R][0.. m] of Boolean := {false ,..., false}
3
4     bitmap[r][i] := true
5     snap := snapshot(bitmap[r])
6     return {p | snap[p] = true} is a survivor set for A'
```

FIGURE 7.4

Testing whether a simulated survivor set has reached a layer.

This computation is simulated by $n+1$ processes $P_0, \ldots, P_n$. Each P_i starts the protocol by proposing its own input value as the input initially written to memory by each Q_j (Line 8). Because the task is colorless, the simulation is correct even if simulated inputs are duplicated or omitted. Thus, if σ is the (colorless) input simplex of the $n + 1$ processes, then each simulated Q_j will take a value from σ as input, and altogether the simplex defined by the $m + 1$ processes' inputs will be a face of σ.

In the main loop (Line 10), P_i tries to complete a step on behalf of each Q_j in round-robin order. For each Q_j, P_i tries to resolve the value Q_j wrote to memory during its previous layer (Line 13). If the resolution is successful, P_i writes the resolved value on Q_j's behalf to the simulated memory (Line 15). Although multiple processes may write to the same location on Q_j's behalf, they all write the same value. When P_i observes that all R simulated layers have been written by simulated survivor sets (Line 16), then P_i returns the final state of some Q_j.

Otherwise, if P_i did not return, P_i checks (Line 18) whether a survivor set of simulated processes for A' has written values for that layer (Figure 7.4). If so, it takes a snapshot of those values and proposes that snapshot (after discarding process names, since the simulated protocol is colorless) as Q_j's state at the start of the next layer. Recall that adversaries A, A' have minimum core size $t + 1$. Thus, when P_i takes a snapshot in Line 19, at least $m + 1 - t$ entries in $mem[r][*]$ have been written, and hence the simulated execution is A'-resilient.

Theorem 7.4.4.   *The BG simulation protocol is correct if s, the maximum survivor set size for the adversaries* A, A' *is less than or equal to* $m + 1 - t$.

Proof.   At most t of the $n + 1$ processors can fail in the unsafe zone of the safe agreement protocol, blocking at most t out of the $m + 1$ simulated processes, leaving $m + 1 - t$ simulated processes capable of taking steps. If $s \leq m + 1 - t$, there are always enough unblocked simulated processes to form a survivor set, ensuring that eventually some process completes each simulated layer. $\square$

7.5 Conclusions

In this chapter we have once more seen the two faces of distributed computing: algorithmic and combinatorial. If we know a task is unsolvable in a certain model $\mathbb{R}$ and we want to show it is unsolvable in another model $\mathbb{V}$, then it is natural to try to reduce model $\mathbb{R}$ to model $\mathbb{V}$ instead of proving the impossibility result from scratch in $\mathbb{V}$, especially if model $\mathbb{V}$ seems more difficult to analyze than model $\mathbb{R}$. The end result of a reduction is a simplicial map from protocols in $\mathbb{R}$ to protocols in $\mathbb{V}$. We can produce such a simplicial map operationally using a protocol in $\mathbb{R}$, or we can show it exists, reasoning about the topological properties of the two models.

The first reduction studied was for k-set agreement. It was known that it is unsolvable in a (real) wait-free model $\mathbb{M}_{WF}^n$ even when $k = n$ for $n + 1$ processes. Proving directly that k-set agreement is unsolvable in a (virtual) t-resilient model, $\mathbb{M}_t^n$, when $k \leq t$ seemed more complicated. Operationally, one assumes (for contradiction) that there is a k-set agreement protocol in $\mathbb{M}_t^n$. Then a generic protocol in $\mathbb{M}_{WF}^n$ is used to simulate one by one the instructions of the protocol to obtain a solution for k-set agreement in $\mathbb{M}_{WF}^n$.

This operational approach has several benefits, including the algorithmic insights discovered while designing a simulation protocol, and its potential applicability for transforming solutions from one model of computation to another. However, to understand the possible reductions among a set of N models of computation, we would have to devise $O(N^2)$ explicit pair-wise simulations, each simulation intimately connected with the detailed structure of two models. Each simulation is likely to be a protocol of nontrivial complexity requiring a nontrivial operational proof.

By contrast, the combinatorial approach described in this chapter requires analyzing the topological properties of the protocol complexes for each of the N models. Each such computation is a combinatorial exercise of the kind that has already been undertaken for many different models of computation. This approach is more systematic and, arguably, reveals more about the underlying structure of the models than explicit simulation algorithms. Indeed, in the operational approach, once a simulation is found, we also learn *why* it existed, but this new knowledge is not easy to formalize; it is hidden inside the correctness proof of the simulation protocol.

We note that the definitions and constructions of this chapter, both the combinatorial and the operational, work only for colorless tasks. For arbitrary tasks, we can also define simulation in terms of maps between protocol complexes, but these maps require additional structure (they must be *color-preserving*, mapping real to virtual processes in a one-to-one way). See Chapter 14.

7.6 Chapter notes

Borowsky and Gafni [23] introduced the BG simulation to extend the wait-free set agreement impossibility result to the t-resilient case. Later, Borowsky, Gafni, Lynch, and Rajsbaum [27] formalized and studied the simulation in more detail.

Borowsky, Gafni, Lynch, and Rajsbaum [27] identified the tasks for which the BG simulation can be used as the colorless tasks. This class of tasks was introduced in Herlihy and Rajsbaum [80,81], under the name *convergence* tasks, to study questions of decidability.

Borowsky and Gafni [25] and later Chaudhuri and Reiners [41] used the BG simulation to define and study the set agreement partial order [79]. Gafni and Kuznetsov [65] used the simulation to reduce solvability of colorless tasks under adversaries to wait-free solvability (see Exercise 7.3). Imbs and Raynal [97] consider a variant of the BG simulation where processes communicate through objects that can be used by at most x processes to solve consensus as well as read-write registers. More broadly, BG simulation can be used to relate the power of different models to solve colorless tasks (see Exercise 7.10). Gafni, Guerraoui, and Pochon [60] use BG simulation to derive a lower bound on the round complexity of k-set agreement in synchronous message-passing systems.

Gafni [62] extends the BG simulation to certain colored tasks, and Imbs and Raynal [96] discuss this simulation further.

The BG-simulation protocol we described is not layered (though the simulated protocol *is* layered). This protocol can be transformed into a layered protocol (see Chapter 14 and the next paragraph). Herlihy, Rajsbaum, and Raynal [87] present a layered safe agreement protocol (see Exercise 7.6).

Other simulations [26,67] address the computational power of layered models, where each shared object can be accessed only once. In Chapter 14 we consider such simulations between models with the same sets of processes, but different communication mechanisms.

Chandra [35] uses a simulation argument to prove the equivalence of t-resilient and wait-free consensus protocols using shared objects.

Exercise 7.1 is based on Afek, Gafni, Rajsbaum, Raynal, and Travers [4], where reductions between simultaneous consensus and set agreement are described.

7.7 Exercises

Exercise 7.1.   In the *k-simultaneous consensus* task a process has an input value for k independent instances of the consensus problem and is required to decide in at least one of them. A process decides a pair (c, d), where c is an integer between 1 and k, and if two processes decide pairs (c, d) and (c', d'), with $c = c'$, then $d = d'$, and d was proposed by some process to consensus instance c and c'. State formally the k-simultaneous consensus problem as a colorless task, and draw the input and output complex for $k = 2$. Show that k-set agreement and k-simultaneous consensus (both with sets of possible input values of the same size) are wait-free equivalent (there is a read-write layered protocol to solve one using objects that implement the other).

Exercise 7.2.   Prove that if there is no protocol for a task using immediate snapshots, then there is no protocol using simple snapshots.

Exercise 7.3.   Using the BG simulation, show that a colorless task is solvable by an A-resilient layered snapshot protocol if and only if it is solvable by a t-resilient layered immediate snapshot protocol, where t is the size of the minimum core of A (and in particular by a $t + 1$ process wait-free layered immediate snapshot protocol).

Exercise 7.4.   Explain why the wait-free safe agreement protocol does not contradict the claim that consensus is impossible in the wait-free layered immediate snapshot memory.

Exercise 7.5.   The BG simulation uses safe agreement objects that are not wait-free. Suppose consensus objects are available. What would be the simulated executions if the BG-simulation used consensus objects instead of safe agreement objects?

Exercise 7.6.   Describe an implementation of safe agreement using two layers of wait-free immediate snapshots. Explain why your protocol is not colorless.

Exercise 7.7.   Prove Lemma 7.3.1.

Exercise 7.8.   In the BG simulation, what is the maximum number of snapshots a process can take to simulate an R-round layered protocol?

Exercise 7.9. For the BG simulation, show that the map ϕ, carrying final views of the simulating protocol to final views of the simulated protocol, is *onto*: every simulated execution is produced by some simulating execution.

Exercise 7.10. Consider Exercise 5.6, where we are given a "black box" object that solves k-set agreement for $m + 1$ processes. Define a wait-free layered model that has access to any number of such boxes as well as read-write registers. Use simulations to find to which of the models considered in this chapter it is equivalent in the sense that the same colorless tasks can be solved.

Exercise 7.11. We have seen that it is undecidable whether a colorless task has a t-resilient layered snapshot protocol for $t \geq 2$ (Corollary 5.6.12). Use simulations to conclude undecidability results in other models. More generally, suppose that in a virtual model $\mathbb{V}$ colorless task solvability is undecidable. State a theorem that allows us to conclude undecidability in a real model $\mathbb{R}$.

General Tasks

PART

III

This page intentionally left blank

Read-Write Protocols
for General Tasks

8

CHAPTER OUTLINE HEAD

So far we have focused on protocols for *colorless* tasks—tasks in which we care only about the tasks' sets of input and output values, not which processes are associated with which values. Whereas many important tasks are colorless, not all of them are. Here is a simple example of a "colored" task: In the *get-and-increment* task, if $n + 1$ processes participate, then each must choose a unique integer in the range $0, \ldots, n$. (If a single process participates, it chooses 0, if two participate, one chooses 0 and the other chooses 1, and so on.) This task is not colorless, because it matters which process takes which value. In this chapter we will see that the basic framework for tasks and protocols extends easily to study general tasks. However, we will have to defer the computability analysis to later chapters. Although we have been able to analyze colorless tasks using simple tools from combinatorial topology, we will see that understanding more general kinds of tasks will require more sophisticated concepts and techniques.

8.1 Overview

The underlying operational model is the same as the one described in Chapter 4. The notions of processes, configurations, and executions are all unchanged.

Distributed Computing Through Combinatorial Topology. http://dx.doi.org/10.1016/B978-0-12-404578-1.00008-5

```
// There are N  layers
shared mem: array[0..N−1][0..n] of Value
protocol ColorlessLayered (input: Value): Value
    view: Value := input          // initial  view is input value
    for ℓ := 0 to N − 1 do
      immediate
        mem[ℓ][i] := view
        snap := snapshot(mem[ℓ][*])
      view :=  snap
    return δ(view)                // apply decision map to final  view
```

FIGURE 8.1

Layered immediate snapshot protocol: Pseudo-code for P_i.

As with colorless protocols, computation is split into two parts: a task-independent, full-information protocol and a task-dependent decision. In the task-independent part, each process repeatedly communicates its view to the others, receives their views in return, and updates its own state to reflect what it has learned. When enough communication layers have occurred, each process chooses an output value by applying a task-dependent decision map to its final view. In contrast to colorless protocols, each process keeps track not only of the set of views it has received but also which process sent which view.

In more detail, each process executes *a layered immediate snapshot protocol* of Figure 8.1. This protocol is similar to the one in Figure 4.1, except that a process does not discard the process names when it constructs its view. Initially, P_i's view is its input value. During layer ℓ, P_i performs an immediate snapshot; it writes its current view to mem$[\ell][i]$ and in the very next step takes a snapshot of that layer's row, mem$[\ell][*]$. mem$[\ell][*]$. Instead of discarding process names, P_i takes as its new view its most recent immediate snapshot. As before, after completing all layers, P_i chooses a decision value by applying a deterministic decision map δ to its final view. An execution produced by a layered immediate snapshot protocol is called a *layered execution*.

Now that processes may behave differently according to which process has which input value, we can consider task specifications that encompass process names, as in the *get-and-increment* example we saw earlier. We first extend colorless tasks to general tasks, and then we extend the combinatorial notions of protocols and protocol complexes to match.

8.2 Tasks

Recall that there are $n + 1$ processes with names taken from Π, V^{in} is a domain of input values and V^{out} a domain of output values.

The principal difference between colorless tasks and general tasks is that for general tasks, the vertices of the input and output complexes are labeled with process names, and the task carrier map preserves process names.

Definition 8.2.1.   A (general) *task* is a triple $(\mathcal{I}, \mathcal{O}, \Delta)$, where

- $\mathcal{I}$ is a pure chromatic *input complex*, colored by Π and labeled by V^{in} such that each vertex is uniquely identified by its color together with its label;
- $\mathcal{O}$ is a pure chromatic *output complex*, colored by Π and labeled by V^{out} such that each vertex is uniquely identified by its color together with its label;
- Δ is a name-preserving (chromatic) carrier map from $\mathcal{I}$ to $\mathcal{O}$.

If v is a vertex, we let name(v) denote its color (usually a process name) and view(v) its label (usually that process's view). The first two conditions of Definition 8.2.1 are equivalent to requiring that the functions (name, view) : $V(\mathcal{I}) \to \Pi \times V^{in}$ and (name, view) : $V(\mathcal{O}) \to \Pi \times V^{out}$ be injective.

Here is how to use these notions to define a specific task: In the *get-and-increment* task described at the start of this chapter, one imagines the processes share a counter, initially set to zero. Each participating process increments the counter, and each process returns as output the counter's value immediately prior to the increment. If $k + 1$ processes participate, each process chooses a unique value in the range $[k]$.

This task is an example of a *fixed-input* task, where V^{in} contains only one element, $\perp$. If in addition the process names are $\Pi = [n]$, the input complex consists of a single n-simplex and all its faces, whose i^{th} vertex is labeled with $(i, \perp)$. Figure 8.2 shows the output complex for the three-process get-and-increment task. The output complex $\mathcal{O}$ consists of six triangles representing the distinct ways one can assign 0, 1, and 2 to three processes. The color of a vertex (white, gray, or black) represents its name. Note that for ease of presentation, some of the vertices drawn as distinct are actually the same.

In general, the facets of the output complex of this task are indexed by all permutations of the set $[n]$, and the carrier map Δ is given by

$$\Delta(\sigma) = \{\tau \in \mathcal{O} \mid \text{name}(\tau) \subseteq \text{name}(\sigma), \text{ and value}(\tau) \subseteq \{0, \ldots, \dim \sigma\}\}.$$

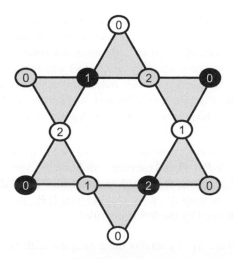

FIGURE 8.2

Output complex for the three-process get-and-increment task. Note that some vertices depicted as distinct are actually the same.

We will review many more examples of tasks in Section 8.3.

> **Mathematical Note 8.2.2.**   The output complex for the *get-and-increment* task is a well-known simplicial complex, which we call a *rook complex*. In the rook complex Rook $(n + 1, N + 1)$, simplices correspond to all rook placements on an $(n + 1) \times (N + 1)$ chessboard so that no two rooks can capture each other (using $n + 1$ rooks). For *get-and-increment*, $\mathcal{O} = \text{Rook} (n + 1, n + 1)$. The topology of these complexes is complicated and not generally known.

8.3 Examples of tasks

In this section, we describe a number of tasks, some of which will be familiar from earlier chapters on colorless tasks. Expressing these tasks as general tasks casts new light on their structures. Some of these tasks cannot be expressed as colorless tasks and could not be analyzed using our earlier concepts and mechanisms.

8.3.1 Consensus

Recall that in the *consensus* task, each process starts with an input value. All processes must agree on a common output value, which must be some process's input value. In the *binary consensus* task, the input values can be either 0 or 1. Formally, there are $n + 1$ processes. The input complex $\mathcal{I}$ has vertices labeled (P, v), where $P \in \Pi$, $v \in \{0, 1\}$. Furthermore, for any subset $S \subseteq \Pi$, $S = \{P_0, \dots, P_\ell\}$, and any collection of values $\{v_0, \dots, v_\ell\}$ from $\{0, 1\}$, the vertices $(P_0, v_0) \cdots (P_\ell, v_\ell)$ form an ℓ-simplex of $\mathcal{I}$, and such simplices are precisely all the simplices of $\mathcal{I}$. Figure 8.3 shows two examples.

> **Mathematical Note 8.3.1.**   In binary consensus, the input complex $\mathcal{I}$ is the *join* of $n + 1$ simplicial complexes $\mathcal{I}_P$, for $P \in \Pi$. Each $\mathcal{I}_P$ consists of two vertices, $(P, 0)$ and $(P, 1)$, and no edges. This complex is homeomorphic to an n-dimensional sphere, and we sometimes call it a *combinatorial sphere*. There is a geometrically descriptive way to view this complex embedded in $(n + 1)$-dimensional Euclidean space, with axes indexed by names from Π. For every $P \in \Pi$, we place the vertex $(P, 1)$ on the P's axis at coordinate 1, and vertex $(P, 0)$ on at coordinate -1. The simplices fit together to form a boundary of a polytope known as a *crosspolytope*.

The output complex $\mathcal{O}$ for binary consensus consists of two disjoint n-simplices. One simplex has $n + 1$ vertices labeled $(P, 0)$, for $P \in \Pi$ and the other one has $n + 1$ vertices labeled $(P, 1)$, for $P \in \Pi$. This complex is disconnected, with two connected components—a fact that will be crucial later.

Finally, we describe the carrier map $\Delta : \mathcal{I} \to 2^{\mathcal{O}}$. Let $\sigma = \{(P_0, v_0), \dots, (P_\ell, v_\ell)\}$ be a simplex of $\mathcal{I}$. The subcomplex $\Delta(\sigma)$ is defined by the following rules:

1. If $v_0 = \cdots = v_\ell = 0$, then $\Delta(\sigma)$ contains the ℓ-simplex with vertices labeled by $(P_0, 0), \dots,$ $(P_\ell, 0)$, and all its faces.
2. If $v_0 = \cdots = v_\ell = 1$, then $\Delta(\sigma)$ contains the ℓ-simplex with vertices labeled by $(P_0, 1), \dots,$ $(P_\ell, 1)$, and all its faces.

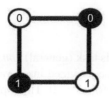

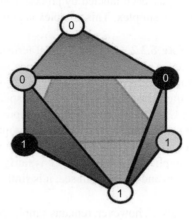

FIGURE 8.3

Input complexes for two and three processes with binary inputs. Here and elsewhere, vertex colors indicate process names, and numbers indicate input values.

3. if $\{v_0, \ldots, v_\ell\}$ contains both 0 and 1, then $\Delta(\sigma)$ contains the two disjoint ℓ-simplices: one has vertices $(P_0, 0), \ldots, (P_\ell, 0)$, and the other has vertices labeled $(P_0, 1), \ldots, (P_\ell, 1)$, together with all their faces.

It is easy to check that Δ is a carrier map. It is clearly rigid and name-preserving. To see that it satisfies monotonicity, note that if $\sigma \subset \tau$, then the set of process names in σ is contained in the set of process names of τ, and similarly for their sets of values. Adding vertices to σ can only increase the set of simplices in $\Delta(\sigma)$, implying that $\Delta(\sigma) \subset \Delta(\tau)$.

Although the carrier map Δ is monotonic, it is not strict. For example, if $\sigma = \{(0, 0), (1, 1)\}$, and $\tau = \{(1, 1), (2, 0)\}$. then $\sigma \cap \tau = \{(1, 1)\}$, and

$$\Delta(\sigma \cap \tau) = \Delta(\{(1, 1)\}) = \{(1, 1)\}.$$

But $\Delta(\sigma)$ has facets

$$\{(0, 0), (1, 0)\} \text{ and } \{(0, 1), (1, 1)\},$$

and $\Delta(\tau)$ has facets

$$\{(1, 0), (2, 0)\} \text{ and } \{(1, 1), (2, 1)\}.$$

It follows that

$$\Delta(\sigma) \cap \Delta(\tau) = \{\{(1, 0)\}, \{(1, 1)\}\},$$

and so

$$\Delta(\sigma \cap \tau) \subset \Delta(\sigma) \cap \Delta(\tau).$$

If there can be more than two possible input values, we call this task (general) *consensus*. As before, there are $n + 1$ processes with names taken from $[n]$ that can be assigned input values from a finite set, which we can assume without loss of generality to be $[m]$, for $m > 0$. The input complex $\mathcal{I}$ has $(m + 1)(n + 1)$ vertices, each labeled by process name and value. Any set of vertices having different process names forms a simplex. This complex is pure of dimension n.

> **Mathematical Note 8.3.2.**   In topological terms, the input complex for consensus with $m + 1$ possible input values is a join of $n + 1$ copies of simplicial complexes $\mathcal{I}_P$, for $P \in \Pi$. Each $\mathcal{I}_P$ consists of the $m + 1$ vertices $(P, 0), \ldots, (P, m)$, and no higher-dimensional simplices. This complex arises often enough that we give it a special name: It is a *pseudosphere*. It is discussed in more detail in Chapter 13. Recall that the input complex for binary consensus is a topological n-sphere, which is a manifold (every $(n - 1)$-simplex is a face of exactly two n-simplices). In the general case, the input complex is not a manifold, since an $(n - 1)$-dimensional simplex is a face of exactly m n-simplices. Nevertheless, $\mathcal{I}$ is fairly standard and its topology (meaning homotopy type) is well known, and as we shall see, it is similar to that of an n-dimensional sphere.

The output complex $\mathcal{O}$, however, remains simple, consisting of $m + 1$ disjoint simplices of dimension n, each corresponding to a possible common output value. The carrier map Δ is defined as follows: Let $\sigma = \{(P_0, v_0), \ldots, (P_\ell, v_\ell)\}$ be an ℓ-simplex of $\mathcal{I}$. The subcomplex $\Delta(\sigma)$ is the union of the simplices $\tau_0 \cup \cdots \cup \tau_\ell$, where $\tau_i = \{(P_0, v_i), \ldots, (P_\ell, v_i)\} \in \mathcal{O}$ for all $i = 0, \ldots, \ell$. Note that two simplices in this union are either disjoint or identical. Again, the carrier map Δ is rigid and name-preserving. Furthermore, monotonicity is satisfied, since growing the simplex can only increase the number of simplices in the union.

8.3.2 Approximate agreement

In the *binary approximate agreement* task, each process is again assigned input 0 or 1. If all processes start with the same value, they must all decide that value; otherwise they must decide values that lie between 0 and 1, all within ϵ of each other, for a given $\epsilon > 0$.

As in Section 4.2.2, we assume for simplicity that $t = \frac{1}{\epsilon}$ is a natural number and allows values $0, \frac{1}{t}, \frac{2}{t}, \ldots, \frac{t-1}{t}, 1$ as output values for the $n+1$ processes. The input complex $\mathcal{I}$ is the same as in the case of binary consensus, namely, the combinatorial n-sphere. The output complex $\mathcal{O}$ here is a bit more interesting. It consists of $(n + 1)(t + 1)$ vertices, indexed by pairs $(P, \frac{v}{t})$, where $P \in \Pi$, and $v \in [t]$. A set of vertices $\{(P_0, \frac{v_0}{t}), \ldots, (P_\ell, \frac{v_\ell}{t})\}$ forms a simplex if and only if the following two conditions are satisfied:

- The $P_0, \ldots, P_\ell$ are distinct.
- For all $0 \le i < j \le \ell$, we have $|v_i - v_j| \le 1$.

> **Mathematical Note 8.3.3.**   It is easy to describe the topology of the output complex for approximate agreement. For $i = 0, \ldots, t - 1$, let $\mathcal{O}_i$ denote the subcomplex of $\mathcal{O}$ consisting of all simplices

spanned by vertices $(0, \frac{i}{t}), \ldots, (n, \frac{i}{t}), (0, \frac{i+1}{t}), \ldots, (n, \frac{i+1}{t})$. As noted before, each $\mathcal{O}_i$ is a combinatorial n-sphere. By the definition of $\mathcal{O}$ we have $\mathcal{O} = \mathcal{O}_0 \cup \cdots \cup \mathcal{O}_{t-1}$. So $\mathcal{O}$ is a union of t copies of combinatorial n-spheres. Clearly, the spheres $\mathcal{O}_i$ and $\mathcal{O}_j$ share an n-simplex if $|i - j| = 1$ and are disjoint otherwise. The concrete geometric visualization of the complex $\mathcal{O}$ is as follows: Start with t disjoint copies of combinatorial n-spheres $\mathcal{O}_0, \ldots, \mathcal{O}_{t-1}$. Glue $\mathcal{O}_0$ with $\mathcal{O}_1$ along the simplex $((0, \frac{1}{t}), \ldots, (n, \frac{1}{t}))$, glue $\mathcal{O}_1$ with $\mathcal{O}_2$ along the simplex $((0, \frac{2}{t}), \ldots, (n, \frac{2}{t}))$, and so on. Note that for each $i = 1, \ldots, t - 1$, the simplices $\sigma_i = ((0, \frac{i}{t}), \ldots, (n, \frac{i}{t}))$ and $\sigma_{i+1} = ((0, \frac{i+1}{t}), \ldots, (n, \frac{i+1}{t}))$ are opposite inside of $\mathcal{O}_i$. So one can view $\mathcal{O}$ as a stretched n-sphere whose inside is further subdivided by $t - 1$ n-disks into t chambers. If we are interested in homotopy type only, we can do a further simplification. It is a well-known fact in topology that shrinking a simplex to a point inside of a simplicial complex does not change its homotopy type. Accordingly, we can also shrink any number of disjoint simplices. In particular, we can shrink the simplices $\sigma_1, \ldots, \sigma_{t-1}$ to points. The result is a chain of n-spheres attached to each other sequentially at opposite points. One can then let the attachment points slide on the spheres. This does not change the homotopy type, and in the end we arrive at a space obtained from t copies of an n-sphere by picking a point on each sphere and then gluing them all together. We obtain what is called a *wedge* of t copies of n-spheres.

Finally, we describe the carrier map Δ. Take a simplex σ in $\mathcal{I}$, $\sigma = ((P_0, v_0), \ldots, (P_\ell, v_\ell))$. We distinguish two different cases:

1. If $v_0 = \cdots = v_\ell = v$, then $\Delta(\sigma)$ is the simplex spanned by the vertices $(P_0, v), \ldots, (P_\ell, v)$, together with all its faces.
2. If the set $\{v_0, \ldots, v_\ell\}$ contains two different values, then $\Delta(\sigma)$ is the subcomplex of $\mathcal{O}$ consisting of all the vertices whose *name* label is in the set $\{P_0, \ldots, P_\ell\}$, together with all the simplices of $\mathcal{O}$ spanned by these vertices.

8.3.3 Set agreement

Approximate agreement is one way of relaxing the requirements of the consensus task. Another natural relaxation is the *k-set agreement* task. Like consensus, each process's output value must be some process's input value. Unlike consensus, which requires that all processes agree, k-set agreement imposes the more relaxed requirement that that no more than k distinct output values be chosen. Consensus is 1-set agreement.

In an input n-simplex, each vertex can be labeled arbitrarily with a value from $[m]$, so the input complex is the same pseudosphere as for general consensus. In an output n-simplex, each vertex is labeled with a value from $[m]$, but the simplex can be labeled with no more than k distinct values. The carrier map Δ is defined by the following rule: For $\sigma \in \mathcal{I}$, $\Delta(\sigma)$ is the subcomplex of $\mathcal{O}$ consisting of all $\tau \in \mathcal{O}$, such that

- $\text{value}(\tau) \subseteq \text{value}(\sigma)$
- $\text{name}(\tau) \subseteq \text{name}(\sigma)$
- $|\text{value}(\tau)| \leq k$

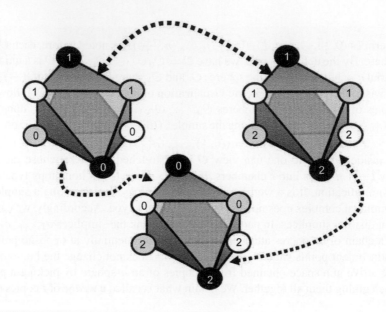

FIGURE 8.4

Output complex for 3-process, 2-set agreement.

Figure 8.4 shows the output complex for three-process 2-set agreement. This complex consists of three combinatorial spheres "glued together" in a ring. It represents all the ways one can assign values to three processes so that all three processes are not assigned distinct values.

8.3.4 Chromatic agreement

We will find that one of the most useful tasks is the *chromatic agreement* task. Here processes start on the vertices of a simplex σ in an arbitrary input complex, $\mathcal{I}$, and they decide on the vertices of a single simplex in the standard chromatic subdivision Chσ (as defined in Section 3.6.3). Formally, the *chromatic agreement* task with input complex $\mathcal{I}$ is the task $(\mathcal{I}, \text{Ch } \mathcal{I}, \text{Ch})$, where, in the triple's last element, the chromatic subdivision operator Ch is interpreted as a carrier map.

8.3.5 Weak symmetry breaking

In the *weak symmetry breaking* task, each process is assigned a unique input name from Π, and the participating processes must sort themselves into two groups by choosing as output either 0 or 1. In any final configuration in which all $n + 1$ processes participate, at least one process must choose 0, and at least one must choose 1. That is, the output complex $\mathcal{O}$ consists of all simplices with at most $n + 1$ vertices of the form $\{(P_0, v_0) \ldots, (P_\ell, v_\ell)\}$, with $P_i \in \Pi$, $v_i \in \{0, 1\}$, and if $\ell = n$, then not all v_i are equal. The part of the output complex for 2-process weak symmetry breaking, shown in Figure 8.5, is for three specific names, represented with the colors white, gray, and black. It is an *annulus*, a combinatorial disk with a hole in the center where the all-zero simplex is missing. One may also think of

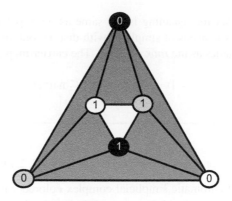

FIGURE 8.5

Output complex: 2-process weak symmetry breaking.

this complex as a combinatorial cylinder in 3-dimensional Euclidean space, where the all-zero and the all-one simplices are missing.

If the names of the processes are taken from a space of names Π with $n + 1$ names, say $\Pi = [n]$, weak symmetry breaking has a trivial protocol: The process with name 0 decides 0; all others decide 1. The task becomes interesting when $|\Pi|$ is large, because no fixed decisions based on input names will work. We study this task in the next chapter.

The weak symmetry-breaking task is formally specified as follows: The input complex $\mathcal{I}$ has $|\Pi|$ vertices, labeled by pairs $(P, \bot)$, where $P \in \Pi$. A set of vertices $\{(P_0, \bot) \ldots, (P_\ell, \bot)\}$ forms a simplex if and only if the P_i are distinct and it contains at most $n + 1$ vertices. We assume $\Pi = [N]$, with $N \gg n$. Each input simplex represents a way of assigning distinct names from $[N]$ to the $n + 1$ processes.

The carrier map $\Delta : \mathcal{I} \to 2^{\mathcal{O}}$ is defined as follows:

$$\Delta(\sigma) = \{\tau \in \mathcal{O} \mid \text{name}(\tau) \subseteq \text{name}(\sigma)\}.$$

Mathematical Note 8.3.4.   The input complex for weak symmetry breaking is the n-skeleton of an N-simplex. The output complex $\mathcal{O}$ is a combinatorial cylinder: the standard combinatorial n-sphere with two opposite n-simplices removed.

8.3.6 Renaming

In the *renaming* task, as in the weak symmetry-breaking task, processes are issued unique input names taken from $[N]$, where $N \gg n$. In this task, however, they must choose unique output names taken from $[M]$, where M is typically much smaller than N. Here too, trivial solutions where processes decide output names based on their input names without communication are not possible when $N \gg n$. We discuss this task, and its relation to weak symmetry breaking, in Chapter 12.

Formally, the input complex for renaming is the same as the input complex for weak symmetry breaking. The output complex consists of simplices with distinct output values taken from $[M]$. This complex is known in mathematics as the *rook complex*. The carrier map Δ is given by

$$\Delta(\sigma) = \{\tau \in \mathcal{O} \mid \text{name}(\tau) \subseteq \text{name}(\sigma)\}.$$

8.4 Protocols

Definition 8.4.1. A *protocol* for $n + 1$ processes is a triple $(\mathcal{I}, \mathcal{P}, \Xi)$ where:

- $\mathcal{I}$ is a pure n-dimensional chromatic simplicial complex colored with names from Π and labeled with values from V^{in} such that each vertex is uniquely identified by its color, together with its label.
- $\mathcal{P}$ is a pure n-dimensional chromatic simplicial complex colored with names from Π and labeled with values from Views such that each vertex is uniquely identified by its color, together with its label.
- $\Xi : \mathcal{I} \to 2^{\mathcal{P}}$ is a chromatic strict carrier map such that $\mathcal{P} = \cup_{\sigma \in \mathcal{I}} \Xi(\sigma)$.

Definition 8.4.2. Assume we are given a task $(\mathcal{I}, \mathcal{O}, \Delta)$ for $n + 1$ processes and a protocol $(\mathcal{I}, \mathcal{P}, \Xi)$. We say that the protocol *solves* the task if there exists a chromatic simplicial map $\delta : \mathcal{P} \to \mathcal{O}$, called the *decision map*, satisfying

$$\delta(\Xi(\sigma)) \subseteq \Delta(\sigma) \tag{8.4.1}$$

for all $\sigma \in \mathcal{I}$.

Treating configurations as simplices gives us an elegant vocabulary for comparing global states. Two configurations σ_0 and σ_1 of a complex are *indistinguishable* to a process if that process has the same view in both. As simplices, σ_0 and σ_1 share a face, $\sigma_0 \cap \sigma_1$, that contains the processes for which σ_0 and σ_1 are indistinguishable. The higher the dimension of this intersection, the more "similar" the configurations.

Just as for colorless tasks, each process executes a layered immediate snapshot protocol with a two-dimensional array $\text{mem}[\ell][i]$, where row ℓ is shared only by the processes participating in layer ℓ, and column i is written only by P_i. Initially, P_i's view is its input value. During layer ℓ, P_i executes an immediate snapshot, writing its current view to $\text{mem}[\ell][i]$, and in the very next step taking a snapshot of that layer's row. Finally, after completing all layers, P_i chooses a decision value by applying a deterministic decision map δ to its final view. Unlike the protocols we used for colorless tasks, the decision map does not operate on colorless configurations. Instead, the decision map may take process names into account.

8.4.1 Single-layer immediate snapshot protocols

Figure 8.6 shows the protocol complex for a three-process, single-layer immediate snapshot protocol. Here, process P (black) has input p, Q (gray) has input q, and R (white) has input r. In the vertex marked at the top, P's immediate snapshot returns only its own value, whereas in the vertex marked on the right edge, Q's immediate snapshot returns both its own value and P's value, whereas at each vertex marked in the center, each immediate snapshot returns all three processes' values.

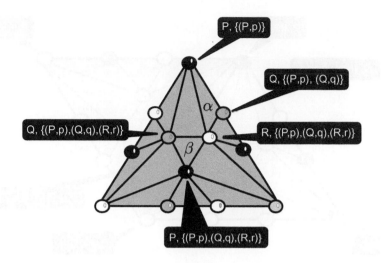

P, {(P,p)}

Q, {(P,p), (Q,q)}

Q, {(P,p),(Q,q),(R,r)}

R, {(P,p),(Q,q),(R,r)}

α

β

P, {(P,p),(Q,q),(R,r)}

FIGURE 8.6

Single-layer immediate snapshot executions for three processes. In this figure and others, we use vertex colors to stand for process names. Here P is black, Q is gray, and R is white. Note that this complex is a standard chromatic subdivision.

The 2-simplex marked α corresponds to the fully sequential execution where P, Q, and R take steps sequentially. It consists of three vertices, each labeled with process state at the end of this execution. The black vertex is labeled with $(P, \{(P, p)\})$, the gray vertex $(Q, \{(P, p), (Q, q)\})$, and the white vertex $(R, \{(P, p), (Q, q), (R, r)\})$.

Similarly, the 2-simplex marked as β corresponds to the fully concurrent execution, and all three processes take steps together. Because the fully sequential and fully concurrent executions are indistinguishable to R, these two simplices share the white vertex labeled $(R, \{(P, p), (Q, q), (R, r)\})$.

Figure 8.6 reveals why we choose to use immediate snapshots as our basic communication pattern: The protocol complex for a single-input simplex σ is a *subdivision* of σ. In fact, this complex is none other than Ch σ, the *standard chromatic subdivision* of σ, defined in Section 3.6.3. Also it is clear that the protocol complex for a single-input simplex σ is a *manifold*: Each $(n-1)$-dimensional simplex is contained in either one or two n-dimensional simplices. We will show in Chapter 16 that protocol complexes for layered immediate snapshot executions are always subdivisions of the input complex.

What happens if we add one more initial configuration? Suppose R can have two possible inputs, r and s. The input complex consists of two 2-simplices (triangles) that share an edge. Figure 8.7 shows the resulting protocol complex, where some of the vertices are labeled with processes' final views. (As one would expect, the protocol complex is the chromatic subdivision of two 2-simplices that share an edge.) R has two vertices corresponding to solo executions, where it completes the protocol before any other process has taken a step. In each vertex, it sees only its own value at one vertex r and the other s.

The vertices along the subdivided edge between the subdivided triangles correspond to executions in which P and Q finish the protocol before R takes a step. These final states are the same whether R starts with input r or s, which is the formal way of saying that "P and Q never learn R's input."

We are ready to define precisely the one-layer immediate snapshot protocol.

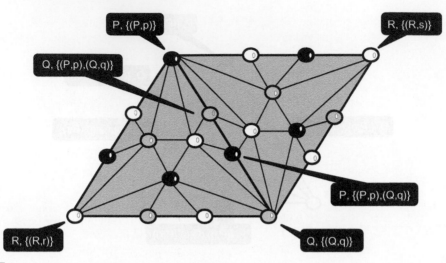

FIGURE 8.7

Protocol complex for two inputs and one layer (selected final views are labeled).

Definition 8.4.3.   The *1-layer layered immediate snapshot protocol* $(\mathcal{I}, \mathcal{P}, \Xi)$ for $n + 1$ processes is:

- The input complex $\mathcal{I}$ can be any pure n-dimensional chromatic simplicial complex colored with names from Π and labeled with V^{in}.
- The carrier map $\Xi : \mathcal{I} \to 2^{\mathcal{P}}$ sends each simplex $\sigma \in \mathcal{I}$ to the subcomplex of final configurations of single-layer immediate snapshot executions where all and only the processes in σ participate.
- The protocol complex $\mathcal{P}$, is the union of $\Xi(\sigma)$, over all $\sigma \in \mathcal{I}$.

It is easy to check that the 1-layer immediate snapshot protocol is indeed a protocol, according to Definition 8.4.1. One needs to verify the three items in this definition; the most interesting is stated as a lemma:

Lemma 8.4.4. Ξ *is a chromatic strict carrier map, with* $\mathcal{P} = \cup_{\sigma \in \mathcal{I}} \Xi(\sigma)$.

The carrier map Ξ describes the structure of $\mathcal{P}$, identifying parts of $\mathcal{P}$ where processes run without hearing from other processes. For example, for a vertex $q \in \mathcal{I}$, $\Xi(q)$ is a vertex $q' \in \mathcal{P}$, with name$(q) =$ name$(q') = P$, and view$(q') = \{q\}$. The final state q' is at the end of the execution

$$C_0, \{P\}, C_1$$

for one process, P, where both C_0 and C_1 contain the state of only P. That is, the $(n + 1)$-process protocol encompasses a solo protocol for each process.

Similarly, a one-layer protocol for two processes, say P, Q, can be obtained by focusing on the executions where only P and Q participate, considering the subcomplex $\mathcal{I}'$ of $\mathcal{I}$ of initial states for P and Q and applying Ξ to $\mathcal{I}'$. See, for example, Figure 8.7, where the solo protocol and the 2-process protocol correspond to the "boundary" of the subdivision. (We will formalize this notion later.) Meanwhile, notice that a boundary is where some processes do not see others, and hence the inputs of those processes can change without the boundary processes noticing. In the figure, the final states corresponding to executions where only P and Q participate are a boundary 1-dimensional complex (a line) that is contained in two subdivisions for three processes, one for the case where the input of R is r and the other where its input is s. Also, the boundary of this 1-dimensional complex contains two vertices, one for the solo execution of P, the other for the solo execution of Q.

In general, take any subcomplex $\mathcal{I}'$ of the input complex $\mathcal{I}$, where $\mathcal{I}'$ is pure and k-dimensional, colored with names Π', where $\Pi' \subseteq \Pi, |\Pi'| \geq k + 1$. Let Ξ' be the restriction of Ξ to $\mathcal{I}'$. Then $(\mathcal{I}', \mathcal{P}', \Xi')$ is a protocol for $k + 1$ processes, where $\mathcal{P}'$ is the image of $\mathcal{I}'$ under Ξ'. This protocol corresponds to executions where $k + 1$ processes participate, and the others crash before taking a step.

8.4.2 Multilayer protocols

In a one-layer protocol, each process participates in exactly one step, communicating exactly once with the others. There are several ways to generalize this model to allow processes to take multiple steps. One approach is to allow processes to take more than one step in a layer. We will consider this extension in later chapters. For now, however, we will construct *layered* protocols using composition. Recall from Definition 4.2.3 that in the composition of two protocols, each view from the first protocol serves as the input to the second protocol. We define an (r-layer) *layered execution* protocol to be the r-fold composition of 1-layer protocols. (Operationally, the corresponding execution is just the concatenation of each layer's execution.)

For example, Figure 8.8 shows a single-input, 2-process protocol complex for one and two layers. Here we assume that each process has its name as its input value. Each input vertex is labeled with that process's name, and each protocol complex vertex is labeled with the values received from each process, or $\emptyset$ if no value was received. Each process communicates its initial state in the first layer, and its state at the end of the layer becomes its input to the second layer.

For three processes, Figure 8.9 shows part of the construction of a 2-layer protocol complex, where all three processes are active in each layer. As before, assume that processes P, Q, R start with respective inputs p, q, r. The simplex marked α in the single-layer complex corresponds to the single-layer execution where P takes an immediate snapshot, and then Q and R take concurrent immediate snapshots. If the protocol now runs for another layer, the input simplex for the second layer is labeled with views $(P, \{p\})$ for P and $\{(P, p), (Q, q), (R, r)\}$ for Q and R. The set of all possible 1-layer executions defines a subdivision of α.

Similarly, the simplex marked β corresponds to the execution where P, Q, and R take their immediate snapshots sequentially, one after the other. Here, P and Q have the same views in α and β. Namely, α and β share an edge, and R has different views in the two simplices (R has view $\{p, q\}$ in σ_2). The input simplex for the second layer is labeled with views $\{(P, p)\}$ for P, $\{(P, p), (Q, q)\}$ for R, and $\{(P, p), (Q, q), (R, r)\}$ for Q. The set of all possible 1-layer executions defines a subdivision of σ_0. Continuing in this way, the 2-layer protocol complex for an input n-simplex σ is the two-fold subdivision $\mathrm{Ch}^2\sigma$.

The k-layered layered immediate snapshot protocol $(\mathcal{I}, \mathcal{P}, \Xi)$ for $n + 1$ processes is the composition of k 1-layer protocols. As in the 1-layer case, if $\mathcal{I}'$ is a subcomplex of the input complex $\mathcal{I}$, where $\mathcal{I}'$ is pure, and k-dimensional, colored with names Π', where $\Pi' \subseteq \Pi, |\Pi'| \geq k+1$, and Ξ' is the restriction of Ξ to $\mathcal{I}'$, then $(\mathcal{I}', \mathcal{P}', \Xi')$ is a protocol for $k + 1$ processes, where $\mathcal{P}'$ is the image of $\mathcal{I}'$ under Ξ'.

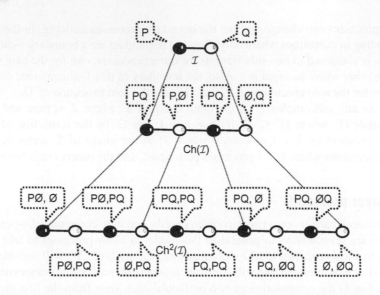

FIGURE 8.8

Input and protocol complexes for two processes: zero, one, and two layers. Each input vertex is labeled with that process's name, and each protocol complex vertex is labeled with the values received from each process, or Ø if no value was received.

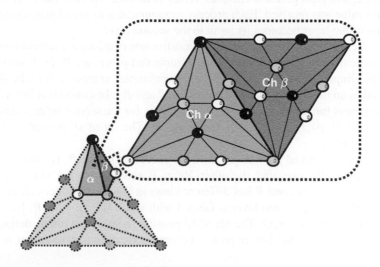

FIGURE 8.9

The complex at the bottom is the protocol complex for a single-layer immediate snapshot protocol, and the complex at the top is a detail of the 2-layer protocol complex, which is a subdivision of the single-layer complex.

It corresponds to executions where $k + 1$ processes participate, and they never see the other processes; the others crash initially.

Protocol complexes for layered immediate snapshot have the following "manifold" property.

Lemma 8.4.5. *If $(\mathcal{I}, \mathcal{P}, \Xi)$ is a layered immediate snapshot protocol complex for $n + 1$ processes, and σ is an $(n − 1)$-dimensional simplex of $\Xi(\tau)$, for some n-simplex τ, then σ is contained either in one or in two n-dimensional simplices of $\Xi(\tau)$.*

The proof of this important property is discussed in Chapter 9.

8.4.3 Protocol composition

Recall from (4.2.3) that the composition of two protocols $(\mathcal{I}, \mathcal{P}, \Xi)$ and $(\mathcal{I}', \mathcal{P}', \Xi')$, where $\mathcal{P} \subseteq \mathcal{I}'$, is the protocol $(\mathcal{I}, \mathcal{P}', \Xi' \circ \Xi)$, where $(\Xi' \circ \Xi)(\sigma) = \Xi'(\Xi(\sigma))$. (This definition applies to protocols for both colored and general tasks.)

8.5 Chapter notes

The first formal treatment of the consensus task is due to Fischer, Lynch, and Paterson [55], who proved that this task is not solvable in a message passing system even if only one process may crash and processes have direct communication channels with each other. The result was later extended to shared memory by Loui and Abu-Amara in [110] and by Herlihy in [78].

Chaudhuri [37] was the first to investigate k-set agreement, where a partial impossibility result was shown. In 1993, three papers [23,90,134] were published together at the same conference showing that there is no wait-free protocol for set agreement using shared read-write memory or message passing. Herlihy and Shavit [90] introduced the use of simplicial complexes to model distributed computations. Borowsky and Gafni [23] and Saks and Zaharoughu [134] introduced layered read-write executions. The first paper called them "immediate snapshot executions"; the second called them "block executions."

Attiya and Rajsbaum [16] later used layered read-write executions in a combinatorial model to show the impossibility of k-set agreement. They explain that the crucial properties of layered executions is that (1) they are a subset of all possible wait-free executions, and (2) they induce a protocol complex that is a divided image (similar to a subdivision) of the input complex. In our terminology, there is a corresponding strict carrier map on a protocol complex that is an orientable manifold. A proof that layered read-write executions induce a subdivision of the input complex appears in [101]. The standard chromatic subdivision of a simplex has appeared before in the discrete geometry literature under the name *anti-prismatic subdivision* [147].

The renaming task was first proposed by Attiya, Bar-Noy, Dolev, Peleg, and Reischuk [9]. Herlihy and Shavit [91], together with Castañeda and Rajsbaum [34], showed that there is no wait-free shared-memory protocol for certain instances of renaming. Several authors [16,91,102] have used weak symmetry breaking to prove the impossibility of renaming. A *symmetry-breaking* family of fixed-input tasks was studied by Castañeda, Imbs, Rajsbaum, and Raynal [29,95]. For an introductory overview of renaming in shared-memory systems, see Castañeda, Rajsbaum, and Raynal [34].

The get-and-increment task is an adaptive instance of *perfect renaming*, whereby processes have to choose names in the range 0 to n. A renaming algorithm is *adaptive* if the size of the new name space depends only on the number of processes that ask for a new name (and not on the total number of processes). See Exercise 8.1. Gafni et al. [66] show that adaptive $(2k - 1)$-renaming (output name space is $1, \ldots, 2k - 1$, where k is the number of processes that actually participate in the execution) is equivalent to n-set agreement (where $n + 1$ processes agree on at most n input values).

In the *strong symmetry-breaking* task, processes decide binary values, and not all processes decide the same value when all participate, as in weak symmetry breaking. In addition, in every execution (even when fewer than $n + 1$ participate) at least one process decides 0. Borowsky and Gafni [23] show that strong symmetry breaking (which they call *(n,n−1)-set-test-and-set*) is equivalent to n-set agreement and hence is strictly stronger than weak symmetry breaking, as we explain in Chapter 9. See Exercise 8.3.

Borowsky and Gafni [24] introduced the immediate snapshot model, showed how to implement immediate snapshots in the conventional read-write model, and showed us how immediate snapshots can be used to solve renaming. Later they introduced the *iterated immediate snapshot* model [26]. Gafni and Rajsbaum [67] present a simulation that shows that if a task is solvable using read-write registers directly, it can also be solved in the iterated model, where each register is accessed only once.

We use the term *layered executions* for our high-level abstract model (sometimes called *iterated executions* in the literature). In the terminology of Elrad and Francez [51], the layered execution model is a *communication-closed layered model*. Instances of this model include the layered read-write memory model and the layered message-passing model. Rajsbaum [128] gives a survey how layered (iterated) immediate snapshot executions have proved useful. Hoest and Shavit [93] examine their implications for complexity.

Other high-level abstract models have been considered by Gafni [61] using failure detectors notions and by Moses and Rajsbaum [120] for situations where at most one process may fail. Various cases of the message-passing model have been investigated by multiple researchers [3,36,103,120,136,138].

Rook complexes appeared first in Garst's Ph.D. thesis [71] and are also known under the name *chessboard complexes*.

The condition-based consensus task of Exercise 8.11 is taken from Mostefaoui, Rajsbaum, and Raynal [122].

8.6 Exercises

Exercise 8.1. Show that the map Δ defined for the get-and-increment task is indeed a chromatic carrier map from $\mathcal{I}$ to $\mathcal{O}$. Consider other chromatic carrier maps from $\mathcal{I}$ to $\mathcal{O}$ and compare the corresponding variants of get-and-increment they define.

Exercise 8.2. In Exercise 5.1 we considered the colorless complex corresponding to independently assigning values from a set V^{in} to a set of $n + 1$ processes. The colored version of the complex is more complicated. As we shall see in Chapter 13, it is a pseudosphere. Give an informal argument that, when $|V^{in}| = 2$, the complex constructed by assigning all combinations of binary values to $n + 1$ processes is a combinatorial sphere. (*Hint:* Think of equators, north and south poles, and argue by induction on n.)

Exercise 8.3. In the *strong symmetry-breaking* task, processes decide binary values, and not all processes decide the same value when all participate, as in weak symmetry breaking. In addition, in every execution (even when less than $n + 1$ participate) at least one process decides 0. Define the input complex, output complex and carrier map formally. Show that strong symmetry breaking is equivalent to n-set agreement: There is a wait-free read-write layered protocol that can invoke n-set agreement objects and solves strong symmetry breaking, and vice versa.

Exercise 8.4.   Let $(\mathcal{I}, \mathcal{P}, \Xi)$ be a protocol for a task $(\mathcal{I}, \mathcal{O}, \Delta)$. Explain why the decision map $\delta : \mathcal{P} \to \mathcal{O}$ must be a simplicial map.

Exercise 8.5.   Explicitly write out the approximate agreement protocol described in Section 8.3.2 for shared memory and for message-passing. Prove it is correct. (*Hint*: Use induction on the number of layers.)

Exercise 8.6.   Consider the following protocol intended to solve k-set agreement for $k \leq n$. Each process has an *estimate*, initially its input. For r layers, each process communicates its estimate, receives estimates from others, and replaces its estimate with the smallest value it sees.
Prove that this protocol does not work for any value of r.

Exercise 8.7.   Show that both the binary consensus and leader election tasks defined in Section 8.2 are monotonic.

Exercise 8.8.   In the *barycentric agreement task* covered in earlier chapters, processes start on the vertices of a simplex σ in a chromatic complex $\mathcal{I}$ and must decide on the vertices of a simplex in *baryσ*. (More than one process can decide the same vertex.) Explain how the chromatic agreement task of Section 8.3.4 can be adapted to solve this task.

Exercise 8.9.   In the *ϵ-approximate agreement* task covered in earlier chapters, processes are assigned as input points in a high-dimensional Euclidean space $\mathbb{R}^N$ and must decide on points that lie within the convex hull of their inputs and within ϵ of one another for some given $\epsilon > 0$. Explain how the iterated chromatic agreement task of Section 8.3.4 can be adapted to solve this task.

Exercise 8.10.   Prove that the standard chromatic subdivision is mesh-shrinking.

Exercise 8.11.   In a *condition-based* consensus task, processes can start with input values from some set $V \geq 2$ and must agree on the input value of one of the processes, as in consensus. However, not all input assignments with values from V are possible. A *condition* defines which input assignments are possible. In the *C-consensus task,* the condition C states that the largest input value assigned to a process must be assigned to at least $f + 1$ processes. Show that although there is no t-resilient colorless protocol that solves C-consensus, there is a t-resilient C-consensus protocol in the form of Figure 5.1, except that instead of computing view using the set of values with values(M), the view is the multiset of values, hence allowing us to count how many times each value is the input of some process.

This page is intentionally left blank

Manifold Protocols

Theoretical distributed computing is primarily concerned with classifying tasks according to their difficulty. Which tasks can be solved in a given distributed computing model? We consider here two important tasks: set agreement and weak symmetry breaking. It turns out that the immediate snapshot protocols of Chapter 8 cannot solve these tasks. Moreover, we will identify a broader class of protocols called *manifold protocols* that cannot solve *k*-set agreement. (The impossibility proof for weak symmetry breaking is more complicated and is deferred to Chapter 12.)

Given that neither task can be solved by layered immediate snapshots, it is natural to ask which task is *harder*. One way of comparing the difficulty of two tasks, T_1, T_2, is to assume we have access to an "oracle" or "black box" that can solve instances of T_1 and ask whether we can now solve T_2. In this sense, we will show that set agreement is strictly stronger than weak symmetry breaking; we can construct a protocol for weak symmetry breaking if we are given a "black box" that solves set agreement, but not vice versa.

We investigate these particular questions here because they can be addressed with a minimum of mathematical machinery. We will rely on two classical constructs. The first is a class of complexes called *pseudomanifolds*, and the second is a classical result concerning pseudomanifolds, called *Sperner's lemma*.[1] In later chapters, we generalize these techniques to address broader questions.

9.1 Manifold protocols

The single-layer immediate snapshot protocol introduced in Chapter 8 has a simple but interesting property: In any $(n + 1)$-process protocol complex, each $(n - 1)$-simplex is contained in either one or two n-simplices. In the 3-process case, the resulting complex looks like a discrete approximation to a surface.

In this section we define this property formally. A protocol that has this property is called a *manifold protocol*, and we will see that any such protocol is limited in the tasks it can solve. Moreover, we will see that all layered immediate snapshot protocols are manifold protocols.

9.1.1 Subdivisions and manifolds

In Figure 8.6, it is apparent that the single-layer immediate snapshot protocol complex shown is a subdivision of the input complex. Formally, an $(n + 1)$-process protocol $(\mathcal{I}, \mathcal{P}, \Xi)$ is a *subdivision protocol* if $\mathcal{P}$ is a subdivision of $\mathcal{I}$ and the subdivision carrier map Ξ is chromatic (recall Definition 3.4.9). Furthermore, Figure 8.9 suggests that longer executions produce finer subdivisions. A subdivision protocol is a special case of a manifold.

> **Mathematical Note 9.1.1.** In point-set topology, an *n-manifold* is a space where every point has a neighborhood homeomorphic to n-dimensional Euclidean space, whereas an *n-manifold with boundary* is a space where every point has a neighborhood homeomorphic either to n-dimensional Euclidean space or to n-dimensional Euclidean half-space. A torus, for example, is a 2-dimensional manifold or *surface*. A pinched torus, shown in Figure 9.1, is not a manifold, because the "pinch" has no neighborhood homeomorphic to the plane.

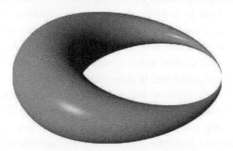

FIGURE 9.1

A pinched torus is not a point-set manifold.

---

[1] We saw a restricted version of Sperner's lemma in Chapter 5.

Definition 9.1.2.   We say that a pure abstract simplicial complex of dimension n is *strongly connected* if any two n-simplices can be connected by a sequence of n-simplices in which each pair of consecutive simplices has a common $(n - 1)$-dimensional face.

For brevity, we sometimes simply say that two such n-simplices can be *linked*, understanding that every n-simplex is linked to itself. Being linked is clearly an equivalence relation. In particular, it is transitive.

Definition 9.1.3.   A pure abstract simplicial complex $\mathcal{M}$ of dimension n is called a *pseudomanifold with boundary* if it is strongly connected, and each $(n - 1)$-simplex in $\mathcal{M}$ is a face of precisely one or two n-simplices.

Because *pseudomanifold with boundary* is such a long and awkward term, we will refer to such complexes simply as *manifolds* in this book, even though, as noted in Remark 9.1.1, this term has a slightly different meaning in other contexts.

An $(n - 1)$-simplex in $\mathcal{M}$ is an *interior* simplex if it is a face of exactly two n-simplices, and it is a *boundary* simplex if it is a face of exactly one. The *boundary* subcomplex of $\mathcal{M}$, denoted $\partial \mathcal{M}$, is the set of simplices contained in its boundary $(n - 1)$-simplices. For an n-dimensional simplex σ, let 2^{σ} be the complex containing σ and all its faces, and $\partial 2^{\sigma}$ the complex of faces of σ of dimension $n - 1$ and lower. (When there is no ambiguity, we will sometimes denote these complexes simply as σ and $\partial \sigma$.)

Manifolds are preserved by subdivisions: If $\mathcal{M}$ is an n-manifold, then any subdivision of $\mathcal{M}$ is again an n-manifold. Figure 9.2 shows a two-dimensional manifold (with an empty boundary complex).

Indeed, the single-layer protocol complex for three processes in Figure 8.6 is a manifold with boundary, as we shall soon prove. Furthermore, the single-layer protocol complex has a recursive structure of manifolds within manifolds, similar to subdivision protocols, with subdivisions within subdivisions. The boundary of a single-layer three-process layered snapshot protocol complex contains the

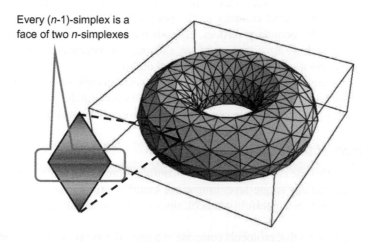

Every (n-1)-simplex is a face of two n-simplexes

FIGURE 9.2

A manifold complex.

executions where only two processes participate and itself consists of the union of three manifolds with boundary. For every two processes, the executions where only they participate again form a manifold with boundary (and in fact, a subdivision) and contain executions where only one process participates. An execution where a single process participates is itself a degenerate manifold, consisting of a single vertex. This structure is conveniently captured using carrier maps.

Definition 9.1.4. An $(n + 1)$-process protocol $(\mathcal{I}, \mathcal{P}, \Xi)$ is a *manifold protocol* if:

- For any simplex σ of $\mathcal{I}$ the subcomplex $\Xi(\sigma)$ is a manifold (automatically it will have the same dimension as σ)
- The protocol map commutes with the boundary operator

$$\partial \Xi(\sigma) = \Xi(\partial \sigma) \tag{9.1.1}$$

for all $\sigma \in \mathcal{I}$.

We say Ξ is a *manifold protocol map* and $\mathcal{P}$ is a *manifold protocol complex*.

Note that this definition applies to arbitrary protocols, not just layered immediate snapshot protocols.

Here is the operational intuition behind Property 9.1.1. Let σ be an input simplex, and σ^{n-1} an $(n - 1)$-face of σ where the vertex labeled with process P is discarded. Recall from Chapter 8 that $\Xi(\sigma^{n-1})$ is the complex generated by executions starting from σ where P does not participate. Consider the following execution: The processes other than P execute by themselves, halting on the vertices of an $(n - 1)$-simplex $\tau \in \Xi(\sigma^{n-1})$. After that, P starts running deterministically by itself until it halts. Because there is only one such execution, there is only one n-simplex τ' containing τ.

For layered immediate snapshot protocols, the protocol complexes are subdivisions of the input complex. However, the manifold protocol definition is more general. Consider the manifold protocol shown in Figure 9.3. The input complex $\mathcal{I}$ is a 2-dimensional simplex with all its faces. The protocol complex is a 2-dimensional "punctured torus," which is a torus with one 2-simplex removed. The map Ξ sends the boundary of the input complex to the boundary of the punctured torus and sends the input complex vertices to the boundary vertices. Ξ sends the input complex's 2-simplex to the entire protocol complex. (Although we are not aware of any existing computer architecture that supports such a protocol, it is nevertheless a well-defined mathematical object.)

Except for layered immediate snapshots, few of the protocol complexes that arise naturally in the study of distributed computing are manifolds. Nevertheless, we start with the study of manifold protocols because the insights they provide will ease our approach to more complicated models.

9.1.2 Composition of manifold protocols

In this section we prove that the composition of two manifold protocols is again a manifold protocol. In Section 9.2 we will see that any single-layer immediate snapshot protocol is a manifold protocol. Any multilayer protocol is therefore a manifold protocol, since it is the composition of single-layer manifold protocols.

We saw in subsection 4.2.4 that protocols compose in a natural way: If $(\mathcal{I}, \mathcal{P}, \Xi)$ and $(\mathcal{P}, \mathcal{P}', \Xi')$ are two protocols where the protocol complex for the first is contained in the input complex for the second, then their composition is the protocol $(\mathcal{I}, \mathcal{P}', \Xi' \circ \Xi)$, where $(\Xi' \circ \Xi)(\sigma) = \Xi'(\Xi(\sigma))$. Informally, the

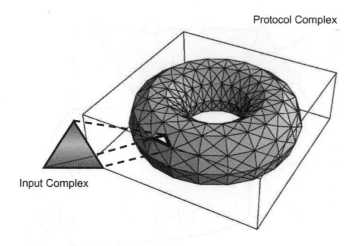

Protocol Complex

Input Complex

FIGURE 9.3

This 3-process manifold protocol complex is not a subdivision.

processes first participate in the first protocol, and then they participate in the second, using their final views from the first as inputs to the second.

We now proceed with the proof that $(\mathcal{I}, \mathcal{P}', \Xi' \circ \Xi)$ is a manifold protocol whenever both $(\mathcal{I}, \mathcal{P}, \Xi)$ and $(\mathcal{P}, \mathcal{P}', \Xi')$ are manifold protocols. Following Definition 9.1.4, we must show that

- For any simplex σ of $\mathcal{I}$, the subcomplex $(\Xi' \circ \Xi)(\sigma)$ is a manifold.
- The protocol map commutes with the boundary operator

$$\partial(\Xi' \circ \Xi)(\sigma) = (\Xi' \circ \Xi)(\partial\sigma)$$

for all $\sigma \in \mathcal{I}$.

To prove that for any m-simplex σ of $\mathcal{I}$ the subcomplex $(\Xi' \circ \Xi)(\sigma)$ is a manifold, we first need to prove that $(\Xi' \circ \Xi)(\sigma)$ is strongly connected: Any two n-simplices can be connected by a sequence of n-simplices in which each pair of consecutive simplices has a common $(n-1)$-dimensional face.

Lemma 9.1.5. *For any simplex σ of $\mathcal{I}$ the subcomplex $(\Xi' \circ \Xi)(\sigma)$ is strongly connected.*

Proof. Assume without loss of generality that $\dim(\sigma) = n$. Thus, $(\Xi' \circ \Xi)(\sigma)$ is a pure simplicial complex of dimension n. Let α^n and β^n be n-simplices of $(\Xi' \circ \Xi)(\sigma)$. If α^n and β^n are in $\Xi'(\sigma')$ for some $\sigma' \in \Xi(\sigma)$, we are done, because by assumption $(\mathcal{P}, \mathcal{P}', \Xi')$ is a manifold protocol, and hence $\Xi'(\sigma')$ is strongly connected.

Assume that α^n is in $\Xi'(\sigma_\alpha)$ and β^n is in $\Xi'(\sigma_\beta)$ for some $\sigma_\alpha, \sigma_\beta$ in $\Xi(\sigma)$. Moreover, we can assume that $\sigma_\alpha \cap \sigma_\beta = \sigma^{n-1}$ for some $(n-1)$-dimensional face because by assumption $(\mathcal{I}, \mathcal{P}, \Xi)$ is a manifold protocol, and hence $\Xi(\sigma)$ is strongly connected. See Figure 9.4.

Because the carrier Ξ' is strict, $\Xi'(\sigma_\alpha \cap \sigma_\beta) = \Xi'(\sigma_\alpha) \cap \Xi'(\sigma_\beta)$. Let γ_0^{n-1} be an $(n-1)$-dimensional simplex such that $\gamma_0^{n-1} \in \Xi'(\sigma_\alpha) \cap \Xi'(\sigma_\beta)$. Now, because Ξ' is a manifold protocol map, there is a

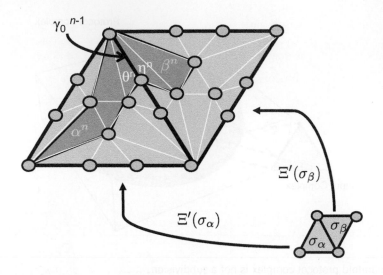

FIGURE 9.4

Lemma 9.1.5, showing strong connectivity of the composition of manifold protocols.

unique n-dimensional simplex θ^n that contains γ_0^{n-1} and such that $\theta^n \in \Xi'(\sigma_\alpha)$. Similarly, there is a unique n-dimensional simplex η^n that contains γ_0^{n-1} and such that $\eta^n \in \Xi'(\sigma_\beta)$.

Finally, because $\Xi'(\sigma_\alpha)$ is strongly connected, the two simplices α^n and θ^n can be linked, and because $\Xi'(\sigma_\beta)$ is strongly connected, the two simplices β^n and η^n can be linked. To complete the proof, observe that θ^n and η^n are linked, because $\gamma_0^{n-1} = \theta^n \cap \eta^n$.   □

Now that we have seen that $(\Xi' \circ \Xi)(\sigma)$ is strongly connected, we need to check the status of the complex's $(n-1)$-simplices.

Lemma 9.1.6.   *If $(\mathcal{I}, \mathcal{P}, \Xi)$ is manifold protocol where $\mathcal{I}$ is an n-manifold, then every $(n-1)$-simplex of $\mathcal{P}$ belongs to one or to two n-simplices.*

Proof.   Let γ^{n-1} be an arbitrary $(n-1)$-simplex of $\mathcal{P}$. Let $\sigma_1^n, \ldots, \sigma_k^n$ be the complete list of those n-simplices of $\mathcal{I}$ for which $\gamma^{n-1} \in \Xi(\sigma_i^n)$. The simplicial complex $\mathcal{P}$ is a union of pure n-dimensional complexes, so it itself is pure and n-dimensional as well. Therefore, $k \geq 1$.

We have $\gamma^{n-1} \in \cap_{i=1}^k \Xi(\sigma_i^n) = \Xi(\cap_{i=1}^k \sigma_i^n)$. Hence $\cap_{i=1}^k \sigma_i^n$ is an $(n-1)$-simplex, which we denote ρ^{n-1}. Since $\mathcal{I}$ is an n-manifold, we must have $k \leq 2$. Now we consider two cases.

Case 1: $k = 1$. All n-simplices containing γ^{n-1} are contained in $\Xi(\sigma_1^n)$, which is an n-manifold. Thus γ^{n-1} is contained in one or two n-simplices.

Case 2: $k = 2$. In this case, each n-simplex of $\mathcal{P}$ containing γ^{n-1} is contained either in $\Xi(\sigma_1^n)$ or in $\Xi(\sigma_2^n)$. On the other hand, we have

$$\gamma^{n-1} \subseteq \Xi(\rho^{n-1}) \subseteq \Xi(\partial\sigma_1^n) = \partial\Xi(\sigma_1^n),$$

implying that γ belongs to precisely one n-simplex from $\Xi(\sigma_1^n)$. The analogous argument for σ_2^n, together with the fact that $\Xi(\sigma_1^n)$ and $\Xi(\sigma_2^n)$ have no common n-simplices because their intersection is pure $(n-1)$-dimensional, yields that γ^{n-1} is contained in exactly two n-simplices.   □

It remains to show that the protocol map commutes with the boundary operator

$$\partial(\Xi' \circ \Xi)(\sigma) = (\Xi' \circ \Xi)(\partial\sigma)$$

for all $\sigma \in \mathcal{I}$.

Theorem 9.1.7.   *If $(\mathcal{I}, \mathcal{P}, \Xi)$ is a manifold protocol such that $\mathcal{I}$ is a manifold, then the simplicial complex $\mathcal{P}$ is also a manifold, and furthermore, $\partial\mathcal{P} = \Xi(\partial\mathcal{I})$.*

Proof.   The first part of the statement is the content of Lemmas 9.1.5 and 9.1.6; hence we just need to show that $\partial\mathcal{P} = \Xi(\partial\mathcal{I})$.

First, we show that $\partial\mathcal{P} \subseteq \Xi(\partial\mathcal{I})$. Let τ^{n-1} be an $(n-1)$-simplex in $\partial\mathcal{P}$. There exists a unique n-simplex α^n such that $\tau^{n-1} \subset \alpha^n$. Furthermore, there exists a unique n-simplex σ^n in $\mathcal{I}$ such that α^n is in $\Xi(\sigma^n)$. We have $\tau^{n-1} \in \partial\Xi(\sigma^n) = \Xi(\partial\sigma^n)$. Hence there exists $\gamma^{n-1} \in \partial\sigma^n$ such that $\alpha^n \in \Xi(\gamma^{n-1})$. We just need to show that $\gamma^{n-1} \in \partial\mathcal{I}$. If this is not the case, there exists an n-simplex $\tilde{\sigma}^n \neq \sigma^n$ such that $\gamma^{n-1} \subset \tilde{\sigma}^n$. But then $\tau^{n-1} \in \Xi(\tilde{\sigma}^n)$, and there will exist an n-simplex in $\Xi(\tilde{\sigma}^n)$ (hence different from α^n), which contains τ^{n-1}, contradicting our assumption that $\tau^{n-1} \in \partial\mathcal{P}$.

Next we show that $\Xi(\partial\mathcal{I}) \subseteq \partial\mathcal{P}$. Let τ^{n-1} be an $(n-1)$-simplex of $\Xi(\partial\mathcal{I})$. Assume γ^{n-1} is the unique $(n-1)$-simplex in $\partial\mathcal{I}$ such that $\tau^{n-1} \in \Xi(\gamma^{n-1})$. Let σ^n be the unique n-simplex in $\mathcal{I}$ such that $\gamma^{n-1} \subset \sigma^n$. Since $\tau^{n-1} \in \Xi(\partial\sigma^n) = \partial\Xi(\sigma^n)$, there will be precisely one n-simplex in $\Xi(\sigma^n)$ containing τ^{n-1}. On the other hand, assume there exists an n-simplex $\tilde{\sigma}^n$ other than σ^n, such that $\tau^{n-1} \in \Xi(\tilde{\sigma}^n)$. We have $\tau^{n-1} \in \Xi(\tilde{\sigma}^n) \cap \Xi(\gamma^{n-1}) = \Xi(\tilde{\sigma}^n \cap \gamma^{n-1})$, but $\dim(\tilde{\sigma}^n \cap \gamma^{n-1}) \leq n-2$, which yields a contradiction.   □

A simple inductive argument yields:

Corollary 9.1.8.   *The composition of any number of manifold protocols is itself a manifold protocol.*

9.2 Layered immediate snapshot protocols

We will show that the any single-layer immediate snapshot protocol is a manifold protocol. Since manifold protocols compose, multilayered immediate snapshot protocols are also manifold protocols.

9.2.1 Properties of single-layer protocol complexes

A single-layer immediate snapshot execution is a sequence

$$C_0, S_0, C_1, S_1, \ldots, S_r, C_{r+1},$$

where C_0 is the initial configuration, step S_i is a set of active processes that execute concurrent immediate snapshots, and each process appears at most once in the schedule $S_0, S_1, \ldots, S_r$.

When a process P participates in step S_j, its immediate snapshot returns the states of processes that participated in S_j and in earlier steps. The process states in the final configuration of an execution satisfy the following properties, where q_i is P_i's final state.

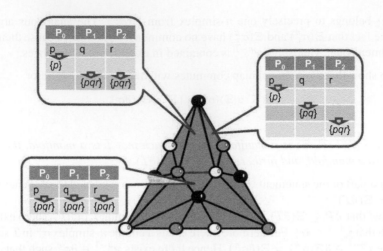

FIGURE 9.5

Protocol complex for 3-process single-layer executions.

Property 9.2.1. Each process's initial state appears in its view, and $P_i \in$ names(q_i).

Property 9.2.2. Because processes in the same step see the same initial states, final states are ordered: For $0 \leq i, j \leq n$, either view$(q_i) \subseteq$ view(q_j) or vice versa.

Property 9.2.3. For $0 \leq i, j \leq n$, if $P_i \in$ names(q_j), then P_i is active in the same or earlier step; hence view$(q_i) \subseteq$ view(q_j).

Consider all 1-layer executions starting in an initial configuration C_0, where every processes appears exactly once in the schedule. Figure 9.5 combines Figures 4.4 and 8.6. It shows a 3-process example with initial process states p, q, and r, respectively, for P_0, P_1, and P_2. Thus, $C_0 = \{(P_0, p), (P_1, q), (P_2, r)\}$, which we write as pqr to avoid clutter. Steps are shown as arrows, and the new value is shown only when a process state changes. Recall from Chapter 8 that the execution at the top right is the execution where P_0, P_1, and P_2 take steps sequentially:

$$C_0, \{P_0\}, C_1, \{P_1\}, C_2, \{P_2\}, C_3$$

where

$$C_0 = pqr$$
$$C_1 = \{p\} qr$$
$$C_2 = \{p\} \{pq\} r$$
$$C_3 = \{p\} \{pq\} \{pqr\}$$

At the bottom left is the fully concurrent execution where all three processes take steps together:

$$C_0, \{P_0, P_1, P_2\}, C_1$$

where $C_1 = \{pqr\}\{pqr\}\{pqr\}$. At the top left is the execution where P_0 takes a step, followed by a step by P_1, P_2:

$$C_0, \{P_0\}, C_1, \{P_1, P_2\}, C_2$$

where

$$C_0 = pqr$$
$$C_1 = \{p\}qr$$
$$C_2 = \{p\}\{pqr\}\{pqr\}$$

When do the final configurations of two executions differ only in the state of a single process? Consider the preceding fully sequential execution, where P_1 is alone in a step. If we want to change its final state without modifying the state of any other process, the only choice is to move it to the next step, resulting in the top-left execution:

$$C_0, \{P_0\}, C_1, \{P_1, P_2\}, C_2.$$

We cannot move P_1 to an earlier step because doing so would change the final states of that step's processes.

What if we want to modify the final state of P_2 in the fully concurrent execution? That process is alone in the last step, so no other process sees its initial state. P_2 cannot be moved because there is no other execution where P_0, P_1 have the same final states. Indeed, as far as P_0, P_1 are concerned, the execution could have ended without P_2 participating.

In summary, if in an execution the state of process P is seen by some other process, then either P appears alone in a step, which is not the last one, or else P appears together with other processes in a step. In either case, we can modify the final state of P without modifying the final states of the others. In the first case, P is moved to the next step; in the second case, P is removed from its step and placed alone in a new step immediately before its old one. Finally, if P is not seen by other processes in an execution, it is alone in the last step, and P's state cannot be changed without affecting the others. The next lemma states this property formally.

Lemma 9.2.4. *Consider these two one-layer executions:*

$$\alpha = C_0, S_0, C_1, S_1, \ldots, S_r, C_{r+1},$$
$$\alpha' = C_0, S_0', C_1', S_1', \ldots, S_t', C_{t+1}',$$

and their final configurations C_{r+1} and C_{t+1}'.

1. *The two configurations C_{r+1} and C_{t+1}' differ in exactly the state of one process, P, if and only if for some $i, i < r, S_i = \{P\}$*

$$\alpha = C_0, S_0, C_1, S_1, \ldots, S_i = \{P\}, C_{i+1}, S_{i+1}, C_{i+2}, \ldots, S_r, C_{r+1},$$
and
$$\alpha' = C_0, S_0, C_1, S_1, \ldots, S_i', C_{i+2}', S_{i+2}, \ldots, S_{r-1}, C_r',$$

with $S_i' = S_i \cup S_{i+1}$, and $S_j = S_j', C_{j+1} = C_{j+1}'$, for all $j < i$. In this case, for all $j \geq i+2, C_j$ and C_j' differ in exactly the state of P and $S_j = S_j$ (or symmetrically for the other execution).

2. *If $S_r = \{P\}$ (or symmetrically for the other execution), then if C_{r+1} and C'_{t+1} differ in the state of P, they differ in the state of at least one other process.*

9.2.2 One-layer protocol complexes are manifolds

When a process P takes an immediate snapshot in step S_i, P's view is the face of the input simplex whose vertices are colored by the processes that participated in the same or earlier steps. For input n-simplex σ, the set of layered executions defines a subdivision of σ, the *standard chromatic subdivision* Ch σ (see Figure 9.5). Each vertex in this subdivision is a pair (P_i, σ_i), where P_i is the name of process taking the steps, and σ_i, the result of its snapshot, is a face of the input simplex σ. In this chapter, we will not prove that Ch σ is a subdivision, only that it is a manifold. A proof that Ch σ is actually a subdivision requires more advanced tools and is postponed to Chapter 16.

Figure 9.5 shows the standard chromatic subdivision of an input simplex for three processes, highlighting the simplices corresponding to certain schedules. Informally, we can see that this complex is a manifold.

First we show that Ch σ is strongly connected. Each simplex in Ch σ corresponds to a particular layered execution. We proceed by "perturbing" executions so that only one process's view is changed by each perturbation. First we show that any execution can be linked to a *sequential* execution in which only one process is scheduled during each step. Next we show that any sequential execution can be linked to the unique *fully concurrent* execution in which all processes are scheduled in a single step. In this way, any simplex can be linked to the fully concurrent simplex, and any two simplices can be linked to each other.

Lemma 9.2.5. *Any simplex $\tau \in$ Ch σ can be linked to a simplex $\hat\tau$ corresponding to a sequential execution.*

Proof. Suppose σ corresponds to the execution $S_0, S_1, \ldots, S_k$. If each $|S_i| = 1$, the execution is already sequential, and we are done. Otherwise, let ℓ be any index such that $|S_\ell| > 1$, and let P_ℓ be a process in S_ℓ. We now "perturb" the execution by moving P_ℓ to a new step immediately before the steps of the other processes in S_ℓ. See Figure 9.6.

Formally, we construct the schedule $S'_0, \ldots, S'_{k+1}$, where

$$
S'_i = \begin{cases}
S_i & \text{if } i < \ell, \\
\{P_\ell\} & \text{if } i = \ell, \\
S_{i-1} \setminus \{P_\ell\} & \text{if } i = \ell+1, \\
S_{i-1} & \text{if } i > \ell+1.
\end{cases}
$$

It is easy to check that the view of every process other than P_ℓ is unchanged by this move, implying that for τ', the simplex generated by this schedule, $\dim(\tau \cap \tau') = n - 1$.

Continuing in this way, we can repeatedly reduce the number of processes scheduled during each step, eventually reaching a sequential schedule. □

Lemma 9.2.6. *Any simplex $\tau \in$ Ch σ can be linked to a simplex $\tilde\tau$ corresponding to a fully concurrent execution.*

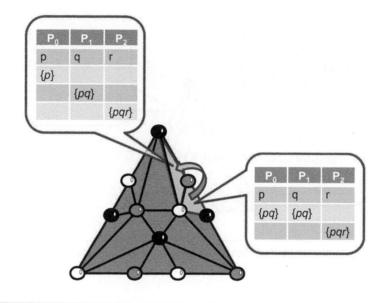

FIGURE 9.6

Linking an execution to a sequential execution, changing one view at a time.

Proof.   We will prove something slightly stronger. An immediate snapshot execution with schedule $S_0, S_1, \ldots, S_k$ is *tail-concurrent* if all steps except possibly the last are sequential: $|S_i| = 1$ for $0 \leq i < k$. Both sequential executions and the fully-concurrent execution are tail-concurrent.

We claim that any tail-concurrent execution can be shortened as follows. Let $S_{k-1} = \{P_{k-1}\}$. If we merge P_{k-1} into S_k, then only the view of P_{k-1} changes. Figure 9.7 shows an example of such a transformation.

Formally, we construct the schedule $S_0', \ldots, S_{k-1}'$, where

$$F_i' = \begin{cases} S_i & \text{if } i < k, \\ \{P_{k-1}\} \cup S_k & \text{if } i = k - 1. \end{cases}$$

Continuing in this way, we can repeatedly reduce the number of layers in any tail-concurrent schedule, eventually reaching the fully concurrent schedule.   □

Lemmas 9.2.5 and 9.2.6 imply the following:

Corollary 9.2.7.   *The simplicial complex* $\mathrm{Ch}\,\sigma$ *is strongly connected.*

Finally, this corollary and Lemma 9.2.4 imply the main result of this section.

Theorem 9.2.8.   *For any simplex* σ, *the simplicial complex* $\mathrm{Ch}\,\sigma$ *is a manifold.*

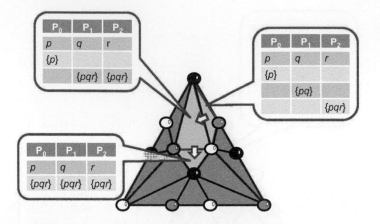

FIGURE 9.7

Linking a sequential execution to the fully concurrent execution in Ch σ, changing one view at a time.

9.3 No set agreement from manifold protocols

Recall that the *k-set agreement* task (Section 8.3.3) is often described in the literature using three (informal) requirements. Each process starts with a private input value and communicates with the others, every process must decide on some process's input, and no more than k distinct inputs can be chosen. For brevity we use *set agreement* as shorthand for $(n + 1)$-process n-set agreement, where the processes agree to discard a single value. We now demonstrate that no manifold protocol can solve set agreement. We will prove a slightly more general result that any protocol that satisfies the termination and validity properties must violate the agreement property in an odd number of distinct executions.

9.3.1 Sperner's lemma

Before we turn our attention to set agreement, we provide a statement of the classical Sperner's lemma for manifolds. We provide a proof for completeness and because this lemma is so important. The proof consists of a simple counting argument that perfectly illustrates the beauty of combinatorial topology, as argued in Chapter 3: Deep, powerful properties of spaces made up of simple pieces can be characterized by counting. Readers uninterested in the proof may read the statement of the lemma and skip to the next subsection.

Recall that an $(n + 1)$-*labeling* of a complex $\mathcal{K}$ is a simplicial map $\chi : \mathcal{K} \to \Delta^n$, where Δ^n is an n-simplex. (We use the same name for a simplex and the complex that consists of the simplex and all its faces.) We say that χ sends a simplex σ *onto* Δ^n if every vertex in Δ^n is the image of a vertex in σ. If $\sigma \in \mathcal{K}$ and Δ^n have the same dimension and χ maps σ onto Δ^n so that each vertex of σ is assigned a distinct color, then we say that σ is *properly colored*.

To state Sperner's lemma in our notation, let Δ^n be equal to the n-simplex $\{(P, P)|P \in \Pi\}$ for a set of $n + 1$ names Π. We then take Δ^n and its faces to be an input complex. Sperner's lemma is usually stated in terms of a subdivision of Δ^n, but the combinatorial proof requires only the manifold property. Thus, instead of a subdivision of Δ^n, consider a manifold protocol, $(\Delta^n, \mathcal{P}, \Xi)$.

A Sperner coloring of $\mathcal{P}$ is a labeling $\delta : \mathcal{P} \to \Delta^n$ that satisfies the properties illustrated in the left-hand complex of Figure 9.4. Here n is 2. Choose three colors, say, the names of Π. The three "corners" of the subdivision are colored with distinct colors. In a Sperner coloring, the interior vertices on each boundary connecting any two corners are colored arbitrarily using only the colors from those two corners, and the interior vertices in each 2-simplex are colored arbitrarily using only colors from those three colors. Sperner's lemma states that no matter how the arbitrary coloring choices are made, there must be an odd number of 2-simplices that are properly colored (with all three colors). In particular, there must be at least one.

More formally, consider Ξ^i, the identity carrier map from Δ^n to itself: For each $\sigma \in \Delta^n$, $\Xi^i(\sigma)$ is equal to the complex 2^σ, consisting of σ and all its faces. The labeling is a *Sperner coloring* if δ is carried by Ξ^i. Namely, for each $\sigma \in \Delta^n$,

$$\delta(\Xi(\sigma)) \subseteq \Xi^i(\sigma).$$

Lemma 9.3.1 (Sperner's Lemma).   *For any manifold protocol, $(\Delta^n, \mathcal{P}, \Xi)$, and any Sperner's coloring $\delta : \mathcal{P} \to \Delta^n$, δ sends an odd number of n-simplices of $\mathcal{P}$ onto Δ^n.*

Sperner's lemma says, in particular, that there exists no Sperner's coloring $\delta : \mathcal{P} \to \partial \Delta^n$.

The proof follows from an inductive application of a rather surprising property: For an n-dimensional manifold, the number of properly colored $(n-1)$-simplices on the boundary can reveal something about the number of properly colored n-simplices in the interior.

First, we recall a simple lemma from graph theory. Recall that a graph is a 1-dimensional complex given by a set of vertices V and a set of edges E. The *degree* of a vertex, $\deg(v)$, is the number of edges that contain v.

Lemma 9.3.2.   *In any graph $G = (V, E)$, the sum of the degrees of the vertices is twice the number of edges:*

$$2|E| = \sum_{v \in V} \deg(v).$$

Proof.   Each edge $e = \{v_0, v_1\}$ adds one to the degree of v_0 and one to the degree of v_1, contributing two to the sum of the degrees.   □

Corollary 9.3.3.   *Any graph has an even number of vertices of odd degree.*

Lemma 9.3.4 (Sperner's Lemma for Manifolds).   *Let $\mathcal{M}$ be an n-dimensional manifold,[2] and let $\chi : \mathcal{M} \to \Delta^n$ be an $(n+1)$-labeling. If χ sends an odd number of $(n-1)$-simplices of $\partial \mathcal{M}$ onto $\mathrm{Face}_n \Delta^n$, then χ sends an odd number of n-simplices of $\mathcal{M}$ onto Δ^n.*

Proof.   Define G to be the *dual* graph whose vertices are indexed by the n-simplices of $\mathcal{M}$, with the addition of one more "external" vertex e. There is an edge between two vertices if their corresponding simplices share a common $(n-1)$-face colored with all colors except n; that is, χ sends that $(n-1)$-face onto $\mathrm{Face}_n \Delta^n$. There is also an edge from the external vertex e to every n-simplex σ with a boundary

---

[2]We do not need strong connectivity here.

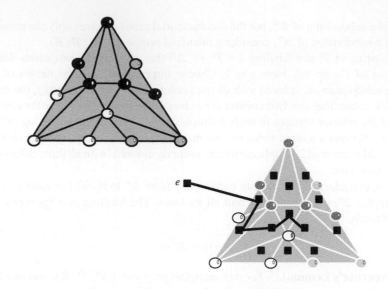

FIGURE 9.8

A colored manifold (top) and its dual graph (bottom) linking triangles that share black-and-white faces.

face colored with every color except n; that is, σ has an $(n-1)$-face in $\partial\mathcal{M}$, and χ sends that face onto $\text{Face}_n\Delta^n$. As an example, Figure 9.8 shows a manifold, in fact a subdivided triangle, where each vertex is colored black, white, or gray, along with its dual graph whose edges cross black-and-white faces.

For an n-simplex σ we let v_σ denote the dual graph vertex corresponding to σ, and we let $\chi(\sigma)$ denote the set of colors of the vertices of σ. We claim that v_σ has an odd degree if and only if $\chi(\sigma) = [n]$. There are three cases to consider.

Case 1. Assume $\chi(\sigma) = [n]$. In this case each color from $[n]$ occurs among the vertices of σ exactly once. In particular, precisely one of the boundary $(n-1)$-simplices has $[n-1]$ as the set of colors, and hence the degree of v_σ is equal to 1.

Case 2. Assume $\chi(\sigma) = [n-1]$. In this case there exists one color that occurs on two vertices of σ, say, a and b, whereas each other color from $[n]$ occurs among the vertices of σ exactly once. This means that there are exactly two $(n-1)$-simplices on the boundary of σ; specifically these are $\sigma \setminus \{a\}$ and $\sigma \setminus \{b\}$, which are mapped onto $[n-1]$ by χ. Hence in this case the degree of $v_\sigma = 2$.

Case 3. Finally, assume $\chi(\sigma) \not\supseteq [n-1]$. Then χ does not map any $(n-1)$-face of σ onto $[n-1]$, so the vertex v_σ has degree 0.

Moreover, the vertex e has odd degree, since by our assumptions, χ sends an odd number of boundary $(n-1)$-simplices onto $[n-1]$, producing an odd number of edges at e.

According to Lemma 9.3.2, the graph G has an even number of vertices of odd degree. Since the external node e has an odd degree, the dual graph must include an odd number of other vertices v_σ with odd degrees. Each of these vertices corresponds to an n-simplex that χ maps onto Δ^n. □

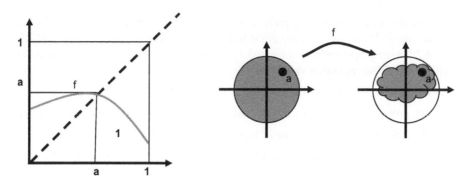

FIGURE 9.9

Brouwer's fixed-point theorem in dimensions 1 and 2.

Mathematical Note 9.3.5.   Sperner's lemma is equivalent to the celebrated Brouwer fixed-point theorem, used across numerous fields of mathematics; we could say it is its discrete version. In its simplest form, Brouwer's fixed-point theorem states that for any continuous function $f : D \to D$, mapping an n-dimensional unit disk D into itself there is a point x_0 such that $f(x_0) = x_0$. This is a generalization of the simple intermediate value theorem, which says that every continuous function $f : [0, 1] \to [0, 1]$ has a fixed point (when the function crosses the diagonal of the unit square). See Figure 9.9. There are many proofs of Brouwer's fixed-point theorem; an elegant one uses Sperner's lemma.

9.3.2 Application to set agreement

The *set validity* task is set agreement without the requirement that at most, n distinct values may be decided. The validity requirement is maintained: Any value decided was some process's input. Thus, any protocol that solves set agreement also solves set validity. We will prove that any layered protocol solving set validity has an execution whereby $n + 1$ different values are decided; hence no set agreement protocol is possible in the layered execution model.

In the *set validity* task, $(\mathcal{I}, \mathcal{O}, \Delta)$, each process has one possible input value: its own name. Processes are required to halt with the name of some participating process (perhaps their own). Formally, there is a single input n-simplex $\sigma = \{(P, P)|P \in \Pi\}$, and the input complex $\mathcal{I}$ is 2^σ, the complex consisting of σ and its faces. The output complex $\mathcal{O}$ has vertices of the form (P, Q) for $P, Q \in \Pi$, and a set of vertices is an output simplex if the process names in the first component are distinct. The validity condition means that

$$\Delta(\sigma) = \{\tau \in \mathcal{O} \mid \text{names}(\tau) \subseteq \text{names}(\sigma) \text{ and value}(\tau) \subseteq \text{names}(\sigma)\}.$$

Let $(\mathcal{I}, \mathcal{P}, \Xi)$ be a manifold protocol that solves set validity with decision map δ. The decision map $\delta : \mathcal{P} \to \mathcal{O}$ induces a map $\chi : \mathcal{P} \to \mathcal{I}$ by projecting onto the output vertex's value. Suppose that in a vertex v of $\mathcal{P}$, the decision value is v, namely, $v = \text{value}(\delta(v))$. Then $\chi(v) = (v, v)$. Notice that χ is a Sperner's coloring of $\mathcal{P}$ because to solve the validity task, for each input simplex σ, $\chi(\Xi(\sigma))$ is sent to a simplex of 2^σ. Using Sperner's Lemma (9.3.1), we obtain that χ sends $\mathcal{P}$ onto an odd number of simplices with $n + 1$ different output values.

Theorem 9.3.6.   *There is no manifold protocol for set agreement.*

Because every protocol complex in a layered execution model is a manifold complex, we have:

Corollary 9.3.7.   *No set agreement protocol is possible in a layered execution model.*

We will discuss again this impossibility result in Chapter 10, where we will consider the connectivity of the protocol complex.

9.4 Set agreement vs. weak symmetry breaking

In the weak symmetry-breaking task of Section 8.3.5, each process is assigned a distinct *input name* taken from Π and chooses a binary output so that if all $n + 1$ processes participate, at least one chooses 0 and at least one chooses 1. We saw that the number of possible names $|\Pi|$ is important when we are considering the difficulty of this task. For impossibility results, the size of the name space is unimportant; any task that cannot be solved if names are taken from a small name space also cannot be solved if names are taken from a larger name space. For algorithms, however, it may be possible to abuse the small name-space assumption to derive trivial protocols. If $\Pi = [n]$, then weak symmetry breaking can be solved with no communication at all: The process with name 0 decides 0 and all others decide 1. Lower bounds are discussed in Chapter 12.

One way of comparing the difficulty of two tasks $\mathcal{T}_1, \mathcal{T}_2$, as in classical (sequential) computability theory, is to assume that the layered execution model has access to an "oracle" or "black box" that can solve instances of $\mathcal{T}_1$ and ask whether it can now solve $\mathcal{T}_2$. Real multicore systems use this approach by including a hardware implementation of the desired black box.

In this section, we compare the "computational power" of weak symmetry breaking and set agreement. Given a "black-box" protocol for set agreement, we will show that we can implement weak symmetry breaking but not vice versa. It follows that weak symmetry breaking is weaker than set agreement, an example of a *separation* result.

9.4.1 Comparing the powers of tasks

There are various ways of comparing the power of tasks. Here we consider a setting that, although not the most general, is particularly elegant. We say a task $\mathcal{T}$ *implements* a task $\mathcal{S}$ if one can construct a protocol for $\mathcal{S}$ by composing one or more instances of protocols for $\mathcal{T}$, along with one or more layered immediate snapshot protocols. If $\mathcal{T}$ implements $\mathcal{S}$ but not vice versa, then we say that $\mathcal{S}$ is *weaker* than $\mathcal{T}$. Otherwise, they are *equivalent*.

Recall from subsection 4.2.4 that given two protocols $(\mathcal{I}, \mathcal{P}, \Xi)$ and $(\mathcal{P}, \mathcal{P}', \Xi')$ such that the first's protocol complex is contained in the second's input complex, their composition is the protocol $(\mathcal{I}, \mathcal{P}', \Xi' \circ \Xi)$, where $(\Xi' \circ \Xi)(\sigma) = \Xi'(\Xi(\sigma))$, which we denote

$$(\mathcal{I}, \mathcal{P}, \Xi) \circ (\mathcal{P}, \mathcal{P}', \Xi').$$

Now consider tasks $\mathcal{T} = (\mathcal{I}, \mathcal{O}, \Delta)$ and $\mathcal{S} = (\mathcal{I}', \mathcal{O}', \Delta')$. If their carrier maps are strict, then the tasks can be treated like protocols. Then, task $\mathcal{T}$ implements task $\mathcal{S}$ if there exists a protocol $(\mathcal{P}_0, \mathcal{P}_k, \Xi)$ equal to the composition

$$(\mathcal{P}_0, \mathcal{P}_1, \Xi_1) \circ (\mathcal{P}_1, \mathcal{P}_2, \Xi_2) \circ \cdots \circ (\mathcal{P}_{k-1}, \mathcal{P}_k, \Xi_k)$$

consisting of a sequence of (consecutively compatible) protocols $(\mathcal{P}_{i-1}, \mathcal{P}_i, \Xi_i)$, $1 \le i \le k$, where each is either an immediate snapshot protocol or else it is $\mathcal{T} = (\mathcal{I}, \mathcal{O}, \Delta)$, and furthermore, the composed protocol $(\mathcal{P}_0, \mathcal{P}_k, \Xi)$ solves $\mathcal{S} = (\mathcal{I}', \mathcal{O}', \Delta')$. Operationally, the processes go through the protocols $(\mathcal{P}_{i-1}, \mathcal{P}_i, \Xi_i)$ in the same order, asynchronously. The processes execute the first protocol, and once a process finishes, it starts the next without waiting for other processes to finish the previous protocol. Each process uses its final view from each protocol as its input value for the next.

Recall that $(\mathcal{P}_0, \mathcal{P}_k, \Xi)$ solves $\mathcal{S} = (\mathcal{I}', \mathcal{O}', \Delta')$ if $\mathcal{P}_0 = \mathcal{I}'$ and there exists a chromatic simplicial decision map $\delta : \mathcal{P}_k \to \mathcal{O}'$, satisfying

$$\delta(\Xi(\sigma)) \subseteq \Delta'(\sigma),$$

for all $\sigma \in \mathcal{I}'$.

9.4.2 Weak symmetry breaking from set agreement

Here we show that one can use a set agreement protocol to implement weak symmetry breaking. Formally, we construct a two-layer protocol, where the first layer is a set agreement protocol and the second an immediate snapshot. The "program logic" resides in the decision map.

For readability, we describe this protocol in terms of a program and flowcharts, but of course this program is just a readable way to specify a protocol complex.

Figures 9.10 and 9.11 shows the control structure and pseudo-code to implement weak symmetry breaking using set agreement. The processes share an $(n+1)$-element array of input names, chosen[·], whose entries are initially $\perp$ (Line 3). The processes also share a set agreement protocol instance (Line 4). Each process P_i calls the set agreement object's decide() method, using its own input name as input, and stores the result in chosen[i] (Line 6). The process then takes a snapshot and returns the value 1 if and only if its own input is in the set of inputs chosen by the set agreement protocol (Line 7).

Lemma 9.4.1.   *If all $n + 1$ processes participate, some process decides 1.*

Proof.   Among the processes that were selected by the set agreement protocol, the last process to take a step will observe its own name and return 1.   □

Lemma 9.4.2.   *If all $(n + 1)$ processes participate, some process decides 0.*

Proof.   If all $n + 1$ processes decide 1, then $n + 1$ distinct inputs were chosen by the set agreement protocol, violating the set agreement specification.   □

Thus, the protocol $(\mathcal{I}, \mathcal{O}, \Delta) \circ (\mathcal{O}, \mathcal{P}_2, \Xi_2)$ with decision map δ, which corresponds to the code in Figure 9.11, solves weak symmetry breaking.

Theorem 9.4.3.   *Set agreement implements weak symmetry breaking.*

Notice that for this result, the size of the name space Π is immaterial.

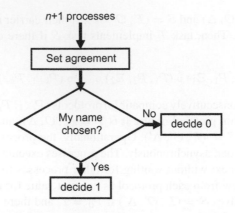

FIGURE 9.10

Flowchart for weak symmetry breaking from set agreement.

```
1   // code for P_i
2   protocol WSB
3     shared chosen: array [0..n] of int  // initially  null
4     shared setAgree: SetAgreeProtocol;   // set agreement
5     Boolean decide (input: int)
6       chosen[i] := setAgree.decide(input)
7       return input ∈ snapshot(chosen)
```

FIGURE 9.11

Pseudo-code for weak symmetry breaking from set agreement.

9.4.3 Weak symmetry breaking does not implement set agreement

For the other direction, we want to show that weak symmetry breaking cannot implement set agreement. We will prove this claim indirectly by constructing a manifold protocol that implements weak symmetry breaking. If weak symmetry breaking could implement set agreement, we could replace the weak symmetry-breaking objects with their manifold task implementations, yielding a manifold protocol for set agreement and contradicting Theorem 9.3.6.

We introduce a new task, $(\mathcal{I}, \mathcal{M}, \Delta)$, which we call the *Moebius* task. First we construct the 2-dimensional Moebius task. The input complex is the same as for weak symmetry breaking: Each process starts with a distinct input name.

The task's output complex $\mathcal{M}$ is illustrated in Figure 9.12. Take three 2-simplices, $\sigma_0, \sigma_1, \sigma_2$, each colored by process names, and define $\xi_i := \text{Ch}\,\sigma_i$, for $i = 0, 1, 2$, a chromatic subdivision. Abusing notation, let $\text{Face}_j\xi_i := \text{Ch}\,\text{Face}_j\sigma_i \subset \xi_i$. We call $\text{Face}_i\xi_i$ the *external face* of ξ_i, (even though it is technically a complex) and $Face_j\xi_i$, for $i \neq j$, the *internal faces*. We then identify (that is, "glue together") $\text{Face}_0\xi_1$ and $\text{Face}_0\xi_2$, $\text{Face}_1\xi_0$ and $\text{Face}_1\xi_2$, and $\text{Face}_2\xi_0$ and $\text{Face}_2\xi_1$. The resulting complex is a manifold the boundary complex of which consists of the external faces of the ξ_i.

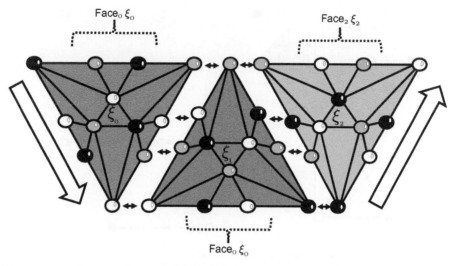

FIGURE 9.12

The Moebius task output complex for three processes. The edges on the sides are identified (glued together) in the direction of the arrows.

Figure 9.13 illustrates the task's carrier map Δ for the 2-dimensional case. Each process chooses an output vertex of matching color. If a proper subset of the processes participates, they choose to the vertices of a simplex in an external face. If they all participate, they converge to the vertices of any simplex.

Although we have defined the Moebius task $(\mathcal{I}, \mathcal{M}, \Delta)$ as a task, we can also treat it as a protocol, where $\mathcal{I}$ is the protocol's input complex, $\mathcal{M}$ is its protocol complex, and Δ is its (strict) execution carrier map. It is easy to check that the Moebius protocol is a manifold protocol. As such, the Moebius protocol cannot solve 2-set agreement.

As illustrated in Figure 9.14, however, the Moebius protocol can solve weak symmetry breaking. We color each vertex with black and white "pebbles" (that is, 0 or 1 values) as follows: For each central simplex of ξ_i, color each node black except for the one labeled P_i. For the central simplex of each external face $\text{Face}_i\xi_i$, color the vertices of the central 2-simplex black. The rest of the vertices are colored white. It is easy to check that (1) no 2-simplex is monochromatic, and (2) the protocol is well defined; namely, there is a corresponding decision map δ. To solve 3-process weak symmetry breaking, run the Moebius protocol from each 2-simplex σ in the weak symmetry-breaking input complex $\mathcal{I}$.

It follows that the 2-dimensional Moebius task *separates* weak symmetry breaking and set agreement in the sense that it can implement one but not the other.

Now we generalize this construction to even dimensions. Let $n = 2N$. Start with $n + 1$ n-simplices, $\sigma_0, \ldots, \sigma_n$, colored with process names, and set $\xi_i := \text{Ch } \sigma_i$. As before, we call the complex $\text{Face}_i\xi_i$ the *external face* of ξ_i and $Face_j\xi_i$, for $i \neq j$, the *internal faces*.

The *rank* of P_i's name in a set of process names is the of names in $[n]$ smaller than i in that set. For each j, $0 \leq j \leq n$, let $\pi_j : [n] \setminus \{j\} \to [n] \setminus \{j\}$ be the map sending the name with rank r in $[n] \setminus \{j\}$ to the name with rank $r + N \mod 2N$.

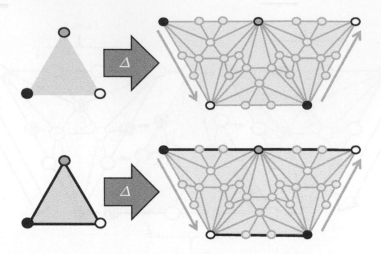

FIGURE 9.13

Carrier map for the Moebius task: one and two-process executions. Note that because the left- and right-hand edges are glued together in the directions of the arrows, some vertices depicted twice are actually the same.

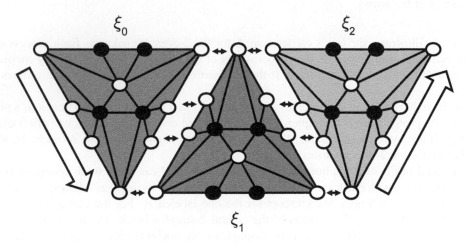

FIGURE 9.14

How the Moebius task solves weak symmetry breaking.

For each i and each $j \neq i, \pi_j(i)$, identify the internal face $\text{Face}_j \xi_i$ with $\text{Face}_j \xi_{\pi_j(i)}$. Because $\pi_j(i) \neq i, \pi_j(i) \neq j$, and $\pi_j(\pi_j(i)) = i$, each $(2N-1)$-simplex in each internal face lies in exactly two $(2N)$-simplices, so the result is a manifold. (This why this construction works only in even dimensions.)

Let σ be an input n-simplex. The Moebius task's carrier map carries each proper face τ of σ to $\text{Ch}\tau$. It carries σ itself to all n-simplices of $\Delta(\sigma)$.

Theorem 9.4.4.   *The Moebius task cannot solve set agreement.*

Proof.   The 1-layer Moebius task is a manifold protocol, so composing the Moebius task with itself, with one-layer protocols, with any other manifold task yields a manifold task. The claim then follows from Theorem 9.3.6.   □

To show that this task solves weak symmetry breaking, we again color the edges with black and white pebbles so that no simplex is monochromatic and the coloring on the boundary is symmetric. For the central simplex of each ξ_i, color each node black except for the one labeled P_i. For the central simplex of each external face ξ_{ii}, color the central $(2N - 2)$-simplex black. The rest are white.

Every $(2N - 1)$-simplex ξ in ξ_i intersects both a face, either internal or external, and a central $(2N - 1)$-simplex. If ξ intersects an internal face, then the vertices on that face are white, but the vertices on the central simplex are black. If ξ intersects an external face, then it intersects the white node of the central simplex of ξ_i and a black node of the central simplex of ξ_{ii}. To solve $(n + 1)$-process weak symmetry breaking, run the Moebius protocol from each n-simplex σ in the weak symmetry-breaking input complex $\mathcal{I}$.

Corollary 9.4.5.   *Set agreement implements weak symmetry breaking but not vice versa.*

The techniques studied here illustrate how combinatorial and algorithmic techniques complement one another: Combinatorial techniques are often effective to prove impossibility, whereas algorithmic techniques are convenient to show that something is possible.

Mathematical Note 9.4.6.   The notion of a pseudomanifold (Definition 9.1.3) can be strengthened as follows.

Definition 9.4.7.   Assume that $\mathcal{M}$ is a pure abstract simplicial complex of dimension n.

(1) $\mathcal{M}$ is called a *simplicial manifold* if the geometric realization of the link of every simplex σ is homeomorphic to a sphere of dimension $n - 1 - \dim \sigma$.
(2) $\mathcal{M}$ is called a *simplicial manifold with boundary* if the geometric realization of the link of every simplex σ is homeomorphic to either a sphere or a closed ball, in each case of dimension $n - 1 - \dim \sigma$.

Note that in the special case when $\dim \sigma = n - 1$, we have $n - 1 - \dim \sigma = 0$. The 0-dimensional sphere consists of two points, whereas the 0-dimensional ball consists of one point, so conditions of (1) and (2) of Definition 9.4.7 specialize precisely to the conditions of Definition 9.1.3.

There is also the following standard topological notion.

Definition 9.4.8.   Assume X is an arbitrary Hausdorff[3] topological space.

(1) X is called a *topological manifold* of dimension n if every point of X has a neighborhood homeomorphic to an open ball of dimension n.

---

[3]This is a technical condition from point-set topology, meaning every two points can be separated by disjoint open neighborhoods; it is needed to avoid all sorts perverse examples, and is always satisfied in the context of this book.

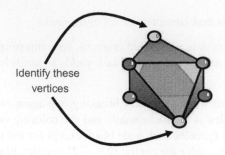

Identify these vertices

FIGURE 9.15

A triangulation of a pinched torus.

(2) X is called a *topological manifold with boundary* of dimension n if every point of X has a neighborhood homeomorphic to an open subset of Euclidean half-space:

$$\mathbb{R}_+^n = \{(x_1, \dots, x_n) \in \mathbb{R}^n : x_n \geq 0\}.$$

The interior of X, denoted $\text{Int}X$, is the set of points in X that have neighborhoods homeomorphic to an open ball of dimension n. The boundary of X, denoted ∂X, is the complement of $\text{Int}X$ in X. The boundary points can be characterized as those points that land on the boundary hyperplane $x_n = 0$ of $\mathbb{R}_+^n$ in their respective neighborhoods. If X is a manifold of dimension n with boundary, then $\text{Int}X$ is a manifold of dimension n, and $\text{Int}X$ is a manifold of dimension $n - 1$.

We note that if $\mathcal{M}$ is a simplicial manifold with boundary, its geometric realization is a topological manifold with boundary of the same dimension; moreover, the geometric realization of the boundary of $\mathcal{M}$ is precisely the boundary of $|\mathcal{M}|$. As you can see in Figure 9.2, a 2-dimensional manifold is a kind of a discrete approximation to a surface.

On the other hand, the geometric realization of the pseudomanifold does not have to be a manifold. Perhaps the simplest example is obtained if we take a simplicial 2-dimensional sphere and glue together the north and south poles,[4] as shown in Figure 9.15. This space is also called the *pinched torus*. Clearly, the condition of being a manifold fails at the glued poles, but the condition of being a pseudomanifold is still satisfied since it is a condition for edges and triangles and is untouched by vertices being glued together.

9.5 Chapter notes

Immediate snapshot executions are due to Borowsky and Gafni [23], and to Saks and Zaharoughu [134], who called them *block executions*. Borowsky and Gafni also showed that the layered execution model is equivalent to the standard read-write memory model.

---

[4]We assume that the poles are vertices of the simplicial complex, and that the mesh is fine enough, so that even after that gluing, we still have a simplicial complex.

Many of the basic properties of one-layered executions presented here were first shown by Attiya and Rajsbaum [16], although in the more general situation where processes repeatedly execute immediate snapshot operations in the same shared memory. The example of a manifold protocol in Figure 9.3 that is not a subdivision is from Attiya and Rajsbaum [16]. Attiya and Castañeda [12] prove the set agreement impossibility by applying Sperner's lemma directly on executions.

Sperner's lemma and its relation with Brouwer's fixed-point theorem has been well studied. See, for example, Bondy and Murty [22] and Henle [77] for a self-contained, elementary proof of Sperner's lemma (the same argument we presented here) and how it is used to prove Brouwer's fixed-point theorem.

The separation between weak symmetry breaking and set agreement is adapted from Gafni, Rajsbaum, and Herlihy [69]. They proved that weak symmetry breaking cannot implement set agreement when the number of processes $n + 1$ is odd. It was shown by Castañeda and Rajsbaum [31,33] that weak symmetry breaking can be solved wait-free, without the help of any tasks (e.g., in the multilayer model) if the number of processes is not a prime power. Thus, in this case too, weak symmetry breaking cannot implement set agreement, because it is known that set agreement is not wait-free solvable [23,91,135]. Therefore, the only case that remains open to prove that weak symmetry-breaking cannot implement set agreement is when the number of processes is at least 4 and a power of 2 (for two processes the tasks are equivalent). Castañeda, Imbs, Rajsbaum, and Raynal [29,30] prove this case in a weaker model and study various definitions of the nondeterminism of the objects involved. More about renaming and its relation to weak symmetry breaking can be found in the survey by Castañeda, Rajsbaum, and Raynal [34].

9.6 **Exercises**

Exercise 9.1.   Show that the following tasks are all equivalent to set agreement in the sense that any protocol for this task can be adapted to solve set agreement (possibly with some extra read-write memory), and vice versa.

a. *Fixed-input set agreement*. Each process has its own name as input, each process decides the name of some participating process, and no more than k distinct names may be decided.

b. *Strong set agreement*. Each process decides some process's input, no more than k distinct inputs may be decided, and at least one process decides its own input.

Exercise 9.2.   Check that the carrier maps for both the Moebius task of Section 9.4.3 and k-set agreement are strict.

Exercise 9.3.   Count the number of simplices in Ch σ for an n-simplex σ.

Exercise 9.4.   Count the number of simplices in the output complex for $(n+1)$-process weak symmetry breaking.

Exercise 9.5.   Compute the Euler characteristic of Ch σ for an n-simplex σ.

Exercise 9.6.   Once upon a time, the city of Königsberg, then in Prussia, included two islands connected to each other and to the city itself by seven bridges, as shown in Figure 9.16. The residents amused

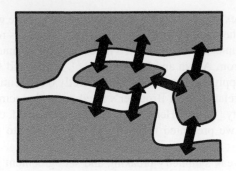

FIGURE 9.16

The bridges of Königsberg.

themselves by trying to find a way through the city by crossing each bridge exactly once. Prove that such a tour is impossible. *Hint*: Use reasoning similar to the proof of Lemma 9.3.2.

Exercise 9.7.   Using read-write memory, implement the Set<Name> object used in Figure 9.11. You may assume that names are integers in the range $[1 : N]$, for some $N > n + 1$. Do not worry about efficiency.

Exercise 9.8.   Show that if $\mathcal{M}$ is a manifold and v a vertex not in $\mathcal{M}$, then

- The cone $v * \mathcal{M}$, and
- The cone $v * \partial \mathcal{M}$

are manifolds.

Exercise 9.9.   Prove that if $(\mathcal{I}, \mathcal{M}, \Xi)$ is a manifold protocol, then

a. For any input simplex σ, $\partial \Xi(\sigma) = \Xi(\partial \sigma)$, and
b. If $\mathcal{I}$ is a manifold, $\Xi(\mathcal{I}) = \Xi(\partial \mathcal{I})$.

Exercise 9.10.   Prove that no manifold protocol can solve the following task: Suppose we want the processes to announce when they have all seen each other. For this purpose, it is sufficient to assume that processes have no inputs (except for their names). The outputs can be anything, but they include a special value "all." The task requirement is that, in at least one execution where all processes see each other (namely, each process sees at least one value written to the shared memory by each other process), all processes output "all." Also, whenever a process does not see another process, it should not output "all."

Exercise 9.11.   Prove that weak symmetry breaking and set agreement are equivalent in the case of two processes.

Connectivity

In Chapter 9, we considered models of computation for which for any protocol Ξ and any input simplex σ, the subcomplex $\Xi(\sigma) \subset \mathcal{P}$ is a manifold. We saw that any such protocol cannot solve k-set agreement for $k \leq \dim \sigma$. In this chapter, we investigate another important topological property of the complex $\Xi(\sigma)$: having no "holes" in dimensions m and below, a property called *m-connectivity*. We will see that if every $\Xi(\sigma)$ is $(k-1)$-connected, then Ξ cannot solve k-set agreement. We will see later that there are natural models of computation for which protocol complexes are not manifolds, but they are m-connected for some $0 \leq m \leq n$. We will also use this notion of connectivity in later chapters to characterize when protocols exist for certain tasks.

10.1 Consensus and path connectivity

We start with the familiar, 1-dimensional notion of connectivity and explore its relation to the consensus task.

Recall from Section 8.3.1 that in the consensus task for $n + 1$ processes, each process starts with a private *input value* and halts with an *output value* such that (1) all processes choose the same output value, and (2) that value was some process's input.

Distributed Computing Through Combinatorial Topology. http://dx.doi.org/10.1016/B978-0-12-404578-1.00010-3
© 2014 Elsevier Inc. All rights reserved.

Here we consider the consensus task $(\mathcal{I}, \mathcal{O}, \Delta)$ with an *arbitrary* input complex. In other words, instead of requiring that the input complex contain all possible assignments of values to processes, we allow $\mathcal{I}$ to consist of an arbitrary collection of initial configurations. There are particular input complexes for which consensus is easily solvable. An input complex is said to be *degenerate for consensus* if every process has the same input in every configuration. Consensus is easy to solve if the input complex is degenerate; each process simply decides its input. We will see that if a protocol's carrier map takes each simplex to a path-connected subcomplex of the protocol complex, then that protocol cannot solve consensus for any nondegenerate input complex.

Informally, consensus requires that all participating processes "commit" to a single value. Expressed as a protocol complex, executions in which they all commit to one value must be distinct, in some sense, from executions in which they commit to another value. We now make this notion more precise.

Recall from Section 3.5.1 that a complex $\mathcal{K}$ is *path-connected* if there is an edge path linking any two vertices of $\mathcal{K}$. In the next theorem, we show that if a protocol carrier map satisfies a local path-connectivity condition, it cannot solve consensus for nondegenerate input complexes.

Theorem 10.1.1. *Let $\mathcal{I}$ be a nondegenerate input complex for consensus. If $(\mathcal{I}, \mathcal{O}, \Delta)$ is an $(n + 1)$-process consensus task, and $(\mathcal{I}, \mathcal{P}, \Xi)$ is a protocol such that $\Xi(\sigma)$ is path-connected for all simplices σ in $\mathcal{I}$, then $(\mathcal{I}, \mathcal{P}, \Xi)$ cannot solve the consensus task $(\mathcal{I}, \mathcal{O}, \Delta)$.*

Proof. Assume otherwise. Because $\mathcal{I}$ is not degenerate, it contains an edge $\{v, w\}$ such that $\mathrm{view}(v) \neq \mathrm{view}(w)$. (That is, there is an initial configuration where two processes have distinct inputs.) By hypothesis, $\Xi(\{v, w\})$ is path-connected, and by Proposition 3.5.3, $\delta(\Xi(\{v, w\}))$ is path-connected as well and lies in a single path-connected components of $\mathcal{O}$. But each path-connected component of the consensus output complex $\mathcal{O}$ is a single simplex whose vertices are all labeled with the same output value, so $\delta(\Xi(\{v, w\}))$ is contained in one of these simplices, τ.

Because Ξ is a carrier map, $\Xi(v) \subset \Xi(\{v, w\})$, $\delta(\Xi(v)) \subset \Delta(\{v, w\}) \subset \tau$. Similarly, $\delta(\Xi(w)) \subset \Delta(\{v, w\}) \subset \tau$. It follows that $\delta(\Xi(v))$ and $\delta(\Xi(w))$ are both vertices of τ; hence they must be labeled with the same value.

Because the protocol $(\mathcal{I}, \mathcal{P}, \Xi)$ solves the task $(\mathcal{I}, \mathcal{O}, \Delta)$, $\delta(\Xi(v))$ is a vertex of $\Delta(v) \in \mathcal{O}$, and $\delta(\Xi(w))$ is a vertex of $\Delta(w) \in \mathcal{O}$. Consensus defines $\Delta(v)$ to be a single vertex labeled with $\mathrm{view}(v)$, and therefore $\delta(\Xi(v))$ is also labeled with $\mathrm{view}(v)$. By a similar argument, $\delta(\Xi(w))$ is labeled with $\mathrm{view}(w)$. It follows that $\delta(\Xi(v))$ and $\delta(\Xi(w))$ must be labeled with distinct values, a contradiction. $\square$

This impossibility result is model-independent: It requires only that each $\Xi(\sigma)$ be path-connected. We will use this theorem and others like it to derive three kinds of lower bounds:

- In asynchronous models, the adversary can typically enforce these conditions for every protocol complex. For these models, we can prove *impossibility*: Consensus cannot be solved by any protocol.
- In synchronous models, the adversary can typically enforce these conditions for r or fewer rounds, where r is a property of the specific model. For these models, we can prove *lower bounds*: Consensus cannot be solved by any protocol that runs in r or fewer rounds.
- In semisynchronous models, the adversary can typically enforce these conditions for every protocol that runs in less than a particular time T, where T is a property of the specific model. For these

models, we can prove *time lower bounds*: Consensus cannot be solved by any protocol that runs in time less than T.

In the next section, we show that layered immediate snapshot protocol complexes are path-connected.

10.2 Immediate snapshot model and connectivity

We now show that if $(\mathcal{I}, \Xi, \mathcal{P})$ is a layered immediate snapshot protocol, then $\Xi(\sigma)$ is path-connected for every simplex $\sigma \in \mathcal{I}$.

10.2.1 Critical configurations

Here we introduce a style of proof that we will use several times, called a *critical configuration* argument. This argument is useful in asynchronous models, in which processes can take steps independently. As noted earlier, we can think of the system as a whole as a state machine where each local process state is a component of the configuration. Each input n-simplex σ encodes a possible initial configuration, the protocol complex $\Xi(\sigma)$ encodes all possible protocol executions starting from σ, and each facet of $\Xi(\sigma)$ encodes a possible final configuration. In the beginning, all interleavings are possible, and the entire protocol complex is reachable. At the end, a complete execution has been chosen, and only a single simplex remains reachable. In between, as the execution unfolds, we can think of the reachable part of the protocol complex as shrinking over time as each step renders certain final configurations inaccessible.

We use simplex notation (such as σ, τ) for initial and final configurations, since they correspond to simplices of the input and protocol complexes. We use Latin letters for transient intermediate configurations (C).

We want to show that a particular property, such as having a path-connected reachable protocol complex, that holds in each final configuration also holds in each initial configuration. We argue by contradiction. We assume that the property does not hold at the start, and we maneuver the protocol into a *critical configuration* where the property still does not hold, but where any further step by any process will make it hold *henceforth* (from that point on). We then do a case analysis of each of the process's possible next steps and use a combination of model-specific reasoning and basic topological properties to show that the property of interest must already hold in the critical configuration, a contradiction.

Let σ be an input m-simplex, $0 \leq m \leq n$, and let C be a configuration reached during an execution of the protocol $(\mathcal{I}, \mathcal{P}, \Xi)$ starting from σ. A simplex τ of $\Xi(\sigma)$ is *reachable* from C if there is an execution starting from configuration C and ending in final configuration τ. The subcomplex of the protocol complex $\mathcal{P}$ consisting of all simplices that are reachable from intermediate configuration C is called the *reachable complex* from C and is denoted $\Xi(C)$.

Definition 10.2.1.   Formally, a *property* $\mathbb{P}$ is a predicate on isomorphism classes of simplicial complexes. A property is *eventual* if it holds for any complex consisting of a single n-simplex and its faces.

For brevity, we say that a property $\mathbb{P}$ *holds* in configuration C if $\mathbb{P}$ holds for $\Xi(C)$, the reachable complex from C.

Definition 10.2.2.   A configuration C is *critical* for an eventual property $\mathbb{P}$ if $\mathbb{P}$ does not hold in C but does hold for every configuration reachable from C.

Informally, a critical configuration is a *last* configuration where $\mathbb{P}$ fails to hold.

Lemma 10.2.3.   *Every eventual property $\mathbb{P}$ either holds in every initial configuration or it has a critical configuration.*

Proof.   Starting from an initial configuration where $\mathbb{P}$ does not hold, construct an execution by repeatedly choosing a step that carries the protocol to another configuration where $\mathbb{P}$ does not hold. Because the protocol must eventually terminate in a configuration where $\mathbb{P}$ holds, advancing in this way will eventually lead to a configuration C where $\mathbb{P}$ does not hold, but every possible next step produces a configuration where $\mathbb{P}$ holds. The configuration C is the desired critical configuration.   □

10.2.2 The nerve graph

We need a way to reason about the path connectivity of a complex from the path connectivity of its subcomplexes.

Definition 10.2.4.   Let I be a finite index set. A set of simplicial complexes $\{\mathcal{K}_i | i \in I\}$ is called a *cover* for a simplicial complex $\mathcal{K}$, if $\mathcal{K} = \cup_{i \in I} \mathcal{K}_i$.

Definition 10.2.5.   The *nerve graph* $\mathcal{G}(\mathcal{K}_i | i \in I)$ is the 1-dimensional complex (often called a *graph*) whose vertices are the components $\mathcal{K}_i$ and whose edges are the pairs of components $\{\mathcal{K}_i, \mathcal{K}_j\}$ where $i, j \in I$, which have non-empty intersections.

Note that the nerve graph is defined in terms of the cover, not just the complex $\mathcal{K}$.

The lemma that follows is a special case of the more powerful *nerve lemma* (Lemma 10.4.2) used later to reason about higher-dimensional notions of connectivity.

Lemma 10.2.6.   *If each $\mathcal{K}_i$ is path-connected and the nerve graph $\mathcal{G}(\mathcal{K}_i | i \in I)$ is path-connected, then $\mathcal{K}$ is also path-connected.*

Proof.   We will construct a path between two arbitrary vertices $v_i \in \mathcal{K}_i$ and $v_j \in \mathcal{K}_j$ for $i, j \in I$. By hypothesis, the nerve graph contains a path $\mathcal{K}_i = \mathcal{K}_{i_0}, \ldots, \mathcal{K}_{i_\ell} = \mathcal{K}_j$, for $0 \leq j < \ell$, where $\mathcal{K}_{i_j} \cap \mathcal{K}_{i_{j+1}} \neq \emptyset$.

We argue by induction on ℓ, the number of edges in this path. When $\ell = 0$, v_i, v_j are both in $\mathcal{K}_{i_0}$, and they can be connected by a path because $\mathcal{K}_{i_0}$ is path-connected by hypothesis.

Assume the claim for paths with fewer than ℓ edges, and let $\mathcal{L} = \cup_{j=0}^{\ell-1} \mathcal{K}_{i_j}$. By construction, $\mathcal{L} \cap \mathcal{K}_{i_\ell}$ is non-empty. Pick a vertex v in $\mathcal{L} \cap \mathcal{K}_{i_\ell}$. By the induction hypothesis, $\mathcal{L}$ is path-connected, so there is a path p_0 from v_i to v in $\mathcal{L}$. By hypothesis, $\mathcal{K}_{i_\ell}$ is path-connected, so there is a path p_1 from v to v_j in $\mathcal{K}_{i_\ell}$. Together, p_0 and p_1 form a path linking v_i and v_j.   □

10.2.3 Reasoning about layered executions

To reason about the connectivity of layered protocol complexes, we need some basic lemmas about their structure. Assume C is a configuration, $U \subseteq [n]$ is a subset of process names, and $(\mathcal{I}, \mathcal{P}, \Xi)$ is a protocol. We introduce the following notations:

- Let $C \uparrow U$ denote the configuration obtained from C by running the processes in U in the next layer.
- Let $\Xi(C)$ denote the complex of executions that can be reached starting from C; we call $\Xi(C)$ the *reachable complex* from C.
- Let $(\Xi \downarrow U)(C)$ denote the complex of executions where, starting from C, the processes in U halt without taking further steps, and the rest finish the protocol.

In the special case $U = \emptyset$, $(\Xi \downarrow U)(C) = \Xi(C)$.

These notations may be combined to produce expressions like $(\Xi \downarrow V)(C \uparrow U)$, the complex of executions in which, starting from configuration C, the processes in U simultaneously take immediate snapshots (write then read), the processes in V then halt, and the remaining processes run to completion. For future reference we note that for all $U, V \subseteq \Pi$, and all configurations C, we have

$$((\Xi \downarrow U) \downarrow V)(C) = (\Xi \downarrow U \cup V)(C). \tag{10.2.1}$$

Recall that each configuration, which describes a system state, has two components: the state of the memory and the states of the individual processes. Let U and V be sets of process names, where $|U| \geq |V|$.

Lemma 10.2.7. *If $V \subseteq U$, then configurations $C \uparrow U$ and $(C \uparrow V) \uparrow (U \setminus V)$ agree on the memory state and on the states of processes not in V, but they disagree on the states of processes in V.*

Proof. Starting in C, we reach $C \uparrow U$ by letting the processes in U take immediate snapshot in a single layer. Each process in U reads the values written by the processes in U.

Starting in C, we reach $C \uparrow V$ by letting the processes in V write, then read in the first layer, and we reach $(C \uparrow V) \uparrow (U \setminus V)$ by then letting the processes in U but not in V write, then read in the second layer. Each process in V reads the values written by the processes in V, but each process in $U \setminus V$ reads the values written by U.

Both executions leave the memory in the same state, and both leave each process not in V in the same state, but they leave each process in V in different states.

Figure 10.1 shows an example where there are four processes, P_0, P_1, P_2, and P_3, where $U = \{P_0, P_1\}$ and $V = \{P_0\}$. The initial configuration C is shown on the left. The top part of the figure shows an execution in which P_0 writes 0 to its memory element and then reads the array to reach $C \uparrow V$, and then P_1 writes 1 and reads to reach $(C \uparrow V) \uparrow (U \setminus V)$. The bottom part shows an alternative execution in which P_0 and P_1 write 0 and 1, respectively, and then read the array to reach $C \uparrow U$. $\square$

Lemma 10.2.8. *If $V \not\subseteq U$ and $U \not\subseteq V$, configurations $(C \uparrow U) \uparrow (V \setminus U)$ and $(C \uparrow V) \uparrow (U \setminus V)$ agree on the memory state and on the states of processes not in $U \cup V$, but they disagree on the states of processes in $U \cup V$.*

Proof. Starting in C, we reach $C \uparrow U$ by letting the processes in U write, then read in the first layer, and we reach $(C \uparrow U) \uparrow (V \setminus U)$ by letting the process in $V \setminus U$ write, then read in the second layer. Each process in the first layer reads the states written by U, and each process in the second layer reads the states written by $U \cup V$. Similarly, starting in C, we reach $C \uparrow V$ by first running V, then $U \setminus V$. Each process in the first layer reads the states written by V, and each process in the second layer reads the states written by $U \cup V$. Both configurations agree on the memory state and on states of processes not in $U \cup V$, but they disagree on the states of processes in $U \cup V$.

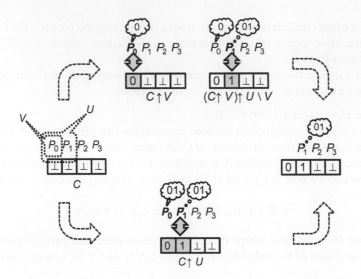

FIGURE 10.1

Proof of Lemma 10.2.7: The starting configuration C is shown on the left, where $U = \{P_0, P_1\}$, $V = \{P_1\}$, and each memory element is initialized to $\perp$. Two alternative executions appear at the top and bottom of the figure. The top shows an execution where $V = \{P_0\}$ writes and reads first, followed by $V \setminus U = \{P_1\}$. The bottom shows an execution where $U = \{P_0, P_1\}$ writes and reads first. In both executions, if we halt the processes in V, then we end up at the same configuration shown on the right.

Figure 10.2 shows an example where there are four processes, P_0, P_1, P_2, and P_3, where $U = \{P_1, P_2\}$ and $V = \{P_0, P_1\}$. The initial configuration C is shown on the left. The top part of the figure shows an execution in which P_0, P_1 write 0 and 1, respectively, read the array to reach $C \uparrow V$, and then P_2 writes 2 and reads to reach $(C \uparrow V) \uparrow (U \setminus V)$. The bottom part shows an alternative execution in which P_1, P_2 write 0 and 1, respectively, read the array to reach $C \uparrow U$, and then P_0 writes 0 and reads to reach $(C \uparrow U) \uparrow (V \setminus U)$.   □

Proposition 10.2.9.   *Assume that C is a configuration and $U, V \subseteq \Pi$; then we have*

$$\Xi(C \uparrow V) \cap \Xi(C \uparrow U) = (\Xi \downarrow W)(C \uparrow U \cup V),$$

where W, the set of processes that take no further steps, satisfies

$$W = \begin{cases} V, & \text{if } V \subseteq U; \\ U, & \text{if } U \subseteq V; \\ U \cup V, & \text{otherwise.} \end{cases}$$

Proof.   There are two cases. For the first case, suppose $V \subseteq U$. For inclusion in one direction, Lemma 10.2.7 states that configurations $C \uparrow U$ and $(C \uparrow V) \uparrow (U \setminus V)$ disagree on the states of processes in

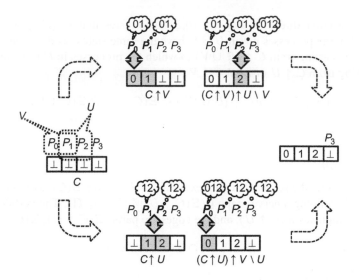

FIGURE 10.2

Proof of Lemma 10.2.8: The starting configuration C is shown on the left, where $U = \{P_1, P_2\}$, $V = \{P_0, P_1\}$, and each memory element is initialized to an arbitrary value $\perp$. Two alternative executions appear at the top and bottom of the figure. The top shows an execution where $V = \{P_0, P_1\}$ writes and reads first, followed by $V \setminus U = \{P_2\}$. The bottom shows an execution where $U = \{P_1, P_2\}$ writes and reads first, followed by $U \setminus V = \{P_0\}$. In both executions, if we halt the processes in $U \cup V$, then we end up at the same configuration, shown on the right.

V, implying that every execution in $\Xi(C \uparrow V) \cap \Xi(C \uparrow U)$ is an execution in $\Xi(C \uparrow U)$ where no process in V takes a step:

$$\Xi(C \uparrow V) \cap \Xi(C \uparrow U) \subseteq (\Xi \downarrow V)(C \uparrow U).$$

For inclusion in the other direction, Lemma 10.2.7 also states that configurations $C \uparrow U$ and $(C \uparrow V) \uparrow (U \setminus V)$ agree on the memory and on states of processes not in V, implying that every execution starting from $C \uparrow U$ in which the processes in V take no steps is also an execution starting from $C \uparrow V$:

$$(\Xi \downarrow V)(C \uparrow U) \subseteq \Xi(C \uparrow V) \cap \Xi(C \uparrow U).$$

The case $U \subseteq V$ is settled analogously.

For the second case, suppose $V \not\subseteq U$ and $U \not\subseteq V$. For inclusion in one direction, Lemma 10.2.8 states that in $(C \uparrow U) \uparrow (V \setminus U)$ and $(C \uparrow V) \uparrow (U \setminus V)$, the processes in $U \cup V$ have distinct states, implying that every execution in $\Xi(C \uparrow V) \cap \Xi(C \uparrow U)$ is an execution in $\Xi(C \uparrow U)$ where no process in $U \cup V$ takes a step:

$$\Xi(C \uparrow V) \cap \Xi(C \uparrow U) \subseteq (\Xi \downarrow U \cup V)(C \uparrow \{U \cup V\}).$$

For inclusion in the other direction, Lemma 10.2.8 also states that in $(C \uparrow U) \uparrow (V \setminus U)$ and $(C \uparrow V) \uparrow (U \setminus V)$, the processes not in $U \cup V$ have the same states, as does the memory, implying that every execution starting from $C \uparrow (U \cup V)$ in which the processes in $U \cup V$ take no steps is also an execution starting from $C \uparrow U$ or from $C \uparrow V$:

$$(\Xi \downarrow U \cup V)(C \uparrow \{U \cup V\}) \subseteq \Xi(C \uparrow V) \cap \Xi(C \uparrow U).$$

$\square$

10.2.4 Application

For each configuration C, the reachable complexes $\Xi(C \uparrow U)$ cover $\Xi(C)$, as U ranges over the non-empty subsets of Π, defining a nerve graph $\mathcal{G}(\Xi(C \uparrow U) | \emptyset \subsetneq U \subseteq \Pi)$. The vertices of this complex are the reachable complexes $\Xi(C \uparrow U)$, and the edges are pairs $\{\Xi(C \uparrow U), \Xi(C \uparrow V)\}$, where

$$\Xi(C \uparrow U) \cap \Xi(C \uparrow V) \neq \emptyset.$$

We know from Proposition 10.2.9 that

$$\Xi(C \uparrow U) \cap \Xi(C \uparrow V) = (\Xi \downarrow W)(C \uparrow U \cup V),$$

which is non-empty if and only if we do not halt every process: $W \neq \Pi$.

Lemma 10.2.10.   *The nerve graph $\mathcal{G}(\Xi(C \uparrow U) | \emptyset \subsetneq U \subseteq \Pi)$ is path-connected.*

Proof.   We claim there is an edge from every nerve graph vertex to the vertex $\Xi(C \uparrow \Pi)$. By Proposition 10.2.9,

$$\Xi(C \uparrow \Pi) \cup \Xi(C \uparrow U) = (\Xi \downarrow U)(C \uparrow \Pi).$$

Because $U \subset \Pi$, this intersection is non-empty, implying that the nerve graph has an edge from every vertex to $\Xi(C \uparrow \Pi)$. It follows that the nerve graph is path-connected. $\square$

Theorem 10.2.11.   *For every wait-free layered immediate snapshot protocol and every input simplex σ, the subcomplex $\Xi(\sigma)$ is path-connected.*

Proof.   We argue by induction on n. For the base case, when $n = 0$, the complex $\Xi(\sigma)$ is a single vertex, which is trivially path-connected.

For the induction step, assume the claim for n processes. Consider $\Xi(\sigma)$, where $\dim \sigma = n$. Being path-connected is an eventual property, so it has a critical configuration C such that $\Xi(C)$ is not path-connected, but $\Xi(C')$ is path-connected for every configuration C' reachable from C. In particular, for each set of process names $U \subseteq \Pi$, each $\Xi(C \uparrow U)$ is path connected.

Moreover, the subcomplexes $\Xi(C \uparrow U)$ cover the simplicial complex $\Xi(C)$, and Lemma 10.2.10 states that the nerve graph of this covering is path-connected. Finally, Lemma 10.2.6 states that these conditions ensure that $\Xi(C)$ is itself path-connected, contradicting the hypothesis that C is a critical state for path connectivity. $\square$

Theorem 10.2.11 provides an alternate, more general proof that consensus is impossible in asynchronous read-write memory.

10.3 *k*-Set agreement and (*k* − 1)-connectivity

We consider the *k*-set agreement task $(\mathcal{I}, \mathcal{O}, \Delta)$ with arbitrary inputs, meaning we allow $\mathcal{I}$ to consist of an arbitrary collection of initial configurations. An input complex is said to be *degenerate* for *k*-set agreement if, in every input configuration, at most *k* distinct values are assigned to processes. Clearly, *k*-set agreement has a trivial solution if the input complex is degenerate. We will see that if a protocol's carrier map satisfies a topological property called (*k* − 1)-connectivity, then that protocol cannot solve *k*-set agreement for any nondegenerate input complex.

Theorem 10.3.1.   *Let $\mathcal{I}$ be a nondegenerate input complex for k-set agreement. If $(\mathcal{I}, \mathcal{O}, \Delta)$ is an $(n + 1)$-process k-set agreement task, and $(\mathcal{I}, \mathcal{P}, \Xi)$ is a protocol such that $\Xi(\sigma)$ is $(k - 1)$-connected for all simplices σ in $\mathcal{I}$, then $(\mathcal{I}, \mathcal{P}, \Xi)$ cannot solve the k-set agreement task $(\mathcal{I}, \mathcal{O}, \Delta)$.*

Proof.   Because $\mathcal{I}$ is not degenerate, it contains a *k*-simplex σ labeled with $k + 1$ distinct values. Let Δ^k denote the *k*-simplex whose vertices are labeled with the input values from σ, and let $\partial \Delta^k$ be its $(k - 1)$-skeleton. Let $c : \Xi(\sigma) \to \partial \Delta^k$ denote the simplicial map that takes every vertex $v \in \Xi(\sigma)$ to its value in $\partial \Delta^k$. Since each vertex of $\Xi(\sigma)$ is labeled with a value from a vertex of σ and since the protocol $(\mathcal{I}, \mathcal{P}, \Xi)$ solves *k*-set agreement, the simplicial map c is well-defined.

Since the subcomplexes $\Xi(\tau)$ are *n*-connected for all simplices $\tau \subseteq \sigma$, Theorem 3.7.5(2) tells us that the carrier map $\Xi|_\sigma$ has a simplicial approximation. In other words, there exists a subdivision Div of σ, together with a simplicial map $\varphi : \mathrm{Div}\,\sigma \to \Xi(\sigma)$, such that for every simplex $\tau \subseteq \sigma$, we have $\varphi(\mathrm{Div}\tau) \subseteq \Xi(\tau)$.

The composition simplicial map

$$c \circ \varphi : \mathrm{Div}\,\sigma \to \partial \Delta^k$$

can be viewed as a coloring of the vertices of Div σ by the vertex values in $\partial \Delta^k$. Clearly, for every $\tau \subseteq \sigma$, the set of values in $c(\varphi(\mathrm{Div}\ \tau))$ is contained in the set of input values of τ, satisfying the conditions of Sperner's lemma. It follows that there exists a *k*-simplex ρ in Div σ colored with all $k + 1$ colors. This is a contradiction, because ρ is mapped to all of Δ^k, which is not contained in the domain complex $\partial \Delta^k$.   □

10.4 Immediate snapshot model and *k*-connectivity

In this section we show that if $(\mathcal{I}, \Xi, \mathcal{P})$ is a layered immediate snapshot protocol, then $\Xi(\sigma)$ is *n*-connected for every simplex $\sigma \in \mathcal{I}$.

10.4.1 The nerve lemma

To compute the connectivity of a complex, we would like to break it down into simpler components, compute the connectivity of each of the components, and then "glue" those components back together in a way that permits us to deduce the connectivity of the original complex from the connectivity of the components.

Definition 10.4.1.   Assume that $\mathcal{K}$ is a simplicial complex and $(\mathcal{K}_i)_{i \in I}$ is a family of non-empty subcomplexes covering $\mathcal{K}$, i.e., $\mathcal{K} = \cup_{i \in I} \mathcal{K}_i$. The cover's *nerve complex* $\mathcal{N}(\mathcal{K}_i | i \in I)$ is the abstract simplicial complex whose vertices are the components $\mathcal{K}_i$ and whose simplices are sets of components $\{\mathcal{K}_j | j \in J\}$, of which the intersection $\cap_{j \in J} \mathcal{K}_j$ is non-empty.

Informally, the nerve of a cover describes how the elements of the cover "fit together" to form the original complex. Like the nerve graph, the nerve complex is determined by the cover, not the complex. The next lemma is a generalization of Lemma 10.2.6.

Lemma 10.4.2 (Nerve Lemma).   Let $\{\mathcal{K}_i | i \in I\}$ be a cover for a simplicial complex $\mathcal{K}$, and let k be some fixed integer. For any index set $J \subseteq I$, define $\mathcal{K}_J = \cap_{j \in J} \mathcal{K}_j$. Assume that $\mathcal{K}_J$ is either $(k - |J| + 1)$-connected or empty, for all $J \subseteq I$. Then $\mathcal{K}$ is k-connected if and only if the nerve complex $\mathcal{N}(\mathcal{K}_i | i \in I)$ is k-connected.

The following special case of the nerve lemma is often useful:

Corollary 10.4.3.   If $\mathcal{K}$ and $\mathcal{L}$ are k-connected simplicial complexes, such that $\mathcal{K} \cap \mathcal{L}$ is $(k - 1)$-connected, then the simplicial complex $\mathcal{K} \cup \mathcal{L}$ is also k-connected.

10.4.2 Reachable complexes and critical configurations

To compute higher-dimensional connectivity, we need to generalize Proposition 10.2.9 to multiple sets.

Lemma 10.4.4.   *Let* $U_0, \ldots, U_m$ *be sets of process names indexed so that* $|U_i| \geq |U_{i+1}|$.

$$\bigcap_{i=0}^{m} \Xi(C \uparrow U_i) = (\Xi \downarrow W)(C \uparrow \cup_{i=0}^{m} U_i),$$

where W, *the set of processes that take no further steps, satisfies*

$$W = \begin{cases} \cup_{i=1}^{m} U_i & \text{if } \cup_{i=1}^{m} U_i \subseteq U_0 \\ \cup_{i=0}^{m} U_i & \text{otherwise.} \end{cases}$$

Proof.   We argue by induction on m. For the base case, when m is 1, the claim follows from Proposition 10.2.9.

For the induction step, assume the claim for m sets. Because the U_i are indexed so that $|U_i| \geq |U_{i+1}|$, $|\cup_{i=0}^{m-1} U_i| \geq |U_m|$, we can apply the induction hypothesis

$$\bigcap_{i=0}^{m} \Xi(C \uparrow U_i) = \bigcap_{i=0}^{m-1} \Xi(C \uparrow U_i) \cap \Xi(C \uparrow U_m)$$

$$= (\Xi \downarrow W)(C \uparrow \cup_{i=0}^{m-1} U_i) \cap \Xi(C \uparrow U_m)$$

where

$$W = \begin{cases} \cup_{i=1}^{m-1} U_i & \text{if } \cup_{i=1}^{m} U_i \subseteq U_0 \\ \cup_{i=0}^{m-1} U_i & \text{otherwise.} \end{cases}$$

Since no process in W takes a step in the intersection,

$$\bigcap_{i=0}^{m} \Xi(C \uparrow U_i) = (\Xi \downarrow W)(C \uparrow \cup_{i=0}^{m-1} U_i) \cap \Xi(C \uparrow U_m)$$

$$= (\Xi \downarrow W)(C \uparrow \cup_{i=0}^{m-1} U_i) \cap (\Xi \downarrow W)(C \uparrow U_m).$$

Applying Proposition 10.2.9 and Equation (10.2.1) yields

$$\bigcap_{i=0}^{m} \Xi(C \uparrow U_i) = (\Xi \downarrow W)(C \uparrow \cup_{i=0}^{m-1} U_i) \cap (\Xi \downarrow W)(C \uparrow U_m)$$

$$= ((\Xi \downarrow W) \downarrow X)(C \uparrow \cup_{i=0}^{m} U_i)$$

$$= (\Xi \downarrow (W \cup X))(C \uparrow \cup_{i=0}^{m} U_i),$$

where

$$X = \begin{cases} U_m & \text{if } U_m \subseteq \cup_{i=0}^{m-1} U_i \\ \cup_{i=0}^{m} U_i & \text{otherwise.} \end{cases}$$

We now compute $W \cup X$, the combined set of processes to halt. First, suppose that $\cup_{i=1}^{m} U_i \subseteq U_0$. It follows that $W = \cup_{i=1}^{m-1} U_i$, and $X = U_m$, so $W \cup X = \cup_{i=1}^{m} U_i$.

Suppose instead that $\cup_{i=1}^{m} U_i \not\subseteq U_0$. If $\cup_{i=1}^{m} U_i \not\subseteq U_0$, then $W = \cup_{i=0}^{m-1} U_i$, and $W \cup X = \cup_{i=0}^{m} U_i$. If $\cup_{i=1}^{m} U_i \subseteq U_0$, then $U_m \not\subseteq \cup_{i=0}^{m-1} U_i = U_0$, so $X = \cup_{i=0}^{m} U_i$, and $W \cup X = \cup_{i=0}^{m} U_i$. Substituting $Y = W \cup X$ yields

$$\bigcap_{i=0}^{m} \Xi(C \uparrow U_i) = (\Xi \downarrow Y)(C \uparrow X),$$

where W, the set of processes that take no further steps, satisfies

$$Y = \begin{cases} \cup_{i=1}^{m} U_i & \text{if } \cup_{i=1}^{m} U_i \subseteq U_0 \\ \cup_{i=0}^{m} U_i & \text{otherwise.} \end{cases}$$

$\square$

For each configuration C, the reachable complexes $\Xi(C \uparrow U)$ cover $\Xi(C)$. They define a nerve complex $\mathcal{N}(\Xi(C \uparrow U) | U \subseteq \Pi)$. The vertices of this complex are the reachable complexes $\Xi(C \uparrow U)$, and the m-simplices are the sets $\{\Xi(C \uparrow U_i) | i \in [0 : m]\}$ such that

$$\bigcap_{i \in I} \Xi(C \uparrow U_i) \neq \emptyset.$$

We know from Lemma 10.4.4 that

$$\bigcap_{i \in I} \Xi(C \uparrow U_i) = (\Xi \downarrow W)(C \uparrow \cup_{i \in I} U_i),$$

where W, the set of processes that halt, depends on U and V. This complex is non-empty if and only if $W \neq \Pi$.

Lemma 10.4.5. *If $\cup_{i=0}^{m} U_i = \Pi$ but each $U_i \neq \Pi$, then $\cap_{i=0}^{m} \Xi(C \uparrow U_i) = \emptyset$.*

Proof. By hypothesis, $\cup_{i=1}^{m} U_i \not\subseteq U_0$, so by Lemma 10.4.4,

$$\bigcap_{i=0}^{m} \Xi(C \uparrow U_i) = (\Xi \downarrow \cup_{i=0}^{m} U_i)(C \uparrow \cup_{i=0}^{m} U_i)$$

$$= (\Xi \downarrow \Pi)(C \uparrow \cup_{i=0}^{m} U_i),$$

which is empty because every process halts. □

Lemma 10.4.6. *The nerve complex $\mathcal{N}(\Xi(C \uparrow U)|\emptyset \subsetneq U \subseteq \Pi)$ is n-connected.*

Proof. We show that the nerve complex is a cone with an apex $\Xi(C \uparrow \Pi)$; in other words, if ν is an non-empty simplex in the nerve complex, so is $\{\Xi(C \uparrow \Pi)\} \cup \nu$. Let $\nu = \{\Xi(C \uparrow U_i | i \in [0:m]\}$.

If $\Pi = U_i$ for some i in $[0:m]$, there is nothing to prove. Otherwise, assume $U_i \neq \Pi$, for $i \in [0:m]$. The simplex $\{\Xi(C \uparrow \Pi)\} \cup \nu$ is non-empty if

$$\Xi(C \uparrow \Pi) \cap \left(\bigcap_{i=0}^{m} \Xi(C \uparrow U_i) \right) \neq \emptyset.$$

Applying Lemma 10.4.4,

$$\Xi(C \uparrow \Pi) \cap \left(\bigcap_{i=0}^{m} \Xi(C \uparrow U_i) \right) = (\Xi \downarrow \cup_{i=0}^{m} U_i)(C \uparrow \Pi)$$

Because each $U_i \neq \Pi$, and ν is non-empty, Lemma 10.4.5 implies that $\cup_{i=0}^{m} U_i \neq \Pi$, so the simplex $\{\Xi(C \uparrow \Pi)\} \cup \nu$ is non-empty.

It follows that every facet of the nerve complex contains the vertex $\Xi(C \uparrow \Pi)$, so the nerve complex is a cone, which is *n*-connected because it is contractible (see Section 3.5.3). □

Theorem 10.4.7. *For every wait-free layered immediate snapshot protocol and every input simplex σ, the complex $\Xi(\sigma)$ is n-connected.*

Proof. We argue by induction on *n*. For the base case, when $n = 0$, the complex $\Xi(\sigma)$ is a single vertex, which is trivially *n*-connected.

For the induction step, assume the claim for *n* processes. Consider $\Xi(\sigma)$, where dim $\sigma = n$. Being *n*-connected is an eventual property, so it has a critical configuration C such that $\Xi(C)$ is not *n*-connected, but $\Xi(C')$ is *n*-connected for every configuration reachable from C. In particular, for each set of process names $U \subseteq \Pi$, each $\Xi(C \uparrow U)$ is *n*-connected. Moreover, the $\Xi(C \uparrow U)$ cover $\Xi(C)$.

Lemma 10.4.4 states that

$$\bigcap_{i \in I} \Xi(C \uparrow U_i) = (\Xi \downarrow W)(C \uparrow X)$$

for $|W| > 0$, $W \subseteq X \subseteq \cup_{i \in I} U_i$. Because $|W| > 0$, this complex is the wait-free protocol complex for $n - |W| + 1$ processes, which is either empty or *n*-connected by the induction hypothesis.

Lemma 10.4.6 states that the nerve complex is *n*-connected, hence $(n-1)$-connected.

It follows from the nerve lemma that $\Xi(C)$ was already *n*-connected, contradicting the assumption that C was a critical configuration for *n*-connectivity. □

10.5 **Chapter notes**

Fischer, Lynch, and Paterson [55] were the first to prove that consensus is impossible in a message-passing system where a single thread can halt. They introduced the critical configuration style of impossibility argument. Loui and Abu-Amara [110] and Herlihy [78] extended this result to shared memory. Biran, Moran, and Zaks [18] were the first to draw the connection between path connectivity and consensus.

Chaudhuri [37] was the first to study the k-set agreement task. The connection between connectivity and k-set agreement appears in Chaudhuri, Herlihy, Lynch, and Tuttle [39], Saks and Zaharoglou [135], Borowsky and Gafni [23], and Herlihy and Shavit [91].

The critical configuration style of argument to show that a protocol complex is highly connected was used by Herlihy and Shavit [91] in the read-write wait-free model. This style of argument is useful to prove connectivity in models where other communication objects are available in addition to read-write objects, as in Herlihy [78] for path connectivity or Herlihy and Rajsbaum [79] for k-connectivity. The layered style of argument was used in Chapter 9 to prove connectivity invariants on the sets of configurations after some number of steps of a protocol. It is further explored in Chapter 13. Yet another approach to prove connectivity is in Chapter 7, based on distributed simulations.

As we have seen in this chapter, $(k - 1)$-connectivity is sufficient to prove the k-set agreement impossibility result. However, it is not a necessary property. In Chapter 9 we saw that the weaker property of being a manifold protocol is also sufficient. Theorem 5.1 in Herlihy and Rajsbaum [82] is a model-independent condition that implies set agreement impossibility in the style of Theorem 10.3.1. The condition is based on homology groups instead of homotopy groups (as is k-connectivity) and is more combinatorial. In fact, from the manifold protocol property it is quite straightforward to derive the homology condition, as explained by Attiya and Rajsbaum [16].

One of the main ideas in this book is that the power of a distributed computing model is closely related to the connectivity of protocol complexes in the model. For instance, given Theorem 10.3.1, the problem of telling whether set agreement is solvable in a particular model is reduced to the problem of showing that protocol complexes in that model are highly connected. A number of tools exist to show that a space is highly connected, such as subdivisions, homology, the nerve theorem, and others. Matousek [113] describes some of them and discusses their relationship. We refer the interested reader to Kozlov [100, section 15.4] for further information on the nerve lemma; in particular, see [100, Theorem 15.24].

Mostefaoui, Rajsbaum, and Raynal [122] introduced the study of the "condition-based approach" with the aim of characterizing the input complexes for which it is possible to solve consensus in an asynchronous system despite the occurrence of up to t process crashes. It was further developed, e.g., for synchronous systems, in Mostefaoui, Rajsbaum, Raynal, and Travers [123] and set agreement in [121].

Obstructions to wait-free solvability of arbitrary tasks based on homology theory were studied by Havlicek [76]. This result is further discussed in Havlicek [75], where it is proved that the wait-free full-information protocol complex (using atomic snapshot memory) is homotopy equivalent to the underlying input complex. The derivation of the homotopy equivalence is based on Theorem 10.4.7 (proved originally in [91]).

10.6 Exercises

Exercise 10.1.   Prove the following stronger version of Lemma 10.2.6: If each $\mathcal{K}_i$ is path-connected, then $\mathcal{K}$ is path-connected *if and only if* the nerve graph $\mathcal{G}(\mathcal{K}_i | i \in I)$ is path-connected.

Exercise 10.2.   Defend or refute the claim that "without loss of generality," it is enough to prove that k-set agreement is impossible when inputs are taken only from a set of size $k + 1$.

Exercise 10.3.   Use the nerve lemma to prove that if $\mathcal{A}$ and $\mathcal{B}$ are n-connected, and $\mathcal{A} \cap \mathcal{B}$ is $(n - 1)$-connected, then $\mathcal{A} \cup \mathcal{B}$ is n-connected.

Exercise 10.4.   Revise the proof of Theorem 10.2.11 to a model in which asynchronous processes share an array of single-writer, multireader registers. The basic outline should be the same except that the critical configuration case analysis must consider individual reads and writes instead of layers.

Exercise 10.5.   Let the simplicial map $\varphi : \mathcal{A} \to \mathcal{B}$ to be a simplicial approximation to the continuous map $f : |\mathcal{A}| \to |\mathcal{B}|$. Show that the continuous map $|\varphi| : |\mathcal{A}| \to |\mathcal{B}|$ is homotopic to f.

Exercise 10.6.   We have defined a simplicial map $\varphi : \mathcal{A} \to \mathcal{B}$ to be a simplicial approximation to the continuous map $f : |\mathcal{A}| \to |\mathcal{B}|$ if, for every simplex $\alpha \in \mathcal{A}$,

$$f(|\alpha|) \subseteq \bigcap_{a \in \alpha} \mathrm{St}\, \varphi(a).$$

An alternative definition is to require that for every vertex $a \in \mathcal{A}$,

$$f(|\mathrm{St}\, a|) \subseteq \mathrm{St}\, \varphi(a).$$

Show that these definitions are equivalent.

Exercise 10.7.   Let C be a configuration, and define

$$Z(C) = \bigcap_{P \in \Pi} \Xi(C \uparrow P),$$

the complex of final configurations reachable by executions in which exactly one process participates in the next layer after C. Clearly, $Z(C) \subseteq \Xi(C)$.

Show that $\Xi(C) \not\subseteq Z(C)$.

Show that the nerve complex $\mathcal{N}(\Xi(C \uparrow P) | P \in \Pi)$ is isomorphic to $\partial \Delta^n$, the $(n - 1)$-skeleton of $\Delta^n = [n] = \{0, \dots n\}$.

Wait-Free Computability for General Tasks

Although many tasks of interest are colorless, there are "inherently colored" tasks that have no corresponding colorless task. Some are wait-free solvable, but not by any colorless protocol; others are not wait-free solvable. In this chapter we give a characterization of wait-free solvability of general tasks. We will see that general tasks are harder to analyze than colorless tasks. Allowing tasks to depend on process names seems like a minor change, but it will have sweeping consequences.

11.1 Inherently colored tasks: the hourglass task

Not all tasks can be expressed as colorless tasks. For example, the weak symmetry-breaking task discussed in Chapter 9 cannot be expressed as a colorless task, since one process cannot adopt the output value of another.

Theorem 4.3.1 states that a colorless task $(\mathcal{I}, \mathcal{O}, \Delta)$ has an $(n+1)$-process wait-free layered snapshot protocol if and only if there is a continuous map $f : |\mathrm{skel}^n \mathcal{I}| \to |\mathcal{O}|$ carried by Δ. Can we generalize this theorem to colorless tasks? A simple example shows that a straightforward generalization will not work.

Consider the following *Hourglass* task, whose input and output complexes are shown in Figure 11.1. There are three processes: P_0, P_1, and P_2, denoted by black, white, and gray, respectively, and only one input simplex. The carrier map defining this task is shown in tabular form in Figure 11.2 and in schematic form in Figure 11.3. Informally, this task is constructed by taking the standard chromatic

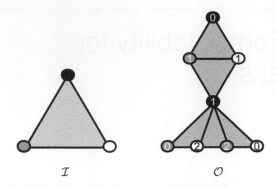

FIGURE 11.1

Input and output complexes for the Hourglass task. If a vertex v is labeled with P_i, then a process that chooses output vertex v in the Hourglass task chooses P_i's input value for the 2-set agreement task. Note that each triangle is labeled with at most two process names.

σ	$\Delta(\sigma)$
$\{P_0\}$	$\{(P_0, 0)\}$
$\{P_1\}$	$\{(P_1, 0)\}$
$\{P_2\}$	$\{(P_2, 0)\}$
$\{P_0, P_1\}$	$\{(P_0, 0), (P_1, 1)\}, \{(P_0, 1), (P_1, 1)\}, \{(P_0, 1), (P_1, 0)\}$
$\{P_0, P_2\}$	$\{(P_0, 0), (P_2, 1)\}, \{(P_0, 1), (P_2, 1)\}, \{(P_0, 1), (P_2, 0)\}$
$\{P_1, P_2\}$	$\{(P_1, 0), (P_2, 2)\}, \{(P_1, 2), (P_2, 2)\}, \{(P_1, 2), (P_2, 0)\}$
$\{P_0, P_1, P_2\}$	$\{(P_0, 0), (P_1, 1), (P_2, 1)\}, \{(P_0, 1), (P_1, 1), (P_2, 1)\}, \{(P_0, 1), (P_1, 0), (P_2, 2)\},$
	$\{(P_0, 1), (P_1, 2), (P_2, 2)\}, \{(P_0, 1), (P_1, 2), (P_2, 0)\}$

FIGURE 11.2

The Hourglass task: Tabular specification.

subdivision and "pinching" it at the waist to identify (that is, "glue together") P_0's vertices on the edges representing its two-process executions.

Note that the Hourglass task satisfies the conditions of Theorem 4.3.1: There is a continuous map $|\mathcal{I}| \to |\mathcal{O}|$ carried by Δ, shown schematically in Figure 11.4.[1]

Nevertheless, even though this task satisfies the conditions of Theorem 4.3.1, it does not have a wait-free layered immediate-snapshot protocol. Perhaps the simplest demonstration is merely to observe that we can solve 2-set agreement by composing a layered immediate snapshot protocol with a protocol for the Hourglass task. It follows that if we had a layered immediate snapshot hourglass protocol, then this composition would yield a layered immediate snapshot protocol for 2-set agreement, which we know to be impossible.

---

[1]This map is a *deformation retraction*, a continuous deformation of the input complex's polyhedron into the output complex's polyhedron that leaves the output complex unchanged.

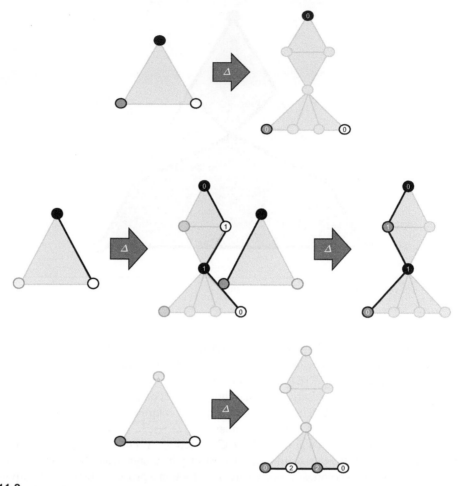

FIGURE 11.3

Carrier map for the Hourglass task. Single-process executions are at the top, executions for P_0 and P_1 on the middle left executions for P_0 and P_2 on the middle right, and executions for P_1 and P_2 on the bottom.

The composite protocol is shown in Figure 11.5. The processes share an array announce[], with one entry for each process, initially null. Process P_i first writes its input value to announce[i] and then calls the layered snapshot Hourglass protocol. If that call returns 0, P_i may be running by itself, so it decides its own input. Otherwise, the processes behave differently. If the Hourglass protocol returns 1 to P_0, then P_0 is running concurrently with either P_1 or P_2, or both, so it decides announce[1] if it is not null; otherwise it decides announce[2]. If the Hourglass protocol returns 1 to P_0 or P_1, it decides announce[0]. If the Hourglass protocol returns 2, the process decides its own input. Figure 11.6 labels each output vertex with its corresponding decision value. It is easy to check that in each execution, the processes decide at most two distinct values.

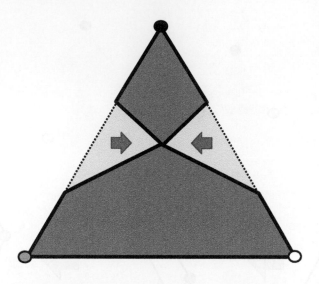

FIGURE 11.4

Continuous map between input and output complexes.

Since there is no wait-free snapshot protocol for k-set agreement, there cannot be a wait-free snapshot protocol for the Hourglass task. Why does Theorem 4.3.1 fail to hold for colored tasks? One direction still works: Given a protocol $(\mathcal{I}, \mathcal{P}, \Xi)$ solving the task $(\mathcal{I}, \mathcal{O}, \Delta)$, it is easy to extend the proof of Theorem 4.3.1 to exploit the connectivity of the snapshot protocol complex to construct a continuous map $|\mathcal{I}| \rightarrow |\mathcal{P}|$. Composing this map with the decision map yields a continuous map $|\mathcal{I}| \rightarrow |\mathcal{O}|$ carried by Δ.

The other direction fails. Given a continuous map $f : |\mathcal{I}| \rightarrow |\mathcal{O}|$ carried by Δ, it is possible to construct a simplicial approximation $\phi : \text{Ch}^N \mathcal{I} \rightarrow \mathcal{O}$ carried by Δ, but that simplicial approximation may not be color-preserving. In other words, one process may be assigned another's output value. Such flexibility is not an issue with colorless tasks, where by definition a process's inputs and outputs do not depend on its identity. By contrast, for tasks such as weak symmetry breaking or Hourglass, an output legal for one process may not be legal for another.

11.2 Solvability for colored tasks

Recall that a simplex $\sigma = \{s_0, \ldots, s_n\}$ is *chromatic* if each vertex is labeled with a distinct color, and a *chromatic subdivision* Div σ is a subdivision of σ where

(1) each simplex of the subdivision is chromatic,
(2) for each $\tau \subset \sigma$, each vertex in Divτ is labeled with a color from τ.

We are now ready to state our main theorem.

```
shared announce: array [0.2] of value
shared hourglass: Hourglass

        decide(input: Value): Value
          announce[0] := input
          select (hourglass.decide(0))
            case 0:
              return announce[0]
            case 1:
              if announce[1] != null then
                return announce[1]
              else
                return announce[2]


  decide(Value input): Value          decide(Value input): Value
    announce[1] := input                announce[2] := input
    select (hourglass.decide(1))        select (hourglass.decide(2))
      case 0:                             case 0:
        return announce[1]                  return announce[2]
      case 1:                             case 1:
        return announce[0]                  return announce[0]
      case 2:                             case 2:
        return announce[1]                  return announce[2]
```

FIGURE 11.5

How to use an Hourglass protocol to solve 2-set agreement: Pseudo-code for P_0, P_1, and P_2.

Theorem 11.2.1. *A task* $(\mathcal{I}, \mathcal{O}, \Delta)$ *has a wait-free layered immediate snapshot protocol if and only if* $\mathcal{I}$ *has a chromatic subdivision* $\mathrm{Div}\,\mathcal{I}$ *and a color-preserving simplicial map*

$$\mu : \mathrm{Div}\,\mathcal{I} \to \mathcal{O}$$

carried by Δ.

Theorem 11.2.1 is depicted schematically in Figure 11.7. The figure's top half shows how a task is specified by a carrier map Δ that takes each simplex σ of the input complex $\mathcal{I}$ to a subcomplex $\Delta(\sigma)$ of the output complex $\mathcal{O}$. The bottom half shows how the simplicial map μ maps each simplex of a chromatic subdivision $\mathrm{Div}\,\mathcal{I}$ to a simplex in the output complex such that every $\tau \in \mathrm{Div}\,\sigma$ is carried to a simplex in $\Delta(\sigma)$.

It is impossible to build such a color-preserving simplicial map for the Hourglass task because the "pinch" in the middle makes it impossible for any simplicial map to be color-preserving.

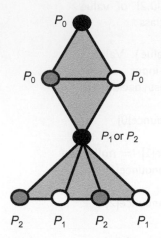

FIGURE 11.6

Hourglass vertices to k-set agreement values.

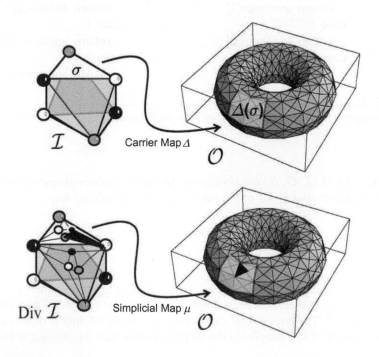

FIGURE 11.7

Fundamental theorem for colored tasks.

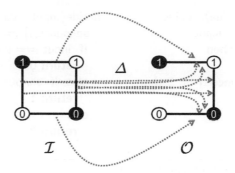

FIGURE 11.8

Input and output complexes for 2-process quasi-consensus.

In Chapter 10 we saw that the (colorless) consensus task has no wait-free layered immediate snapshot protocol. We can illustrate how Theorem 11.2.1 works by relaxing the consensus task's requirements as follows:

> **Quasi-Consensus** Each of P_0 and P_1 is given a binary input. If both have input v, then both must decide v. If they have mixed inputs, then either they agree or P_1 may decide 0 and P_0 may decide 1 (but not vice versa).

Figure 11.8 shows the input and output complexes for the quasi-consensus task. Note that quasi-consensus is not a colorless task. Does it have a wait-free read-write protocol?

It is easy to see that there is no color-preserving simplicial map carried by Δ from the input complex to the output complex. The vertices of input simplex $\{(P_0, 0), (P_1, 1)\}$ map to $(P_0, 0)$ and $(P_1, 1)$, but there is no single output simplex containing both vertices. Nevertheless, there is a map satisfying the conditions of the theorem from a *subdivision* $\mathrm{Div}\,\mathcal{I}$ of the input complex. If input simplex $\{(P_0, 0), (P_1, 1)\}$ is subdivided as shown in Figure 11.9, then it can be "folded" around the output complex, allowing input vertices $(P_0, 0)$ and $(P_1, 1)$ to be mapped to their counterparts in the output complex.

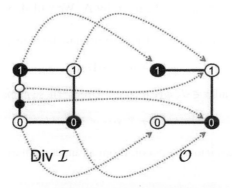

FIGURE 11.9

Subdivided input and output complexes for 2-process quasi-consensus.

```
decide(input: Value): Value          decide(input: Value): Value
    announce[0] := input                 announce[1] := input
    if input = 1 then                    if input == 0 then
      return 1                             return 0
    else if announce[1] != 1 then        else if announce[0] != 0 then
      return 0                             return 1
    else                                 else
      return 1                             return 0
```

FIGURE 11.10

Quasi-consensus protocols for P_0 (left) and P_1 (right).

Figure 11.10 shows a simple protocol for quasi-consensus. If P_0 has input 0 and P_1 has input 1, then this protocol admits three distinct executions: one in which both decide 0, one in which both decide 1, and one in which P_1 decides 0 and P_0 decides 1. These three executions correspond to the three simplices in the subdivision of $\{(P_0, 0), (P_1, 1)\}$, which are carried to the edges $\{(P_0, 0), (P_1, 0)\}$, $\{(P_0, 1), (P_1, 1)\}$, and $\{(P_0, 1), (P_1, 0)\}$.

11.3 Algorithm implies map

One direction of Theorem 11.2.1 is straightforward: If $(\mathcal{I}, \mathcal{P}, \Xi)$ solves $(\mathcal{I}, \mathcal{O}, \Delta)$, then the protocol's simplicial decision map

$$\delta : \mathcal{P} \to \mathcal{O},$$

is color-preserving and carried by Δ. On the other hand, any wait-free layered immediate snapshot protocol complex $\mathcal{P}$ is a chromatic subdivision of the input complex $\mathcal{I}$.

11.4 Map implies algorithm

Assume we are given a task $(\mathcal{I}, \mathcal{O}, \Delta)$, a chromatic subdivision $\mathrm{Div}\,\mathcal{I}$ of the input complex, and a color-preserving simplicial map $\mu : \mathrm{Div}\,\mathcal{I} \to \mathcal{O}$ carried by Δ. We will show that this task has a wait-free read-write protocol.

Our strategy is to show there exists a color-preserving simplicial map

$$\phi : \mathrm{Ch}^N\mathcal{I} \to \mathrm{Div}\,\mathcal{I}$$

for some $N > 0$ such that for all $\sigma \in \mathcal{I}$, $\phi(\mathrm{Ch}^N\sigma) \subseteq \mathrm{Div}\,\sigma$. These maps compose as follows:

$$\mathrm{Ch}^N\mathcal{I} \xrightarrow{\phi} \mathrm{Div}\,\mathcal{I} \xrightarrow{\mu} \mathcal{O}.$$

Here is how to turn these maps into a protocol. From an input simplex σ, each process performs the following three steps:

step 1. execute an N-layer immediate snapshot protocol, halt on a vertex x of the simplicial complex $\mathrm{Ch}^N\sigma$,

step 2. compute $y = \phi(x)$, yielding a vertex in $\text{Div}\,\sigma$,
step 3. compute $z = \mu(y)$, yielding an output vertex.

It is easy to check that all processes halt on the vertices of a single simplex in $\Delta(\sigma)$. Moreover, because all maps are color-preserving, each process halts on an output vertex of matching color.

Because the identity map $|\text{Ch}^N \mathcal{I}| \to |\text{Div}\,\mathcal{I}|$ is continuous, it has a simplicial approximation $\psi :$ $\text{Ch}^N \mathcal{I} \to \text{Div}\,\mathcal{I}$ carried by Δ (Theorem 3.7.5). Unfortunately, there is no guarantee that this map is color-preserving. To provide such a guarantee, we will prove the following generalization of the Simplicial Approximation theorem:

Theorem 11.4.1.   *If $\mathcal{I}$ is a chromatic complex, and $\text{Div}\,\mathcal{I}$ is a chromatic subdivision of $\mathcal{I}$, then there exists a color-preserving simplicial map*

$$\phi : \text{Ch}^N \mathcal{I} \to \text{Div}\,\mathcal{I}$$

such that for all $\sigma \in \mathcal{I}$, $\phi(\text{Ch}^N \sigma) \subseteq \text{Div}\,\sigma$.

Both $\text{Div}\,\mathcal{I}$ and $\text{Ch}^N \mathcal{I}$ are abstract complexes, defined in purely combinatorial terms. Nevertheless, we do not know how to prove the existence of this map in a combinatorial way. Instead, we will work with *geometric complexes*, embedded in high-dimensional Euclidean space, where we can exploit tools provided by point-set topology. Henceforth, all simplices and complexes will be geometric unless explicitly stated otherwise, so we will not always distinguish between a simplex or complex and its polyhedron. The exact meaning should be clear from context.

11.4.1 Basic concepts from point-set topology

Recall that geometric simplices "live" in a a Euclidean space of high but finite dimension. Any such space is a metric space, where the distance between points x and y is denoted $|x - y|$. The *open ϵ-ball* $B(x, \epsilon)$ around a point x is the set of points y such that $|x - y| < \epsilon$ for some $\epsilon > 0$. An ϵ-ball is an open set.

A *Cauchy sequence* is an infinite sequence of points $x_0, x_1, \ldots$ with the property that the distance between successive points $|x_i - x_{i+1}|$ limits to zero.

Fact 11.4.2.   In Euclidean space, every Cauchy sequence $x_0, x_1, \ldots$ *converges* to a point x^*, meaning that for any $\epsilon > 0$, there is an integer $N > 0$ such that $|x^* - x_i| < \epsilon$ for all $i > N$.

The set of points that can be expressed as affine combinations of points $x_0, x_1, \ldots, x_m$,

$$y = \sum_{i=0}^{m} c_i \cdot x_i$$

where $\sum_i c_i = 1$, is called the *hyperplane* defined by those points. If a hyperplane is generated by $m+1$ affinely-independent points, then it has *dimension m*. A set of points need not be affinely independent to define a hyperplane, so a hyperplane generated by $m + 1$ arbitrary points has dimension at most m.

As long as a finite set of hyperplanes does not fill up the entire space, any point has arbitrarily small neighborhoods that include points not on any hyperplane.

Fact 11.4.3. If x is a point and $\mathbb{H}$ a finite set of hyperplanes, each of dimension less than n, then there is an $\varepsilon > 0$ such that for every $\epsilon < \varepsilon$, $B(x, \epsilon)$ contains a point not on any hyperplane in $\mathbb{H}$.

Moreover, any point not on any hyperplane has arbitrarily small neighborhoods that do not intersect the hyperplane.

Fact 11.4.4. If x is a point and $\mathbb{H}$ a finite set of hyperplanes, each of dimension less than n, none of which contains v, then there is an $\varepsilon > 0$ such that $B(x, \varepsilon)$ does not intersect any hyperplane in $\mathbb{H}$.

Definition 11.4.5. An *open cover* $\mathbb{U}$ for a simplicial complex $\mathcal{K}$ is a finite collection of open sets $U_0, \ldots, U_k$ such that $\mathcal{K} \subseteq \cup_{i=0}^{k} U_i$.

The following fact is the basis for the Simplicial Approximation theorem used in earlier chapters.

Fact 11.4.6. If $U_0, \ldots, U_k$ an open cover for a finite simplicial complex $\mathcal{K}$, there exists a real number $\lambda > 0$, called the *Lebesgue number*, such that any set of diameter less than λ lies in a single U_i.

11.4.2 Geometric complexes

In Chapter 9, we defined the standard chromatic subdivision $\mathrm{Ch}\,\sigma$ in a purely combinatorial way, as an abstract simplicial complex. Now we give an equivalent definition of $\mathrm{Ch}\,\sigma$ as a geometric complex.

Definition 11.4.7. Recall that $\mathrm{Ch}\,\sigma$, as an abstract complex, is defined as follows. First, each vertex is a pair (P_i, σ_i), where $\sigma_i \subseteq \sigma$, and $P_i \in \mathrm{names}(\sigma_i)$. Second, if (P_i, σ_i) and (P_j, σ_j) are vertices of $\mathrm{Ch}\,\sigma$, then $\sigma_i \subseteq \sigma_j$ or vice versa. Finally, if $P_i \in \mathrm{names}(\sigma_j)$, then $\sigma_i \subseteq \sigma_j$.

We need to assign a point within $|\sigma|$ to each (P_i, σ_i). Recall that $b = \sum_{i=0}^{n} \frac{s_i}{n+1}$ is the barycenter of σ. Let δ be any real value such that $0 < \delta < \frac{1}{n+1}$. For each P_i and simplex τ, define

$$|(P_i, \tau)| = (1 + \delta)b - \delta|s_i|$$

See Figure 11.11.

For any sufficiently small value of δ, it can be shown that this definition gives a geometric construction for the chromatic subdivision. We will use this construction for the remainder of this section. Since all simplices and complexes in this section are geometric, we will not distinguish between an abstract vertex v and the point $|v|$ or an abstract simplex σ and its polyhedron $|\sigma|$.

Recall (Definition 3.6.7) that the *mesh* of a complex is the maximum diameter of any simplex.

Fact 11.4.8. For an n-simplex σ, $\mathrm{mesh}(\mathrm{Bary}\,\sigma) \leq \frac{n}{n+1}\mathrm{diam}(\sigma)$.

By taking sufficiently large N, $\mathrm{mesh}(\mathrm{Ch}^N \mathcal{I})$ can be made arbitrarily small.

11.4.3 Colors and covers

Fact 11.4.9. A set of vertices $\{v_0, \ldots, v_q\}$ of a complex $\mathcal{K}$ forms a simplex if and only if the intersection of their open stars is non-empty:

$$\bigcap_{i=0}^{q} \mathrm{St}^{\circ}(v_i, \mathcal{K}) \neq \emptyset.$$

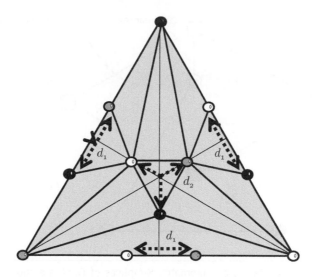

FIGURE 11.11

Geometric representation for $\operatorname{Ch}\sigma$.

To construct a color-preserving simplicial map from $\operatorname{Ch}^N\mathcal{I}$ to $\operatorname{Div}\mathcal{I}$, we will need the vertex colors to "align" nicely. More specifically, the open stars of the vertices in a $\operatorname{Div}\mathcal{I}$ form an open cover for $\mathcal{I}$. This *open-star cover* has a natural coloring inherited from the coloring of $\mathcal{I}$: $\operatorname{name}(\operatorname{St}^\circ(v, \mathcal{I})) = \operatorname{name}(v)$. The open-star cover of $\operatorname{Div}\mathcal{I}$ is *chromatic* on $\operatorname{Ch}^N\mathcal{I}$ if every simplex of $\operatorname{Ch}^N\mathcal{I}$ is covered by open stars of vertices of a simplex in $\operatorname{Div}\mathcal{I}$ of matching color:

$$\text{for all } \sigma \in \operatorname{Ch}^N\mathcal{I}, \sigma \subseteq \bigcup_{t \in \tau} \operatorname{St}^\circ(t, \operatorname{Div}\mathcal{I}),$$

for some $\tau \in \operatorname{Div}\mathcal{I}$ where $\operatorname{names}(\sigma) \subseteq \operatorname{names}(\tau)$.

We will show that without loss of generality, the geometric realization of $\operatorname{Div}\mathcal{I}$ can be chosen so that its open-star cover is chromatic on *any* iterated standard chromatic subdivision of $\mathcal{I}$.

Lemma 11.4.10.   *The geometric realization of* $\operatorname{Div}\mathcal{I}$ *can be chosen so that its open-star cover is chromatic on each of the subdivisions*

$$\mathcal{I}, \operatorname{Ch}\mathcal{I}, \operatorname{Ch}^2\mathcal{I}, \operatorname{Ch}^3\mathcal{I}, \ldots.$$

Figure 11.12 shows how an open-star covering can fail to be chromatic. Simplices of $\operatorname{Div}\mathcal{I}$ are shown with dotted lines and square vertices, whereas simplices of $\operatorname{Ch}^n\mathcal{I}$ are shown with solid lines and round vertices. The gray vertex marked v lies on the boundary between two gray open stars, so it is not covered by an open star of the same color.

Note, however, that if we perturb some of the square vertices by an arbitrarily small amount, then v moves off the boundary and into the open star of a vertex of matching color. This observation suggests a strategy: We will pick an arbitrary geometric realization of $\operatorname{Div}\mathcal{I}$, and if its open-star cover fails to

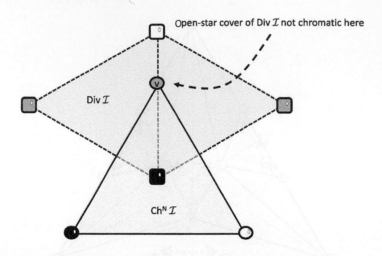

FIGURE 11.12

How an open-star covering can fail to be chromatic. Simplices of $\mathrm{Div}\,\mathcal{I}$ are shown with dotted lines and square vertices; simplices of $\mathrm{Ch}^n\mathcal{I}$ are shown with solid lines and round vertices. The gray vertex marked v lies on the boundary between two gray open stars, so this open-star covering fails to be chromatic.

be chromatic, then we will "perturb" vertex positions by very small amounts until the open-star cover becomes chromatic.

Readers willing to accept Lemma 11.4.10 can skip directly to Section 11.4.4.

We can recast some familiar concepts in terms of convex combinations. Every point x in the polyhedron of a complex $\mathcal{I}$ can be expressed uniquely as the convex combination of the vertices of a simplex σ of $\mathcal{I}$. The open star $\mathrm{St}^\circ(v, \mathcal{I})$ is just the set of points that can be expressed as the convex combination of the vertices of a simplex that includes v.

In this definition and in the subsequent lemmas, $\mathrm{Div}\,\mathcal{I}$ and $\mathrm{Ch}^n\mathcal{I}$ may be replaced by arbitrary chromatic subdivisions. Two simplices *conflict* if they share no colors.

Definition 11.4.11. A *conflict point* for $\mathrm{Div}\,\mathcal{I}$ and $\mathrm{Ch}^N\mathcal{I}$ is a point that can be expressed as the convex combination of vertices of two conflicting simplices, $\sigma = \{s_0, \ldots, s_p\} \in \mathrm{Div}\,\mathcal{I}$ and $\tau = \{t_0, \ldots, t_q\} \in \mathrm{Ch}^N\mathcal{I}$, were $\mathrm{names}(\sigma) \cap \mathrm{names}(\tau) = \emptyset$:

$$x = \sum_{i=0}^{p} a_i \cdot t_i = \sum_{j=0}^{q} b_j \cdot t_j$$

for $0 < a_i, b_j$ and $1 = \sum_i a_i = \sum_j b_j$.

Lemma 11.4.12. *The open-star cover of* $\mathrm{Div}\,\mathcal{I}$ *is chromatic for* $\mathrm{Ch}^N\mathcal{I}$ *if and only if there are no conflict points.*

Proof. As noted, every point x in $\mathcal{I}$ has a unique expression as the convex combination of the vertices of a $\sigma \in \mathrm{Div}\,\mathcal{I}$, and similarly for a $\tau \in \mathrm{Ch}^N\mathcal{I}$.

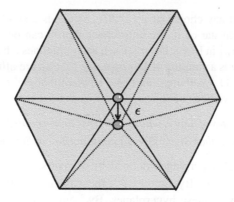

FIGURE 11.13

An ϵ-perturbation.

If x is a conflict point, then it lies in the interior of τ but not in $\mathrm{St}^\circ(v, \mathrm{Div}\,\mathcal{I})$ for any vertex v where $\mathrm{name}(v) \in \mathrm{names}(\tau)$. The open stars of $\mathrm{Div}\,\mathcal{I}$ with colors from $\mathrm{names}(\tau)$ therefore fail to cover τ.

If the open-star cover is chromatic, then every x lies in the open star of some vertex v of $\mathrm{Div}\,\mathcal{I}$ and u of $\mathrm{Ch}^N\mathcal{I}$ such that $\mathrm{name}(v) = \mathrm{name}(u)$, implying that x is not a conflict point. $\qquad\square$

Corollary 11.4.13 follows because the definition of conflict point is symmetric in terms of the two subdivisions.

Corollary 11.4.13.   *If the open-star cover of* $\mathrm{Div}\,\mathcal{I}$ *is chromatic on* $\mathrm{Ch}^N\mathcal{I}$, *then the open-star cover of* $\mathrm{Ch}^N\mathcal{I}$ *is chromatic on* $\mathrm{Div}\,\mathcal{I}$.

We say that $\mathrm{Div}\,\mathcal{I}$ has an ϵ-*perturbation* at vertex v for some $\epsilon > 0$ if there is a point v', $|v - v'| < \epsilon$, such that replacing v with v' in each simplex of $\mathrm{Div}\,\mathcal{I}$ yields a subdivision $\mathrm{Div}'\,\mathcal{I}$ isomorphic to $\mathrm{Div}\,\mathcal{I}$. (This isomorphism means that both complexes are geometric realizations of the same abstract complex.) See Figure 11.13. For brevity, we write such a perturbation as:

$$(v, \mathrm{Div}\,\mathcal{I}) \mapsto_\epsilon (v', \mathrm{Div}'\,\mathcal{I}).$$

If $\mathrm{Div}'\,\mathcal{I}$ is the result of applying ϵ-perturbations at multiple vertices, we write:

$$\mathrm{Div}\,\mathcal{I} \mapsto_\epsilon \mathrm{Div}'\,\mathcal{I}.$$

We will see that there is always an $\epsilon > 0$ such that any vertex of a $\mathrm{Div}\,\mathcal{I}$ can be perturbed to any position within ϵ within its carrier.

Lemma 11.4.14.   *If* v *is a vertex in* $\mathrm{Div}\,\mathcal{I}$ *whose carrier is an* n-*simplex* σ, *then there is an* ε-*perturbation*

$$(v, \mathrm{Div}\,\mathcal{I}) \mapsto_\varepsilon (v', \mathrm{Div}'\,\mathcal{I})$$

for any $v' \in B(v, \varepsilon)$.

Proof. We must check that any choice of $v' \in B(v, \varepsilon)$ yields a subdivision. We can pick ε small enough that v' lies in the open star of v. We must check that v' can be made affinely independent of each simplex $\upsilon = \{u_0, \ldots, u_p\}$ in $\mathrm{Lk}(v, \mathrm{Div}\,\sigma)$. Each such υ defines a hyperplane $\mathbb{H}(\upsilon)$ of dimension $p < n$. Because each $\{v\} \cup \upsilon$ is a simplex of $\mathrm{Div}\,\sigma$, $v, u_0, \ldots, u_p$ are affinely independent, and hence v is not in any $\mathbb{H}(\upsilon)$. By Fact 11.4.4, there is an $\varepsilon > 0$ so that for any $v' \in B(v, \varepsilon)$, $v' \notin \mathbb{H}(\upsilon)$, and thus is affinely independent of the vertices of υ. $\qquad\square$

Lemma 11.4.15. *If v is a vertex of $\mathrm{Div}\,\mathcal{I}$ whose carrier is an n-simplex σ, then there is a perturbation $(v, \mathrm{Div}\,\mathcal{I}) \mapsto_\epsilon (v', \mathrm{Div}'\,\mathcal{I})$ such that $\mathrm{St}(v', \mathrm{Div}'\,\mathcal{I})$ contains no conflict points with $\mathrm{Ch}^N\mathcal{I}$.*

Proof. Let $\mathbb{H}$ be the set of hyperplanes defined by all pairs of conflicting simplices, one from $\mathrm{Lk}(v, \mathrm{Div}\,\sigma)$ and one from $\mathrm{Ch}^N\sigma$. By Fact 11.4.3, there is an $\epsilon > 0$ so that some point v' within ϵ of v does not intersect any of these hyperplanes. By Lemma 11.4.14, this choice of v' defines a perturbation $(v, \mathrm{Div}\,\mathcal{I}) \mapsto_\epsilon (v', \mathrm{Div}'\,\mathcal{I})$.

We claim that $\mathrm{St}^\circ(v', \mathrm{Div}'\,\mathcal{I})$ and $\mathrm{Ch}^N\mathcal{I}$ have no conflict points. Otherwise, if x is a conflict point, then there are conflicting simplices $\rho = \{r_0, \ldots, r_p\}$ in $\mathrm{Lk}(v, \mathrm{Div}\,\sigma)$ and $\tau = \{t_0, \ldots, t_q\}$ in $\mathrm{Ch}^N\sigma$ such that

$$x = \left(\sum_{i=0}^p a_i \cdot r_i\right) + a_{p+1} \cdot v = \sum_{j=0}^q b_j t_j$$

where $1 = \sum_i a_i = \sum_j b_j$, and $0 < a_i, b_j$. And yet, this equation implies that x lies on the hyperplane defined by the vertices of $\rho \cup \tau$, which contradicts the choice of v'. $\qquad\square$

Lemma 11.4.16. *If v is a vertex of $\mathrm{Div}\,\sigma$ whose carrier is an n-simplex σ, then there is a perturbation $(v, \mathrm{Div}\,\mathcal{I}) \mapsto_\epsilon (v^*, \mathrm{Div}^*\,\mathcal{I})$ such that $\mathrm{St}^\circ(v^*, \mathrm{Div}^*\,\mathcal{I})$ has no conflict points with any of the subdivisions*

$$\sigma, \mathrm{Ch}\,\sigma, \mathrm{Ch}^2\sigma, \mathrm{Ch}^3\sigma, \ldots.$$

Proof. By Lemma 11.4.14, there is a $\varepsilon > 0$ such that $(v, \mathrm{Div}\,\mathcal{I}) \mapsto_\varepsilon (v', \mathrm{Div}'\,\mathcal{I})$ is a perturbation for any $v' \in B(v, \varepsilon)$.

We inductively construct a sequence of subdivisions

$$(v^{(i-1)}, \mathrm{Div}^{(i-1)}\sigma) \mapsto_{\epsilon_i} (v^{(i)}, \mathrm{Div}^{(i)}\sigma)$$

such that $\mathrm{St}^\circ(v^{(i)}, \mathrm{Div}^{(i)}\sigma)$ has no conflict points with $\mathrm{Ch}^i\sigma$.

For the base case, the open-star cover of $\mathrm{Ch}^0\sigma = \sigma$ is already a chromatic cover for $\mathrm{Div}\,\sigma$, so let $v^{(0)} = v$ and $\mathrm{Div}^{(0)}\sigma = \sigma$.

For the induction step, Lemma 11.4.15 states that there is a perturbation

$$(v^{(i-1)}, \mathrm{Div}^{(i-1)}\sigma) \mapsto_{\epsilon_i} (v^{(i)}, \mathrm{Div}^{(i)}\sigma)$$

such that $\mathrm{St}^\circ(v^{(i)}, \mathrm{Div}^{(i)}\sigma)$ has no conflict points with $\mathrm{Ch}^i\sigma$.

If we pick each

$$\epsilon_i \le \frac{\varepsilon}{2^i},$$

then $v^{(0)}, v^{(1)}, \ldots$ is a Cauchy sequence that converges to v^*, where

$$|v - v^*| \leq |v^{(0)} - v^{(1)}| + |v^{(1)} - v^{(2)}| + \cdots$$
$$\leq \frac{\varepsilon}{2} + \frac{\varepsilon}{4} + \cdots$$
$$\leq \varepsilon.$$

As a result, we have constructed a perturbation

$$(v, \text{Div}\,\sigma) \mapsto_\varepsilon (v^*, \text{Div}^*\sigma)$$

where $\text{St}^\circ(v^*, \text{Div}^*\sigma)$ and $\text{Ch}^i \sigma$ have no conflict points for all $i \geq 0$. $\square$

Lemma 11.4.17. *Every chromatic subdivision* $\text{Div}\,\mathcal{I}$ *has a perturbation*

$$\text{Div}\,\mathcal{I} \mapsto_\epsilon \text{Div}^*\,\mathcal{I}$$

such that $\text{Div}^*\mathcal{I}$ *has no conflict points with any of the subdivisions*

$$\mathcal{I}, \text{Ch}\,\mathcal{I}, \text{Ch}^2\mathcal{I}, \text{Ch}^3\mathcal{I}, \ldots.$$

Proof. By induction on n. In the base case, when $n = 0$, the claim is trivial because both complexes are discrete sets of vertices.

Inductively assume that the open-star cover of $\text{Div}\,\text{skel}^{n-1}\mathcal{I}$ is chromatic for each of

$$\text{skel}^{n-1}\mathcal{I}, \text{Ch}\,\text{skel}^{n-1}\mathcal{I}, \text{Ch}^2\text{skel}^{n-1}\mathcal{I}, \text{Ch}^3\text{skel}^{n-1}\mathcal{I}, \ldots.$$

For each vertex v in $\text{Div}\,\mathcal{I}$ whose carrier has dimension n, Lemma 11.4.16 states that we can construct a perturbation that eliminates all conflict points from its open star. Successively applying this construction to each such vertex yields a perturbation

$$\text{Div}\,\mathcal{I} \mapsto_\varepsilon \text{Div}^*\mathcal{I}$$

that has no conflict points with any iterated standard chromatic subdivision $\text{Ch}^i\mathcal{I}$. $\square$

Because $\text{Div}\,\mathcal{I}$ and its perturbation $\text{Div}^*\mathcal{I}$ are just different geometric realizations of the same abstract complex, we have completed the proof of Lemma 11.4.10.

11.4.4 Construction

We are now ready to prove Theorem 11.4.1, showing there exists a color-preserving simplicial map

$$\phi : \text{Ch}^N\mathcal{I} \to \text{Div}\,\mathcal{I}$$

such that for all $\sigma \in \mathcal{I}$, $\phi(\text{Ch}^N\sigma) \subseteq \text{Div}\,\sigma$.

We will construct a sequence of chromatic subdivisions, $\text{Ch}^{K_i}\mathcal{I}$, for $i = 0, \ldots, n + 2$, where $K_{n+2} > K_{n+1} > \cdots > K_0 = 0$, along with a sequence of simplicial maps

$$\phi_i : (\text{Ch}^{K_i}\mathcal{I}) \setminus \mathcal{K}_i \to \text{Div}\,\mathcal{I},$$

defined on $\text{Ch}^{K_i}\mathcal{I}$, except for a subcomplex $\mathcal{K}_i \subseteq \text{Ch}^{K_i}\mathcal{I}$ of dimension at most $n - i$.

At the end of the sequence, $\mathcal{K}_{n+2}$ is empty, so

$$\phi_{n+2} : \text{Ch}^{K_{n+2}}\mathcal{I} \to \text{Div}\,\mathcal{I}$$

is the desired map.

This sequence of subdivisions induces a *parent map*,

$$\text{Ch}^{K_{n+2}}\mathcal{I} \xrightarrow{\pi} \text{Ch}^{K_{n+1}}\mathcal{I} \xrightarrow{\pi} \cdots \xrightarrow{\pi} \text{Ch}^{K_0}\mathcal{I},$$

carrying each vertex of $\text{Ch}^{K_i}\mathcal{I}$ to the unique vertex of matching color in its carrier in $\text{Ch}^{K_{i-1}}\mathcal{I}$. The *ancestors* of a vertex $v \in \text{Ch}^{K_i}\mathcal{I}$ are the vertices $v, \pi(v), \pi^2(v), \ldots, \pi^i(v)$.

Subdividing $\text{Ch}^{K_i}\mathcal{I}$ induces subdivisions on its subcomplexes. Given a subcomplex $\mathcal{K}_i \subseteq \text{Ch}^{K_i}\mathcal{I}$, define $\mathcal{K}^i_{i+1}$ to be the maximal subcomplex of $\text{Ch}^{K_{i+1}}\mathcal{I}$ satisfying $\pi(\mathcal{K}^i_{i+1}) = \mathcal{K}_i$. Similarly, for $\ell > i$, define $\mathcal{K}^i_\ell$ to be the maximal subcomplex of $\text{Ch}^{K_\ell}\mathcal{I}$ satisfying $\pi^{\ell-i}(\mathcal{K}^i_\ell) = \mathcal{K}_i$.

Definition 11.4.18. The *extended star* $\text{St}^*(\tau, \mathcal{K})$ of a simplex τ in a complex $\mathcal{K}$ is the union of the stars of its vertices:

$$\text{St}^*(\tau, \mathcal{K}) = \bigcup_{v \in \tau} \text{St}(v, \mathcal{K}).$$

Like the star of a vertex, the extended star of a simplex is a subcomplex of $\mathcal{K}$, and its polyhedron is a closed set. Moreover,

$$\text{diam}(\text{St}^*(\tau, \mathcal{K})) \leq 3 \cdot mesh(\mathcal{K}).$$

For the base case of our construction, let $K_0 = 0$, $\mathcal{K}_0 = \mathcal{I}$, and ϕ_0 is everywhere undefined.

For the inductive step, assume we are given K_i and ϕ_i for all $\ell, 0 \leq \ell < i$.

Lemma 11.4.10 states that we may assume, without loss of generality, that the open-star cover of $\text{Div}\,\mathcal{I}$ is chromatic for $\text{Ch}^{K_i}\mathcal{I}$, including $\mathcal{K}_i \subseteq \text{Ch}^{K_i}\mathcal{I}$, the subcomplex where ϕ_i is undefined. Let λ_i be the Lebesgue number of this cover of $\mathcal{K}_i$. Pick K_{i+1} large enough that for every facet κ of $\mathcal{K}^i_{i+1} \subseteq \text{Ch}^{K_{i+1}}\mathcal{I}$,

$$\text{diam}\,\text{St}^*(\kappa, \mathcal{K}^i_{i+1}) < \lambda_i. \tag{11.4.1}$$

We will use this inequality later.

We define ϕ_{i+1} as follows: Each vertex v in $\text{Ch}^{K_{i+1}}$ not in $\mathcal{K}^i_{i+1}$ "inherits" the map from its parent: $\phi_{i+1}(v) = \phi_i(\pi(v))$. Otherwise, for each vertex in $\mathcal{K}^i_{i+1}$, if there exits a vertex $u \in \text{Div}\,\mathcal{I}$ such that

$$\text{St}(v, \mathcal{K}^i_{i+1}) \subset \text{St}^\circ(u, \text{Div}\,\mathcal{I}), \tag{11.4.2}$$

and $\text{name}(u) = \text{name}(v)$, then $\phi_{i+1}(v) = u$.

The remaining vertices of $\mathcal{K}^i_{i+1}$ define $\mathcal{K}_{i+1}$, the subcomplex of $\text{Ch}^{K_{i+1}}\mathcal{I}$ where ϕ_{i+1} is not defined. Note that this definition implies that for $0 \leq j \leq i$,

$$\pi(\mathcal{K}^j_{i+1}) \subseteq \mathcal{K}^j_i. \tag{11.4.3}$$

Lemma 11.4.19. *Let v_i be a vertex in $(\mathrm{Ch}^{K_i}\mathcal{I}) \setminus \mathcal{K}_i$, and $v_i, v_{i-1}, \ldots, v_0$ its sequence of ancestors. Let j be the least index (earliest ancestor) for which $\phi_j(v_j)$ is defined. We claim that*

$$|\mathrm{St}(v_i, \mathcal{K}_i^j)| \subseteq \mathrm{St}^\circ(\phi_i(v_i), \mathrm{Div}\,\mathcal{I}).$$

Proof. We argue by induction on $i - j$. For the base case, when $i = j$, the claim follows because $\phi_i(v)$ is defined for the first time by Equation 11.4.2.

Assume the result for $0 < i - j < \ell$. By the induction hypothesis,

$$|\mathrm{St}(\pi(v), \mathcal{K}_{i-1}^j)| \subseteq \mathrm{St}^\circ(\phi_{i-1}(\pi(v)), \mathrm{Div}\,\mathcal{I}),$$

where j is the least index for which ϕ_j is defined for an ancestor of v. Note that $\mathcal{K}_i^k$ is a subdivision of $\mathcal{K}_{i-1}^j$, and $\pi(v)$ is a vertex in the carrier of v for this subdivision, so

$$|\mathrm{St}(v, \mathcal{K}_i^j)| \subseteq |\mathrm{St}(\pi(v), \mathcal{K}_{i-1}^j)|.$$

Putting these containments together,

$$
\begin{aligned}
|\mathrm{St}(v, \mathcal{K}_i^j)| &\subseteq |\mathrm{St}(\pi(v), \mathcal{K}_{i-1}^j)| \\
&\subseteq \mathrm{St}^\circ(\phi_{i-1}(\pi(v)), \mathrm{Div}\,\mathcal{I}) \\
&\subseteq \mathrm{St}^\circ(\phi_i(v), \mathrm{Div}\,\mathcal{I}) \qquad\qquad \square
\end{aligned}
$$

Lemma 11.4.20. *Each ϕ_i is a color-preserving simplicial map.*

Proof. The color-preserving property is immediate from Equation 11.4.2. To show that ϕ is simplicial, we argue by induction on i. When $i = 0$, the claim holds vacuously. Let $\kappa = \{k_0, \ldots, k_m\}$ be a simplex of $\mathrm{Ch}^{K_i}\mathcal{I} \setminus \mathcal{K}_i$. We must show that $\{\phi_i(k_0), \ldots, \phi_i(k_m)\}$ is a simplex of $\mathrm{Div}\,\mathcal{I}$.

By Lemma 11.4.19, for each ℓ, $0 \leq \ell \leq m$, there is a j_ℓ such that for each vertex k_ℓ,

$$|\mathrm{St}(k_\ell, \mathcal{K}_i^{j_\ell})| \in \mathrm{St}^\circ(\phi_i(k_\ell), \mathrm{Div}\,\mathcal{I}).$$

Assume without loss of generality that $j_0 \leq j_1 \leq \cdots \leq j_m$, so

$$\mathcal{K}_i^{j_m} \subseteq \mathcal{K}_i^{j_{m-1}} \subseteq \cdots \mathcal{K}_i^{j_0}.$$

In particular, v_k is a vertex of them all.

$$v_k \in |\mathrm{St}(k_\ell, \mathcal{K}_i^{j_\ell})| \subseteq \mathrm{St}^\circ(\phi_i(k_\ell), \mathrm{Div}\,\mathcal{I})$$

for $0 \leq \ell \leq m$, so

$$\bigcap_{\ell=0}^{k} \mathrm{St}^\circ(\phi_i(k_\ell), \mathrm{Div}\,\mathcal{I}) \neq \emptyset.$$

It follows from Fact 11.4.9 that $\{\phi_i(k_0), \ldots, \phi_i(k_m)\}$ is a simplex of $\mathrm{Div}\,\mathcal{I}$, completing the proof that ϕ_i is a simplicial map. $\qquad\square$

Lemma 11.4.21.   *For $0 \leq i \leq n+2$, $\dim \mathcal{K}_{i+1} < \dim \mathcal{K}_k$.*

Proof.   Recall that the open-star cover of $\mathrm{Div}\,\mathcal{I}$ is chromatic on $\mathcal{K}_i$. Moreover, by Equation 11.4.1, the number $\mathcal{K}_{i+1}$ is large enough to ensure that the diameter of the extended star of every facet κ of $\mathcal{K}_{i+1}$ is less than the cover's Lebesgue number:

$$|\mathrm{St}^*(\kappa, \mathcal{K}^i_{i+1})| \in \mathrm{St}^\circ(u, \mathrm{Div}\,\mathcal{I}),$$

for a vertex $u \in \mathrm{Div}\,\mathcal{I}$. Because the cover is chromatic, there is a vertex $v \in \kappa$ of matching color: $\mathrm{name}(v) = \mathrm{name}(u)$. Because $|\mathrm{St}(v, \mathcal{K}^i_{i+1})| \subset |\mathrm{St}^*(\kappa, \mathcal{K}^i_{i+1})|$,

$$|\mathrm{St}(v, \mathcal{K}^i_{i+1})| \subset \mathrm{St}^\circ(u, \mathrm{Div}\,\mathcal{I}),$$

so by Equation 11.4.2, $\phi_{i+1}(v)$ is defined, and v is not a vertex of $\mathcal{K}_{i+1}$. In this way, ϕ_{i+1} is defined on at least one vertex of every facet of $\mathcal{K}^i_{i+1}$, so the dimension of $\mathcal{K}_{i+1}$ drops by at least one.   □

Lemma 11.4.20 states that the vertex map

$$\phi_{n+2} : (\mathrm{Ch}^{K_{n+2}}\mathcal{I}) \setminus \mathcal{K}_{n+2} \to \mathrm{Div}\,\mathcal{I}$$

is color-preserving and simplicial, whereas Lemma 11.4.21 states that $\mathcal{K}_{n+2}$ is empty. Together these imply that

$$\phi_{n+2} : \mathrm{Ch}^{K_{n+2}}\mathcal{I} \to \mathrm{Div}\,\mathcal{I}$$

is a color-preserving simplicial map.

This completes the proof of the "map implies algorithm" direction of Theorem 11.2.1.

It is useful to observe that the "property implies protocol" part of the proof assumes only that the model of computation in which the protocol is constructed is strong enough to solve the immediate snapshot task.

Corollary 11.4.22.   *A task $(\mathcal{I}, \mathcal{O}, \Delta)$ has a protocol in any model of computation that solves immediate snapshot if $\mathcal{I}$ has a chromatic subdivision $\mathrm{Div}(\mathcal{I})$ and a color-preserving simplicial map $\mu : \mathrm{Div}\,\mathcal{I} \to \mathcal{O}$ carried by Δ.*

Of course, the converse of this corollary does not hold in general.

11.5 A sufficient topological condition

We now give a simple topological condition that ensures that a colored task $(\mathcal{I}, \mathcal{O}, \Delta)$ has a wait-free read-write protocol. We will use the following topological property.

Definition 11.5.1.   A pure simplicial complex $\mathcal{O}$ of dimension n is called *link-connected* if for each simplex $\tau \in \mathcal{O}$, $\mathrm{Lk}(\tau, \mathcal{O})$ is $(n - 2 - \dim \tau)$-connected.

The output complex for the Hourglass task shown in Figure 11.1 is *not* link-connected, since the vertex at the hourglass's "waist" is disconnected, that is, not 0-connected.

Theorem 11.5.2.   *The colored task $(\mathcal{I}, \mathcal{O}, \Delta)$ has a wait-free layered immediate snapshot protocol if for each $\sigma \in \mathcal{I}$, $\Delta(\sigma)$ is $(\dim(\sigma) - 1)$-connected, and $\mathcal{O}$ is link-connected.*

This theorem establishes the existence of a protocol in terms of the topological properties of complexes and carrier maps.

By Theorem 11.2.1, it is enough to prove the following lemma.

Lemma 11.5.3. *If for each $\sigma \in \mathcal{I}$, $\Delta(\sigma)$ is $(\dim(\sigma) - 1)$-connected, and $\mathcal{O}$ is link-connected, then there exists a chromatic subdivision $\mathrm{Div}\,\mathcal{I}$ and a color-preserving simplicial map*

$$\mu : \mathrm{Div}\,\mathcal{I} \to \mathcal{O}$$

carried by Δ.

We need the following lemma about link connectivity.

Lemma 11.5.4. *If $\mathcal{O}$ is a pure link-connected simplicial complex, then so is $\mathrm{Lk}(\kappa, \mathcal{O})$ for any simplex $\kappa \in \mathcal{O}$.*

Proof. Assume $\dim \mathcal{O} = n$. Note that $\mathrm{Lk}(\kappa, \mathcal{O})$ is a pure $(n - \dim \kappa - 1)$-complex, and for any λ in $\mathrm{Lk}(\kappa, \mathcal{O})$, $\dim \lambda \cup \kappa = \dim \lambda + \dim \kappa + 1$. We will show that for any λ in $\mathrm{Lk}(\kappa, \mathcal{O})$, $\mathrm{Lk}(\lambda, \mathrm{Lk}(\kappa, \mathcal{O}))$ is $(n - \dim \kappa - \dim \lambda - 3)$-connected.

We claim that

$$\mathrm{Lk}(\lambda, \mathrm{Lk}(\kappa, \mathcal{O})) = \mathrm{Lk}(\lambda \cup \kappa, \mathcal{O}),$$

If $\gamma \in \mathrm{Lk}(\lambda, \mathrm{Lk}(\kappa, \mathcal{O}))$, then $\gamma \cup \lambda \cup \kappa \in \mathcal{O}$, and therefore $\gamma \in \mathrm{Lk}(\lambda \cup \kappa, \mathcal{O})$, so

$$\mathrm{Lk}(\lambda, \mathrm{Lk}(\kappa, \mathcal{O})) \subseteq \mathrm{Lk}(\lambda \cup \kappa, \mathcal{O}).$$

Moreover, if $\gamma \in \mathrm{Lk}(\lambda \cup \kappa, \mathcal{O})$, then $\gamma \cup \lambda \cup \kappa \in \mathcal{O}$, and therefore $\gamma \in \mathrm{Lk}(\lambda, \mathrm{Lk}(\kappa, \mathcal{O}))$, so

$$\mathrm{Lk}(\lambda, \mathrm{Lk}(\kappa, \mathcal{O})) \supseteq \mathrm{Lk}(\lambda \cup \kappa, \mathcal{O}).$$

Because λ and κ are disjoint, $\dim \lambda \cup \kappa = \dim \lambda + \dim \kappa + 1$. The complex $\mathrm{Lk}(\lambda \cup \kappa, \mathcal{O})$ is link-connected by hypothesis, so $\mathrm{Lk}(\lambda \cup \kappa, \mathcal{O})$, and therefore $\mathrm{Lk}(\lambda, \mathrm{Lk}(\kappa, \mathcal{O}))$, is $(n - \dim \lambda - \dim \kappa - 3)$-connected. $\square$

We make use of the following fact, discussed in the chapter notes.

Fact 11.5.5. Let $\mathcal{A}$, $\mathcal{B}$, and $\mathcal{C}$ be complexes such that $\mathcal{B} \subset \mathcal{A}$, and $f : \mathcal{A} \to \mathcal{C}$ a continuous map such that the vertex map induced by f restricted to $\mathcal{B}$ is simplicial. There exists a subdivision Div of $\mathcal{A}$ such that $\mathrm{Div}\,\mathcal{B} = \mathcal{B}$, and a simplicial map $\phi : \mathrm{Div}\,\mathcal{A} \to \mathcal{C}$ that agrees with f on vertices of $\mathcal{B}$.

Lemma 11.5.6. *Let $\mathcal{I}$ and $\mathcal{O}$ be n-complexes such that there is a continuous map*

$$f : \mathcal{I} \to \mathcal{O}$$

such that the restriction of f to $\mathrm{skel}^{n-1}\mathcal{I}$ is a rigid simplicial map. There exist a subdivision $\mathrm{Div}\,\mathcal{I}$ that $\mathrm{Div}\,\mathrm{skel}^{n-1}\mathcal{I} = \mathrm{skel}^{n-1}\mathcal{I}$ and a rigid simplicial map $\phi : \mathrm{Div}\,\mathcal{I} \to \mathcal{O}$ that agrees with f on vertices of $\mathrm{skel}^{n-1}\mathcal{I}$.

Proof. We argue by induction on n. For the base case, when $n = 1$, $\mathcal{I}$ is a graph, and $\mathrm{skel}^0\mathcal{I}$ is a set of discrete points. By way of contradiction, assume that every subdivision $\mathrm{Div}\,\mathcal{I}$ and every simplicial map

$$\phi : \mathrm{Div}\,\mathcal{I} \to \mathcal{O}$$

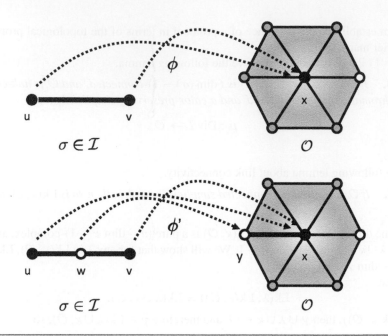

FIGURE 11.14

Eliminating collapse: Base case.

that agrees with f on $\mathrm{skel}^0\,\mathcal{I}$ collapses an edge of $\mathrm{Div}\,\mathcal{I}$.

Pick Div and ϕ to collapse a minimal number of edges. There is an edge $\sigma = \{u, v\}$ in $\mathrm{Div}\,\mathcal{I}$ such that $\phi(u) = \phi(v) = x$, where x is a vertex in $\mathcal{O}$ (see Figure 11.14). Because $\mathcal{O}$ is link-connected, $\mathrm{Lk}(x, \mathcal{O})$ is non-empty and so contains a vertex y. Define a new subdivision $\mathrm{Div}'\,\mathcal{I}$ by taking the stellar subdivision of $\mathcal{I}$ with center w. Define $\phi' : \mathrm{Div}'\,\mathcal{I} \to \mathcal{O}$ to agree with ϕ except that $\phi'(w) = y$. It is easy to check that ϕ' agrees with f on $\mathrm{skel}^{n-1}\mathcal{I}$ but collapses one fewer simplex than ϕ, contradicting our assumption that ϕ collapses a minimum number of simplices.

For the induction step, assume the claim for complexes of dimension less than n. Let $\mathcal{I}$ be an n-complex, and $f : \mathcal{I} \to \mathcal{O}$ a continuous map that is simplicial and rigid on $\mathrm{skel}^{n-1}\mathcal{I}$. Fact 11.5.5 implies there exists a subdivision $\mathrm{Div}\,\mathcal{I}$ such that $\mathrm{Div}\,\mathrm{skel}^{n-1}\mathcal{I} = \mathrm{skel}^{n-1}\mathcal{I}$ and a simplicial map $\phi : \mathrm{Div}\,\mathcal{I} \to \mathcal{O}$ that agrees with f on $\mathrm{skel}^{n-1}\mathcal{I}$.

Suppose, by way of contradiction, that for every such Div and ϕ, ϕ collapses a simplex of $\mathrm{Div}\,\mathcal{I}$. Pick Div and ϕ to minimize the number of collapsed simplices. We will show how to adjust Div and ϕ to collapse one fewer simplex, a contradiction.

Pick σ in $\mathrm{Div}\,\mathcal{I}$ such that ϕ collapses σ to a single vertex $x \in \mathcal{O}$ but does not collapse any simplex strictly containing σ. Because ϕ does not collapse any simplices in $\mathrm{skel}^{n-1}\mathcal{I}$, $\sigma \in \mathrm{Div}\,\tau$ for some n-simplex $\tau \in \mathcal{I}$. Because τ is an n-simplex, $\mathrm{Lk}(\sigma, \mathrm{Div}\,\mathcal{I}) = \mathrm{Lk}(\sigma, \mathrm{Div}\,\tau)$.

As before, pick a point w in the interior of σ, and take the stellar subdivision $\mathrm{stel}\,\mathrm{Div}\,\mathcal{I}$ with center w. Because $\mathrm{Div}\,\tau$ is a manifold, $\mathrm{Lk}(\sigma, \mathrm{Div}\,\tau)$ is an $(n - m - 1)$-sphere, and $\{w\} \cdot \mathrm{Lk}(\sigma, \mathrm{Div}\,\tau)$ is an $(n - m)$-disk with boundary $\mathrm{Lk}(\sigma, \mathrm{Div}\,\tau)$.

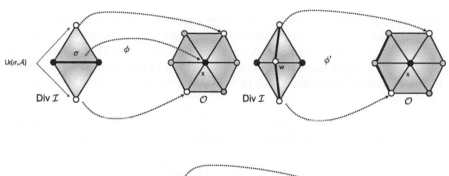

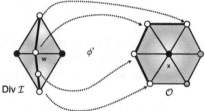

FIGURE 11.15

Eliminating collapse: Induction step.

The subcomplex $\mathrm{Lk}(v, \mathcal{O})$ is $(n-1)$-connected by hypothesis and link-connected by Lemma 11.5.4. Because ϕ does not collapse any simplices strictly containing σ,

$$\phi(\mathrm{Lk}(\sigma, \mathrm{Div}\,\mathcal{I})) \subseteq \mathrm{Lk}(v, \mathcal{O}).$$

Because $\dim\left(\{w\} \cdot \mathrm{Lk}(\sigma, \mathcal{I})\right) = (n-m) < n$, we apply the induction hypothesis as follows: Because $\mathrm{Lk}(x, \mathcal{O})$ is connected and link-connected, there exists a subdivision Div' and simplicial map

$$\phi' : \mathrm{Div}'\,\{t\} \cdot \mathrm{Lk}(\sigma, \mathrm{Div}\,\mathcal{I}) \to \mathrm{Lk}(x, \mathcal{O})$$

such that $\mathrm{Div}'\,\mathrm{Lk}(\sigma, \mathrm{Div}\,\mathcal{I}) = \mathrm{Lk}(\sigma, \mathrm{Div}\,\mathcal{I})$, ϕ' agrees with ϕ on $\mathrm{Lk}(\sigma, \mathrm{Div}\,\mathcal{I})$, and ϕ' does not collapse any simplices of $\mathrm{Div}'\,\{t\} \cdot \mathrm{Lk}(\sigma, \mathrm{Div}\,\mathcal{I})$ (see Figure 11.15).

Because Div' does not subdivide any simplices of $\mathrm{Lk}(\sigma, \mathrm{Div}\,\mathcal{I})$, it extends to a subdivision $\mathrm{Div}'\,\mathcal{I}$ of all of $\mathcal{I}$. Define

$$\psi(u) = \begin{cases} \phi'(u) & \text{if } u \in \mathrm{Div}'(\sigma \cdot \mathrm{Lk}(\sigma, \mathrm{Div}\,\mathcal{I})) \\ \phi(u) & \text{otherwise.} \end{cases}$$

Note that ψ agrees with ϕ on every vertex not in $\mathrm{Div}'(\sigma \cdot \mathrm{Lk}(\sigma, \mathrm{Div}\,\mathcal{I}))$ and on every vertex of $\mathrm{Lk}(\sigma, \mathrm{Div}\,\mathcal{I})$, so ψ cannot collapse more of these simplices than ϕ. Moreover, ψ cannot collapse any simplices of $\mathrm{Div}'\,\{w\} \cdot \mathrm{Lk}(\sigma, \mathcal{I})$ because it is rigid by the induction hypothesis. It follows that ψ collapses one fewer simplex than ϕ, contradicting our assumption that ϕ collapses a minimum number of simplices. $\square$

Lemma 11.5.7. *Let σ be a chromatic n-simplex, $\mathrm{Div}\,\sigma$ a chromatic subdivision, and $\phi : \mathrm{Div}\,\sigma \to \mathcal{L}$ is a rigid simplicial map. If ϕ is color-preserving on $\mathrm{Div}\,\partial\sigma$, then ϕ is color-preserving on $\mathrm{Div}\,\sigma$.*

Proof. Div σ is a manifold with boundary, so for any n-simplex τ, there is a sequence of n-simplices $\tau_0, \ldots, \tau_p$, where τ_0 has an $(n-1)$-face on the boundary, τ_i and τ_{i+1} share an $(n-1)$-face, and $\tau_\ell = \tau$.

We argue by induction on ℓ. When $\ell = 0$, the claim is trivial. Otherwise, assume that ϕ is color-preserving on $\tau_{\ell-1}$. The map ϕ is color-preserving on the $(n-1)$ face shared by $\tau_{\ell-1}$ and τ_ℓ. Because ϕ is rigid, it cannot send the remaining vertex of τ_ℓ to any of the other n colors in the shared face, so it must send it to a vertex of the same color. $\qquad\square$

We are now ready to complete the proof of Lemma 11.5.3 (and hence Theorem 11.5.2). For $0 \le d \le n$, we inductively construct a sequence of chromatic subdivisions Div_d and a color-preserving simplicial map

$$\phi_d : |\mathrm{Div}_d \mathrm{skel}^d \mathcal{I}| \to |\mathcal{O}|$$

carried by Δ.

For the base case, let f_0 send any vertex a of $\mathcal{I}$ to any vertex of $\Delta(a)$. This construction is well-defined because $\Delta(a)$ is (-1)-connected (non-empty) by hypothesis. This map is trivially color-preserving.

For the induction hypothesis, assume we have constructed a chromatic subdivision and color-preserving simplicial map

$$\phi_{d-1} : \mathrm{Div}_{d-1} \mathrm{skel}^{d-1} \mathcal{I} \to \mathcal{O}.$$

This simplicial map induces a continuous map that sends the boundary of each d-simplex σ in $\mathrm{skel}^d \mathcal{I}$ to $\Delta(\sigma)$. By hypothesis, $\Delta(\sigma)$ is $(d-1)$-connected, so this map of the $(d-1)$-sphere $\partial\sigma$ can be extended to a continuous map of the d-disk σ:

$$f_d : |\sigma| \to \Delta(\sigma).$$

These extensions agree on $\mathrm{skel}^{d-1} \mathcal{I}$, so together they define a continuous map,

$$f_d : \mathrm{skel}^d \mathcal{I} \to \mathcal{O},$$

where for each $\sigma \in \mathrm{skel}^d \mathcal{I}$, $f_d(\sigma) \subseteq \Delta(\sigma)$.

Note that the restriction of f_d on the $(d-1)$ skeleton is just ϕ_{d-1}, a color-preserving simplicial map, so by Lemma 11.5.6, there is a subdivision Div_d of $\mathrm{skel}^d \mathcal{I}$ such that $\mathrm{Div}_d \mathrm{Div}_{d-1} \mathrm{skel}^{d-1} \mathcal{I} = \mathrm{Div}_{d-1} \mathrm{skel}^{d-1} \mathcal{I}$ and a rigid simplicial map $\phi_d : \mathrm{Div}_d \mathrm{skel}^d \mathcal{I} \to \mathcal{O}$ extending ϕ_{d-1}. By Lemma 11.5.7, ϕ_d is also color-preserving. These extensions agree on the $(d-1)$-skeleton, so together they define a color-preserving simplicial map:

$$\phi_d : \mathrm{Div}_d \mathrm{skel}^d \mathcal{I} \to \mathcal{O}$$

carried by Δ.

When $n = \dim \mathcal{I}$, ϕ_n is a color-preserving simplicial map carried by Δ, completing the proof.

By analogy with Corollary 11.4.22, the proof of Theorem 11.5.2 requires only that the protocol model support the immediate snapshot.

Corollary 11.5.8. *Assume that for any input complex $\mathcal{I}$ and any $N > 0$, there is a protocol that solves the chromatic agreement task $(\mathcal{I}, \mathrm{Ch}^N \mathcal{I}, \mathrm{Ch}^N)$. Then a task $(\mathcal{I}, \mathcal{O}, \Delta)$ has a protocol if, for each $\sigma \in \mathcal{I}$, $\Delta(\sigma)$ is $(\dim(\sigma) - 1)$-connected, and $\mathcal{O}$ is link-connected.*

One interesting and useful fact that emerges from this discussion is that if two different read-write models with different adversaries have the same minimal core size, then they solve the same set of colorless tasks. In this sense, an adversary's minimum core size completely determines its computational power for colorless tasks.

11.6 Chapter notes

The results in this chapter originally appeared in Herlihy and Shavit [91].

Herlihy, Rajsbaum, and Raynal [87] present an implementation of the safe agreement task in a layered model. See Exercise 11.1.

Mostefaoui, Rajsbaum, and Raynal [122] and Mostefaoui, Rajsbaum, Raynal, and Travers [121] study "condition-based" variations of tasks such as consensus in which the input complexes are restricted to permit t-resilient layered snapshot protocols. Such tasks provide simple, natural examples of colored tasks, where processes can adopt one another's output values but not their input values.

Imbs, Rajsbaum, and Raynal [95] study *generalized symmetry-breaking* tasks (GSB) that include election, renaming, and other tasks that are not colorless. These are fixed-input in the sense that the only input to a process is its name.

Fact 11.5.5 is Theorem IV.2 of Glaser [72].

11.7 Exercises

Exercise 11.1.   In Exercise 7.6 we asked you to describe an implementation of safe agreement using two layers of wait-free immediate snapshots. Show that there is no one-layer implementation.

Exercise 11.2.   Consider the following fixed-input colored task: $n+1$ processes choose distinct values in the range $0, \ldots, n+1$. Prove that the output complex for this task is a manifold. (This task is a special case of the renaming task considered in the next chapter.)

Exercise 11.3.   As a special case of Exercise 11.2, draw the output complex for the fixed-input colored task where three processes choose distinct values in the range $0, 1, 2, 3$. What is this surface called?

Exercise 11.4.   Let $f : |\mathcal{A}| \to |\mathcal{B}|$ be a continuous map and $\mathbb{U}$ be the open-star cover of $|\mathcal{B}|$.

- Show that $\mathbb{V} = \{ f^{-1}(U) | U \in \mathbb{U} \}$ is an open cover of $\mathcal{A}$.
- Suppose $\mathbb{V}$ has Lebesgue number λ and every edge of $\mathcal{A}$ has length 1. For what value of N can we guarantee that for every vertex v in $\text{Bary}^N \mathcal{A}$, diam $\text{St}(v, \text{Bary}^N \mathcal{A}) < \lambda$?
- For each v in $\text{Bary}^N \mathcal{A}$, define $\phi(v) = u$, where u is any vertex such that $\text{St}(v, \text{Bary}^N \mathcal{A}) \subset f^{-1}(\text{St}°(v, \mathcal{B}))$. Prove that ϕ is a simplicial approximation for f.

Exercise 11.5.   For an n-simplex σ, what is the Lebesgue number of the open-star cover of Bary σ? Of $\text{Bary}^N \sigma$?

Exercise 11.6.   Prove Fact 11.4.3: If x is a point in n-dimensional Euclidean space, and $\mathbb{H}$ is a finite set of hyperplanes, each of dimension less than n, then there is an $\varepsilon > 0$ such that for every $\epsilon < \varepsilon$, $B(x, \epsilon)$ contains a point not on any hyperplane in $\mathbb{H}$.

Exercise 11.7.   Prove Fact 11.4.4: If x is a point in n-dimensional Euclidean space, and $\mathbb{H}$ is a finite set of hyperplanes, each of dimension less than n, none of which contains v, then there is an $\varepsilon > 0$ such that $B(x, \varepsilon)$ does not intersect any hyperplane in $\mathbb{H}$.

Exercise 11.8. Prove Fact 11.4.9: A set of vertices $\{v_0, \ldots, v_q\}$ of a complex $\mathcal{K}$ forms a simplex if and only if the intersection of their open stars is non-empty.

Exercise 11.9. Let $\mathcal{K}$ be a geometric complex. Define the *distance* between two simplices σ_0, σ_1 of $\mathcal{K}$ to be $\min (d(x, y) | x \in |\sigma_0|, y \in |\sigma_1|)$. Show that the Lebesgue number of the open-star cover of $\mathcal{K}$ is $\min dist(\sigma_0, \sigma_1)$, taken over all pairs of simplices σ_0, σ_1 in $\mathcal{K}$ such that $\sigma_0 \cap \sigma_1 = \emptyset$.

Advanced Topics

This page intentionally left blank

CHAPTER

Renaming and Oriented Manifolds

12

CHAPTER OUTLINE HEAD

Consider the following air traffic control problem: There are $n + 1$ airplanes flying across the Atlantic in all directions. To avoid collisions, we must assign a distinct altitude to each plane, where an altitude is measured in thousands of feet. This task is easy for a centralized air traffic control system: Simply sort the planes by flight number and assign the ith plane to the ith altitude.

Suppose, instead, that we want to design a protocol so the planes themselves can coordinate to assign altitudes. We call this problem the *renaming* task: Each plane starts out with a unique name taken from a large name space (its flight number) and halts with a unique name taken from a smaller name space (its altitude). We are interested in asynchronous protocols because we do not know in advance which planes are participating, and interference may cause communications to be lost or delayed. In real life, of course, planes would communicate by message passing, but for ease of presentation here we will assume they communicate through some kind of shared read-write memory.

How many altitudes are needed to solve renaming? Consider the following simple layered-execution protocol for two planes, P_0 and P_1. The planes share an array, initially all $\perp$. In each layer, each plane takes an immediate snapshot: It writes its flight number to memory and then takes a snapshot of the memory.

- If a plane sees only its own flight number, it chooses altitude 1,000 feet.
- If a plane sees both flight numbers, it compares them.
  - If its own number is less than the other's, it chooses altitude 2,000.
  - Otherwise, it chooses altitude 3,000.

Distributed Computing Through Combinatorial Topology. http://dx.doi.org/10.1016/B978-0-12-404578-1.00012-7
© 2014 Elsevier Inc. All rights reserved.

Informally, here is why we need three altitudes for two planes. Suppose P_i's flight number is less than P_j's. If P_i sees P_j's flight number, then P_j may or may not have seen P_i's flight number. If P_j did not see P_i's number, then it will choose 1,000 feet, and if it did see the number, it will choose 3,000 feet, so the only safe choice for P_i is to choose 2,000 feet.

Is it possible for two planes to choose between two altitudes? The answer is *no*, because if they could, they could solve two-process consensus, which we know to be impossible using layered immediate snapshot or message-passing protocols. For two planes, three altitudes are both necessary and sufficient.

In the *M-renaming* task, each of $n + 1$ processes is issued a distinct name taken from a large name space, and after coordinating with one another, each chooses a distinct *output name* taken from a (much smaller) name space of size M. We will be interested in *adaptive* protocols, where the range of names chosen depends on the number of participating processes. As usual, processes are *asynchronous* and potentially faulty. Using layered snapshot protocols, for which values of M can we devise a protocol that ensures that all nonfaulty processes choose distinct output names?

To rule out trivial solutions, such as having P_i choose output name i, the renaming task is *index independent*, meaning that each P_i knows its name but not its index, which is not part of its state. In any execution, a process's output name can depend only on its input name, and the way its protocol steps are interleaved with the others', but not on its index.

12.1 An upper bound: renaming with $2n + 1$ names

We now present an $(n + 1)$-process renaming protocol where $p \le n + 1$ participating processes choose output names in the range $[0, \ldots, 2p]$. As noted, processes can compare their names for order and equality, but process indexes are not known to protocols. (That is, P_i "knows" that its name is P_i, but not that its index is i.)

12.1.1 An existence proof

We start by restating this task as a combinatorial problem. Let σ^n denote the n-simplex $\{P_0, P_1, \ldots, P_n\}$, and let Δ^N denote the N-simplex $\{0, 1, \ldots, N\}$. We will construct a subdivision and a rigid simplicial map,

$$\rho : \text{Rename}(\sigma^n) \to \Delta^{2n}.$$

This construction is inductive. For two processes, $\sigma^1 = \{P_0, P_1\}$. $\text{Rename}(\sigma^1)$ is the 1-dimensional chromatic subdivision, consisting of the four vertices

$$e_0 = (P_0, \{P_0\}) \qquad e_1 = (P_1, \{P_1\})$$
$$e_{10} = (P_0, \{P_0, P_1\}) \qquad e_{01} = (P_1, \{P_0, P_1\})$$

and three edges

$$\{e_0, e_{01}\}, \{e_{01}, e_{10}\}, \{e_{10}, e_1\}.$$

The map ρ is defined as follows:

$$\rho((P, \{P\})) = 0$$
$$\rho((P, \{P, Q\})) = \begin{cases} 1 & \text{if } P < Q \\ 2 & \text{otherwise} \end{cases}$$

FIGURE 12.1

The 2-process subdivision Rename(σ^1). Vertex color indicates each process's input name, and vertex number indicates its choice of output name.

It is easy to check that ρ is rigid and that ρ depends on the order of process names, not their indexes. These definitions are illustrated in Figure 12.1.

Inductively, suppose we have constructed a subdivision and a rigid map

$$\rho : \text{Rename}(\sigma^{n-1}) \to \Delta^{2n-2}$$

as illustrated for two processes in Figure 12.1. Next, we use the Rename operator to subdivide each $(n-1)$-face of $\text{skel}^{n-1}\sigma^n$. We extend ρ as follows: Let $\eta : \text{Face}_i(\sigma^n) \to \sigma^{n-1}$ be the unique bijection that preserves order; for vertices v_0, v_1 of $\text{Face}_i(\sigma^n)$, if $v_0 < v_1$, then $\eta(v_0) < \eta(v_1)$. Define ρ on $\text{Rename}(\text{Face}_i(\sigma^n))$ by $\rho(v) = \rho(\eta(v))$. This subdivision is illustrated in the top of Figure 12.2.

We extend this subdivision of the boundary to a subdivision $\text{Div}\,\sigma^n$ by introducing a "central" n-simplex $\kappa = \{k_0, \ldots k_n\}$, just as in the standard chromatic subdivision. The facets of $\text{Div}\,\sigma^n$ are defined as follows: For each simplex τ in $\text{Rename}(\text{skel}^{n-1}\sigma^n)$, let τ' be the face of κ labeled with complementary process names. The facets of $\text{Div}\,\sigma^n$ are κ itself, along with all simplices of the form $\tau \cup \tau'$. See the illustration in the middle of Figure 12.2.

Finally, let k_m be the vertex of κ labeled with the highest process name. Define $\text{Rename}(\sigma^n)$ to be the relative subdivision of $\text{Div}\,\sigma^n$ constructed by replacing $\text{Face}_m\,\kappa$ with $\text{Rename}(\text{Face}_m\,\kappa)$. Define $\rho(k_m) = 2n$, and for each $v \in \text{Rename}(\text{Face}_m\,\kappa)$, assign output names starting from $2n - 1$ and moving "down,"

$$\rho(v) = 2n - 1 - \rho(\eta(v)),$$

where $\eta : \text{Rename}(\text{Face}_m\,\kappa) \to \text{Rename}(\sigma^{n-1})$ is the order-preserving bijection.

Lemma 12.1.1. *Every facet of* $\text{Rename}(\sigma^n)$ *is assigned* $n + 1$ *distinct names in the range* $0, \ldots, 2n$.

Proof. We argue by induction on n. The claim for $n = 1$ holds by inspection.

Assume the claim holds for $\text{Rename}(\sigma^{n-1})$. By the induction hypothesis, for $\ell < n$, ρ assigns each ℓ-simplex in $\text{Rename}(\sigma^\ell)$ $\ell + 1$ distinct names in the range $0, \ldots, 2\ell$.

By construction, ρ assigns n distinct names in the range $1, \ldots, 2n - 1$ to each simplex in $\text{Face}_m\,\kappa$, and it assigns $2n$ to k_m, so it assigns $n + 1$ distinct names in the range $1, \ldots, 2n$ to each simplex in $\text{Rename}(\kappa)$.

Let τ be an ℓ-simplex in $\text{Rename}(\text{skel}^{n-1}\sigma^n)$, and let τ' be an $(n-\ell-1)$-face of $\text{Rename}(\kappa)$ labeled with the complementary process names. By construction, $\tau \cup \tau'$ is a facet of $\text{Rename}(\sigma^n)$, and every facet other than κ can be expressed in this way. By the induction hypothesis, ρ assigns to the vertices of τ distinct names in the range $0, \ldots, 2\ell$. The greatest name that can be assigned to τ' is $2n$, and the least is

$$2n - 2(n - \ell + 1) = 2\ell - 1.$$

Since the ranges assigned to τ and τ' do not overlap, the facet $\tau \cup \tau'$ is assigned $n + 1$ distinct names in the range $0, \ldots, 2n - 1$. $\qquad\square$

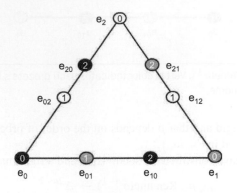

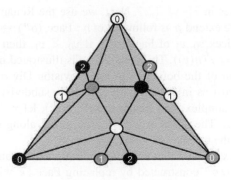

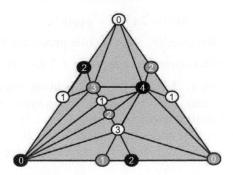

FIGURE 12.2

Construction of the 3-process renaming subdivision. Apply the 2-process subdivision to each edge of a triangle boundary (top). Add a central triangle, as in the standard chromatic subdivision (middle). Finally, assign name $2n - 2$ to the central vertex with the largest name, apply the 2-process subdivision to the opposite face, and recursively assign names from 1 to 3 to the remaining face (bottom).

Theorem 11.2.1, which is proved earlier in the book, implies that there is a color-preserving simplicial map

$$\mu : \text{Ch}^N \sigma^n \rightarrow \text{Rename}(\sigma^n)$$

for some N, which translates directly into an N-round protocol. In the next section, we give an explicit construction for this protocol.

12.1.2 An explicit protocol

We will use as a building block the *participating set* task, where each process P_i has its own name as its only possible input and has a set of process names S_i as output, where:

- $P_i \in S_i$,
- For all i, j, either $S_i \subseteq S_j$ or vice versa, and
- For all i, j, if $P_j \in S_i$, then $S_j \subseteq S_i$.

We leave it as an exercise (12.1) to show how to construct a single-layer immediate snapshot participating set protocol. We will use a slight variation of this task, called *tagged* participating set: Each process has a *tag* value as input, and the sets returned include the names of the processes with matching tags only.

Lemma 12.1.2. *Let $T_0 \subset \cdots \subset T_\ell$ be the distinct sets returned by a participating set protocol, and let S_i denote the set returned to P_i. We claim that P_j observes the first participating set in which it appears: If $j \in T_{i+1} \setminus T_i$, then $S_j = T_{i+1}$.*

Proof.   Left as Exercise 12.3.   □

Figure 12.3 shows the protocol that corresponds to the previous section's subdivision construction. Output names are allocated either *top-down* or *bottom-up*, depending on whether we assign the next output name from the high or low end of the range of unassigned output names. Each recursive call to

```
 1   shared tps: array [0.. n] of tagged  participating  set  objects
 2   // initial  call : rename(⟨∠, 0, true,  0)
 3   rename(tag: sequence of set of names,
 4           first : int,  dir: Boolean, r: int ): int
 5     peers: set of name := tps[r]. participants (tag)
 6     if  direction  then
 7        first  :=  first  + 2 ∗ |peers|
 8     else
 9        first  :=  first  − 2 ∗ |peers|
10     if  P_i = max(peers) then
11       return  first
12     else
13       return rename(append(tag, peers),  first ,  not  dir ,  r+1)
```

FIGURE 12.3

Renaming protocol.

the tagged participating set protocol takes the history of prior participating sets, so processes observe only other processes with the same history.

As before, $T_0 \subset \cdots \subset T_\ell$ denote distinct participating sets, and S_i denotes the set observed by P_i. If P_m, the process with the highest name, observes the largest participating set, that is, if $S_m = T_\ell$, then P_m chooses output name $2n$ and halts. Every other P_i such that $S_i = T_\ell$ proceeds as if P_m had chosen $2n$ and joins a recursively called protocol to choose names bottom-up from the range $0, \ldots, 2n - 1$.

The processes that observe $T_{\ell-1}$, the second-largest participating set, act as if they are the only participants. If the process with the largest name in $T_{\ell-1}$ observes $T_{\ell-1}$, then it chooses $2|T_{\ell-1}|$ and halts. The others proceed as if that process had chosen the output name $2|T_{\ell-1}|$ and recursively call a protocol to choose names bottom-up in the range $0, \ldots, 2|T_{\ell-1}| - 1$. In this way, the range of names is recursively broken up.

Theorem 12.1.3. *The protocol of Figure 12.3 is wait-free.*

Proof. In each round, the active process with the highest name chooses a name and halts, so every process finishes in $n + 1$ rounds. □

Theorem 12.1.4. *The protocol of Figure 12.3 assigns distinct output names in the range $0, \ldots, 2n$ to each of $n + 1$ participants.*

Proof. We argue by induction on n. When $n = 1$, the claim is clear from inspection.

For the induction hypothesis, assume that the protocol assigns top-down distinct names in the range $0, \ldots, 2m - 2$, for all $m < n + 1$. By Lemma 12.1.2, the processes that observe T_ℓ, the largest set, are exactly the processes in $T_\ell \setminus T_{\ell-1}$. There are two cases to consider. First, if all processes observe T_ℓ, then the process with the highest name drops out with $2n$, and the remaining n processes choose names bottom-up, starting at $2n - 1$. By the induction hypothesis, they choose distinct names in the range $1, \ldots, 2n - 1$.

Otherwise, if some processes observe $T_{\ell-1} \subset T_\ell$, then at most $|T_\ell| - |T_{\ell-1}| \leq n$ processes choose names bottom-up starting at $2n - 1$. By the induction hypothesis, if they were choosing top-down, they would choose names in the range $0, \ldots, 2(|T_\ell| - |T_{\ell-1}|) - 2$. But because they are choosing bottom-up starting at $2|T_\ell| - 1$, the least name they can choose is

$$2|T_\ell| - (2(|T_\ell| - |T_{\ell-1}|) - 1) = 2|T_{\ell-1}| + 1.$$

The largest name that can be chosen by the processes not in T_ℓ is $2|T_{\ell-1}|$, so these ranges do not overlap. □

We will now show that for some values of $n + 1$, this $(2n + 1)$-renaming protocol is the best one can do using asynchronous read-write memory.

12.2 Weak symmetry breaking

Recall from Chapter 9 that in the *weak symmetry-breaking* (WSB) task, processes have fixed input values and binary output values. In every execution in which all $n + 1$ processes participate, at least one process decides 0 and at least one process decides 1. Like renaming, any protocol that implements WSB must be index-independent. We now show that weak symmetry breaking is equivalent to (non-adaptive)

$2n$-renaming: Any protocol for one can be transformed to a protocol for the other. Suppose we are given a "black-box" index-independent protocol that solves $2n$-renaming. The processes run a $2n$-renaming protocol, and each process decides the parity (value mod 2) of its output name. If all $n + 1$ processes participate, at least one must decide 0 and at least one must decide 1, because the range $0, \ldots, 2n - 1$ does not contain $n + 1$ even or $n + 1$ odd values.

Here is how to transform a "black-box" solution to weak symmetry breaking into a protocol for non-adaptive $2n$-renaming. First, each process executes the weak symmetry-breaking protocol and uses its binary output to choose between two instances of the $(2n + 1)$-renaming protocol of Figure 12.3. The first instance assigns output names moving up from 0, and the second assigns output names moving down from $2n - 1$. The q processes that decide value 0 join the first renaming protocol, and each is assigned a name in the range $0, \ldots, 2q - 2$. The other $n - q + 1$ processes join the second protocol, and each is assigned an output name in the range $0, \ldots, 2n - 2q$. Each process in the second group subtracts this name from $2n - 1$, yielding an output name in the range $2q - 1, \ldots, 2n - 1$. The ranges of names chosen by the two groups do not overlap.

It follows that instead of deriving a lower bound for adaptive or non-adaptive $2n$-renaming, it is enough to derive a lower bound for weak symmetry breaking. Here is our strategy. If there is a wait-free layered immediate snapshot protocol for weak symmetry breaking, then we know from the proof of Theorem 11.2.1 that there is a layered read-write protocol and a color-preserving simplicial map

$$\delta : \mathrm{Ch}^N \sigma \to \mathcal{O}$$

carried by Δ, the input-output relation defining weak symmetry breaking. The map δ, however, just maps each vertex to a binary value, so we can think if it simply as a binary coloring $\delta : \mathrm{Ch}^N \sigma \to \{0, 1\}$, with no *monochromatic* n-simplices (that is, n-simplices in which all vertices are mapped to the same value). Moreover, the protocol is index-independent, implying that the coloring is symmetric on the boundary of σ.

To show a lower bound, we need an analog to Sperner's lemma for binary colorings. Under what circumstances is it impossible to construct a symmetric binary coloring for a subdivided n-simplex that admits no monochromatic n-simplices?

12.3 **The index lemma**

We now prove a combinatorial lemma that will help us understand weak symmetry breaking. An $(n+1)$-*coloring* c of a complex $\mathcal{K}$ is a map from the vertices of $\mathcal{K}$ to Δ^n:

$$c : V(\mathcal{K}) \to \Delta^n.$$

A simplex $\sigma \in \mathcal{K}$ is *properly colored* by c if c is rigid: It maps each vertex of σ to a distinct vertex of Δ^n. A complex $\mathcal{K}$ is *chromatic* if it has a coloring

$$\mathrm{name} : V(\mathcal{K}) \to \Delta^n$$

that properly colors every simplex of $\mathcal{K}$.

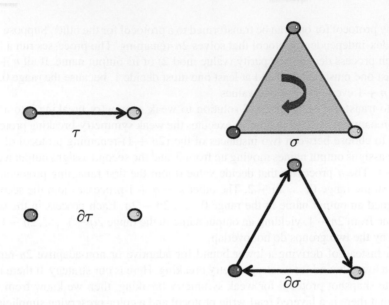

FIGURE 12.4

Oriented simplices and their boundaries.

An *orientation* of a simplex $\sigma = \{s_0, \ldots, s_n\}$ is given by a sequence of its vertices, up to the action of even permutations of that sequence.[1] For example, we use the notation $\langle s_0, s_1, s_2 \rangle$ to denote the orientation given by the sequence s_0, s_1, s_2 and its even permutations, such as s_2, s_0, s_1.

Mathematical Note 12.3.1. Formally, the group of all even permutations acts on the set of all ordered sequences of vertices, there are two orbits, and an orientation is a choice of an orbit.

Any simplex has two possible orientations. An orientation of an n-simplex σ induces an orientation on its $(n-1)$-faces: If σ^n is oriented $\langle s_0, \ldots, s_n \rangle$, then $\text{Face}_i \sigma$ has the induced orientation $\langle s_0, \ldots, \widehat{s_i}, \ldots, s_n \rangle$ if i is even, where a circumflex ($\frown$) denotes omission, and the opposite orientation if i is odd. As illustrated in Figure 12.4, it is natural to think of an oriented edge as an arrow and an oriented triangle as being oriented clockwise or counterclockwise.

A *oriented manifold* $\mathcal{M}$ is an n-manifold with boundary together with an orientation for each n-simplex with the property that every two n-simplices that share an $(n-1)$-face induce opposite orientations on their common face. Not all manifolds are orientable, but we here we will be concerned with subdivided simplices, which are orientable (see Figure 12.5). Henceforth, $\mathcal{M}$ will be an oriented chromatic manifold with boundary.

We say that an n-simplex $\sigma \in \mathcal{M}$ is *positively oriented* if the sequence $\text{name}(s_0), \ldots, \text{name}(s_n)$ is an even permutation of the sequence $0, \ldots, n$, and it is *negatively oriented* otherwise. Positive orientation is denoted by $\text{sgn}\ \sigma = +1$ and negative orientation by $\text{sgn}\ \sigma = -1$. The positive and the negative orientations alternate as we pass through each internal $(n-1)$-simplex. For each $(n-1)$-simplex τ on the boundary of $\mathcal{M}$, there is a unique σ such that $\tau = \text{Face}_i \sigma$. Define $\text{sgn}\ \tau = (-1)^i \text{sgn}\ \sigma$.

---

[1] An *even* permutation is one that can be constructed by exchanging the positions of adjacent elements an even number of times.

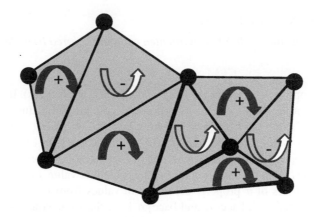

FIGURE 12.5

Oriented manifold with boundary.

We will use the following facts about subdivisions.

Fact 12.3.2.   If Div σ is a chromatic subdivision of σ, then Div σ is an orientable manifold, as is Div σ', for any $\sigma' \subset \sigma$. An orientation of Div σ induces an orientation of Div σ'.

Now consider $\mathcal{M}$ together with a coloring $c : V(\mathcal{M}) \to \Delta^n$, not necessarily proper. In other words, the complex $\mathcal{M}$ now has two colorings: the proper coloring name : $V(\mathcal{M}) \to \Delta^n$, and an arbitrary coloring $c : V(\mathcal{M}) \to \Delta^n$. Henceforth, when we say a simplex of $\mathcal{M}$ is properly colored, we mean properly colored by c. A properly colored simplex $\sigma = \{s_0, \dots, s_n\}$ in $\mathcal{M}$ is *counted by orientation* as follows: Index the vertices of σ in the order of their colors: $\langle s_0, \dots, s_n \rangle$ such that $c(s_i) < c(s_{i+1})$. If $\langle s_0, \dots, s_n \rangle$ belongs to the orientation induced by $\mathcal{M}$, then $C(\sigma, c) = +1$, and otherwise $C(\sigma, c) = -1$. If σ is not properly colored, then $C(\sigma, c) = 0$.

Definition 12.3.3.   The *content* of c in $\mathcal{M}$, denoted $C(\mathcal{M}, c)$, is the number of simplices properly colored by c, counted by orientation:

$$C(\mathcal{M}, c) = \sum_{\substack{\sigma \in \mathcal{M} \\ \dim \sigma = n}} C(\sigma, c).$$

For each $i \in \Delta^n$, consider the set of boundary $(n-1)$-simplices colored properly with values from $\mathrm{Face}_i(\Delta^n)$. These simplices, too, can be counted by orientation, with the definition of the corresponding quantity $I_i(\tau, c)$ repeating verbatim that of $C(\sigma, c)$. Index the vertices of τ in the order of their colors: $\langle t_0, \dots, t_{n-1} \rangle$ such that $c(t_i) < c(t_{i+1})$. If $\langle t_0, \dots, t_{n-1} \rangle$ belongs to the orientation induced by $\mathcal{M}$, then define $I_i(\tau, c) = +1$, and otherwise define $I_i(\tau, c) = -1$. If τ is not properly colored with values from $\mathrm{Face}_i(\Delta^n)$, then $I_i(\tau, c) = 0$.

Definition 12.3.4.   For each $i \in [n]$, the ith *index* of $\mathcal{M}$, denoted by $I_i(\mathcal{M}, c)$, is the number of $(n-1)$-simplices of its boundary complex $\partial \mathcal{M}$, properly colored with values from $\mathrm{Face}_i(\Delta^n)$, counted by orientation.

$$I_i(\mathcal{M}, c) = \sum_{\substack{\tau \in \partial \mathcal{M} \\ \dim \tau = n-1}} I_i(\tau, c).$$

Content and index are related as follows.

Lemma 12.3.5 (Index Lemma).   *If M is an oriented manifold colored by c, then for each $i \in [n]$ we have the identity*

$$C(M, c) = (-1)^i I_i(M, c).$$

Proof.   Fix $i \in [n]$. Suppose M consists of a single simplex σ (and its faces). If σ is properly colored, then $C(\sigma, c) = \text{sgn } \sigma$. For each $i \in \Delta^n$, $\text{Face}_i \sigma$ is properly colored with values from $\text{Face}_i(\Delta^n)$, with induced orientation $(-1)^i \text{sgn } \sigma$, so $I_i(\sigma, c) = (-1)^i \text{sgn } \sigma$, and $C(\sigma, c) = (-1)^i I_i(\sigma, c)$.

Suppose instead that σ is not properly colored, so $C = 0$. Clearly $I_i(\sigma, c) = 0$ if σ has no properly colored $(n-1)$-faces. Suppose instead that at least one $(n-1)$-face of σ is properly colored with values from $\text{Face}_i(\Delta^n)$. Because i does not appear in the coloring, another value j must appear twice, so there are exactly two $(n-1)$-faces of σ properly colored with values from $\text{Face}_i(\Delta^n)$. Index the vertices of σ such that $c(s_0) = c(s_1) = j$. $\text{Face}_0 \sigma$ and $\text{Face}_1 \sigma$ have the same coloring but opposite induced orientations, so their contributions to $I_i(\sigma, c)$ cancel, yielding $C(\sigma, c) = (-1)^i I_i(\sigma, c) = 0$.

Now consider an arbitrary manifold with boundary M. Consider the following sum:

$$S_i := \sum_{\substack{\sigma \in M \\ \dim \sigma = n}} I_i(\sigma, c) = \sum_{\substack{\sigma \in M \\ \dim \sigma = n}} \sum_{\substack{\tau \in \partial\sigma \\ \dim \tau = n-1}} I_i(\tau, c). \tag{12.3.1}$$

Each internal $(n-1)$-simplex τ is counted twice in (12.3.1), and it is counted with opposite orientations, so these contributions cancel out. On the other hand, each boundary $(n-1)$-simplex is counted once, so we have

$$S_i = \sum_{\substack{\tau \in \partial M \\ \dim \tau = n-1}} I_i(\tau, c) = I_i(M, c). \tag{12.3.2}$$

We have seen that $C(\sigma, c) = (-1)^i I_i(\sigma, c)$, so

$$S_i = \sum_{\substack{\sigma \in M \\ \dim \sigma = n}} I_i(\sigma, c) = (-1)^i \sum_{\substack{\sigma \in M \\ \dim \sigma = n}} C(\sigma, c) = (-1)^i C(M, c). \tag{12.3.3}$$

It follows from (12.3.2) and (12.3.3) that $C(M, c) = (-1)^i I_i(M, c)$.   □

Because a manifold's content is determined by its boundary, two colorings that agree on the boundary have the same content.

Corollary 12.3.6.   *If c and d are colorings of M such that $c(v) = d(v)$ for all $v \in \partial M$, then $C(M, c) = C(M, d)$.*

Consider the special case where M is a subdivided simplex. Corollary 12.3.6 implies that when counting properly colored simplices by orientation, we can "recolor" the interior vertices to any convenient color.

12.4 Binary colorings

Although the Index Lemma 12.3.5 and Corollary 12.3.6 are concerned with colorings that take on $n+1$ values, we can extend these results to reason about binary colorings as well.

Let $\mathcal{M}$ be a manifold with boundary with a proper coloring name : $V(\mathcal{M}) \to \Delta^n$ and a binary labeling $b : V(\mathcal{M}) \to \Delta^{[1]}$. Define a new coloring $c : V(\mathcal{M}) \to \Delta^n$ by setting

$$c(v) = \text{name}(v) + b(v) \quad (\text{mod } n+1)$$

for all $v \in V(\mathcal{M})$. The following is immediate:

Lemma 12.4.1.   *An n-simplex of $\mathcal{M}$ is monochromatic under b if and only if it is properly colored under c.*

Because of Lemma 12.4.1, we can use the content $C(\text{Div } \sigma, c)$ to count by orientation monochromatic n-simplices under b. Here we note that if σ is m-monochromatic under b for some $m \in \{0, 1\}$, then it is counted as $(-1)^{m \cdot n} \text{sgn}(\sigma)$ by orientation. The Index Lemma 12.3.5 implies the next statement.

Corollary 12.4.2.   *If $b : V(\text{Div } \sigma) \to \Delta^{[1]}$ and $b' : V(\text{Div } \sigma) \to \Delta^{[1]}$ are binary colorings that agree on $\partial\text{Div } \sigma$, then the number of monochromatic simplices of $\text{Div } \sigma$ counted by orientation is the same in both.*

We can therefore assume, without loss of generality, that $b(v) = 0$ for every v not on the boundary of Div σ. We now show that the content of the n-simplices containing these internal all-0 vertices is $(-1)^n$.

We will use $0(\cdot)$ to denote the coloring that assigns 0 to every vertex, and similarly for $1(\cdot)$.

Let κ, the *central simplex*, be the simplex of $\text{Ch}^N \sigma$ that represents the execution in which all processes execute in lock-step (together) for N rounds. We take $(-1)^{n \cdot N}$ as the orientation of κ and induce uniquely an orientation on the entire $\text{Ch}^N \sigma$. We take that orientation as standard; see Figure 12.6.

Lemma 12.4.3.   *Assume σ is an n-simplex and $\text{Ch}^N \sigma$ is the N-fold standard chromatic subdivision of σ. Then we have $C(\text{Ch}^N \sigma, 0(\cdot)) = 1$ and $C(\text{Ch}^N \sigma, 1(\cdot)) = (-1)^n$.*

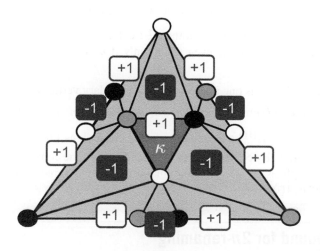

FIGURE 12.6

Standard orientation for Ch σ. Here there are three processes, so $n = 2$, and one round of execution, so $N = 1$. The orientation of the central simplex marked κ is therefore $(-1)^{n \cdot N} = 2 = +1$.

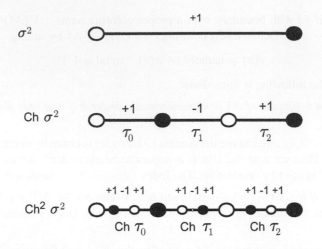

FIGURE 12.7

Orientation for a 1-dimensional subdivision.

Proof.   We first calculate that $C(\mathrm{Ch}^N\sigma, 0(\cdot)) = 1$. For $n = 1$, we note that $\mathrm{Ch}^N\sigma$ is a subdivision of a 1-simplex into an odd number of simplices. These simplices are oriented alternating, and the leftmost simplex is positively oriented. Hence there is one more simplex oriented positively than negatively, and hence the total sum is 1; see Figure 12.7.

Having verified the base, we can now use induction on n. For the induction step, assume that $C(\mathrm{Ch}^N\tau, 0(\cdot)) = 1$ when $\dim \tau < n$. By the Index lemma 12.3.5, we have

$$C(\mathrm{Ch}^N\sigma, 0(\cdot)) = I_0(\mathrm{Ch}^N\sigma, 0(\cdot)) = C(\mathrm{Ch}^N\mathrm{Face}_0\sigma, 0(\cdot)) = 1.$$

To see that $C(\mathrm{Ch}^N\sigma, 1(\cdot)) = (-1)^n$ note that the all-1 coloring "rotates" the process names, sending i to $(i + 1)\bmod n$. This permutation is even if and only if n is even, so orientations are multiplied by $(-1)^n$. $\square$

An *internal simplex* of $\mathrm{Ch}^N\sigma$ is one that contains no boundary vertices.

Lemma 12.4.4.   *If all internal vertices are colored 0, then the content of the internal simplices is* $(-1)^n$.

Proof.   Let $\hat{b} : V(\mathrm{Ch}^N\sigma) \to \Delta^{[1]}$ be the binary coloring that assigns 1 to every vertex on the boundary and 0 to every internal vertex. By construction, $C(\mathrm{Ch}^N\sigma, \hat{b})$ is just the content of the internal simplices, since all the other n-simplices are not monochromatic. Because $\hat{b}$ agrees with $1(\cdot)$ on the boundary, $C(\mathrm{Ch}^N\sigma, \hat{b}) = C(\mathrm{Ch}^N\sigma, 1(\cdot)) = (-1)^n$. $\square$

12.5 A lower bound for $2n$-renaming

We now show that if $n + 1$ is a prime power, weak symmetry breaking, and hence $2n$-renaming, is impossible. We argue by contradiction. Assume we have an index-independent wait-free protocol for weak symmetry breaking.

Assume σ is an n-simplex. For all $I, J \subseteq [n]$, let $\alpha_{I,J} : I \to J$ denote the rank-preserving bijection, and let $\varphi_{I,J} : \sigma^I \to \sigma^J$ denote the affine isomorphism of the corresponding boundary simplices σ^I and σ^J of σ induced by the map $\alpha_{I,J}$.

Definition 12.5.1.   A binary coloring $b : \mathrm{Ch}^N \sigma \to \{0, 1\}$ is called *rank symmetric* if, for any choice of I and J as above, the map $\varphi_{I,J}$ induces an isomorphism of the subdivisions $\varphi_{I,J} : \mathrm{Ch}^N \sigma^I \to \mathrm{Ch}^N \sigma^J$, and, in addition, b is invariant with respect to $\varphi_{I,J}$. Formally, the last condition says that for every $v \in V(\mathrm{Ch}^N \sigma^I)$, we have $b(v) = b(\varphi_{I,J}(v))$.

The following is a useful lemma about rank-symmetric subdivisions.

Lemma 12.5.2.   *Let τ be a simplex in the subdivided boundary $\mathrm{Ch}^N \sigma^I$. For any set $J \subseteq [n]$ such that $|I| = |J|$, there is an orientation-preserving bijection between the following two sets of n-simplices of $\mathrm{Ch}^N \sigma$:*

$$\left\{ \gamma \mid \gamma \cap \mathrm{Ch}^N \partial \sigma = \tau \right\} \text{ and } \left\{ \gamma \mid \gamma \cap \mathrm{Ch}^N \partial \sigma = \varphi_{I,J}(\tau) \right\}.$$

Proof.   Let κ be, as above, the central simplex of $\mathrm{Ch}^N \sigma$. Define a *flip* to be the operation of moving from one n-simplex to another by replacing a single vertex. For example, a flip might go from $\rho_0 = \{r_0, \ldots, r_i, \ldots, r_n\}$ to $\rho_1 = \{r_0, \ldots, r_i', \ldots, r_n\}$, where $\mathrm{name}(r_i) = \mathrm{name}(r_i')$. If we can get from ρ_0 to ρ_1 in a single flip, then ρ_0 and ρ_1 have opposite orientations. We can characterize a sequence of flips by the names of the vertices replaced: A sequence of flips $i_0, \ldots, i_{\ell-1}$ replaces the vertex whose ID is i_j at step j. Two n-simplices have the same orientation if and only if we can get from one to the other through an even sequence of flips.

Assume that τ is a face of an n-simplex τ'. We can get from τ' to κ by a sequence of ℓ flips: $i_0, \ldots, i_{\ell-1}$. Because $\mathrm{Ch}^N \sigma$ is symmetric, we can use a reversed, symmetric sequence of ℓ flips $\pi(i_{\ell-1}), \ldots, \pi(i_0)$ to get from κ to an n-simplex, which we call $\varphi_{I,J}(\tau')$. It takes a total of 2ℓ flips to get from τ' to $\varphi_{I,J}(\tau')$, so they have the same orientation in $\mathrm{Ch}^N \sigma$. Clearly, $\varphi_{I,J}(\tau')$ is an n-simplex such that $\varphi_{I,J}(\tau') \cap \mathrm{Ch}^N \partial \sigma = \varphi_{I,J}(\tau)$, and the above procedure yields a bijection. $\square$

Recall that the Index lemma 12.3.5 ensures that we can assume without loss of generality that $\delta(v) = 0$ for any vertex v not in $\partial \mathrm{Ch}^N \sigma$. Since every n-simplex in $\mathrm{Ch}^N \sigma$ contains at least one such vertex, $\mathrm{Ch}^N \sigma$ has no 1-monochromatic simplices, so we can compute the number by orientation of monochromatic n-simplices in $\mathrm{Ch}^N \sigma$ simply by computing the number by orientation of 0-monochromatic simplices.

Lemma 12.5.3.   *The number of monochromatic n-simplices in $\mathrm{Ch}^N \sigma$, counted by orientation, is*

$$(-1)^n + \sum_{q=0}^{n-1} \binom{n+1}{q+1} k_q$$

for integers $k_0, \ldots, k_{n-1}$.

Proof.   Take $0 \leq q \leq n - 1$. By Lemma 12.4.4, the internal simplices contribute $(-1)^n$ to the content. Let $\sigma^q \subset \sigma$, and suppose $\mathrm{Ch}^N \sigma^q$ has k_q 0-monochromatic q-simplices counted by orientation. There are $\binom{n+1}{q+1}$ such faces, and by Lemma 12.5.2, they have the same contributing $k_q \cdot \binom{n+1}{q+1}$ to the count of 0-monochromatic n-simplices of $\mathrm{Ch}^N \sigma$. $\square$

Now we make use of the following fact from number theory.

Fact 12.5.4. The integers in the set

$$\left\{ \binom{n+1}{1}, \ldots, \binom{n+1}{n} \right\}$$

have a common factor if and only if $n + 1$ is a prime power.

Theorem 12.5.5. *If $n + 1$ is a prime power, then there is no index-independent wait-free read-write protocol for $2n$-renaming.*

Proof. If there is a protocol for weak symmetry breaking, then there is an iterated standard chromatic subdivision $\mathrm{Ch}^N \sigma$ that has a binary coloring with no monochromatic n-simplices. There is therefore a binary coloring for $\mathrm{Ch}^N \sigma$ with no 0-monochromatic n-simplices. Because $\mathrm{Ch}^n \sigma$ is symmetric, however, the number of 0-monochromatic n-simplices it contains is

$$1 + \sum_{q=0}^{n-1} \binom{n+1}{q+1} k_q. \tag{12.5.1}$$

By Fact 12.5.4, if $(n + 1)$ is a prime power, then these binomial coefficients share a common factor, and this number cannot be zero. □

12.6 Chapter notes

The renaming algorithm presented here is due to Borowsky and Gafni [24]. The recursive formulation of Figure 12.3 is due to Gafni and Rajsbaum [68]. An earlier and less efficient algorithm is due to Attiya et al. [9]. The connection between renaming and weak symmetry breaking is due to Gafni, Herlihy and Rajsbaum [69]. The renaming lower bound is due to Castañeda and Rajsbaum [31,32]. Attiya and Paz [15] prove the weak symmetry-breaking impossibility applying the Index Lemma directly on executions. There is an overview of renaming in shared memory systems by Castañeda, Rajsbaum and Raynal [34].

The version of the Index Lemma we used is a restatement of Corollary 2 by Fan [54] using our notation. For a simple description of this lemma in dimension 2 see Henle [77, pp. 46-47].

It turns out that if $n + 1$ is not a prime power, then there is an index-independent wait-free read-write protocol for $2n$-renaming, as shown by Castañeda and Rajsbaum [31,33]. An algebraic topology proof is given by Castañeda, Herlihy, and Rajsbaum [125] and a more direct upper-bound construction by Attiya et al. [11]; see also Kozlov [102].

Exercise 12.6 and 12.7 are inspired by work of Moir and Anderson [117].

12.7 Exercises

Exercise 12.1. Give a single-round immediate snapshot protocol for the tagged participating set task defined in Section 12.1.2.

Exercise 12.2. Prove Fact 12.5.4.

```
 1  // pseudo−code for P
 2  int last = ⊥                 //  initially
 3  goRight: Boolean = false     //  initially
 4  visit (box: MA−Box): Status
 5    box. last := P
 6    if (box.goRight) then
 7      return RIGHT
 8    box.goRight = true
 9    if  last = P then
10      return STOP
11    else
12      return DOWN
```

FIGURE 12.8

Code for MA-Box.

Exercise 12.3.   Prove Lemma 12.1.2.

Exercise 12.4.   Show that if a manifold with boundary $\mathcal{M}$ has an orientation, then there are exactly two possible orientations.

Exercise 12.5.   Recall that the testAndSet() operation atomically swaps *true* with the contents of a memory location. Devise a wait-free M-renaming protocol using testAndSet() objects. How small can you make M?

Exercise 12.6.   An *MA-box* is named after its inventors, Mark Moir and James Anderson. It provides a single method, visit (), which returns one of three values: STOP, DOWN, or RIGHT. Figure 12.8 shows the code for visit (). The MA-Box object has two fields, last, initially ⊥, and goRight, initially *false*. These fields can be read and written by multiple processes. When a process calls visit (), it stores its own name in the last field. If goRight field is *true*, it returns RIGHT. Otherwise, it rereads last and if its name is still present, it returns STOP; otherwise it reruns DOWN.

Prove that if *n* threads call visit (),

- At most, one thread gets the value STOP.,
- At most, $n - 1$ threads get the value DOWN, and
- At most, $n - 1$ threads get the value RIGHT.

Note that the last two proofs are not symmetric.

Exercise 12.7.   Arrange MA-box objects in a triangular matrix, where each MA-Box is given an integer name as shown in Figure 12.9. Each thread starts by visiting MA-Box zero. If it gets STOP, it stops. If it gets RIGHT, it visits 1, and if it gets DOWN, it visits 2. In general, if a thread gets STOP, it stops. If it gets RIGHT, it visits the next MA-Box on that row, and if it gets DOWN, it visits the next MA-Box in that column. Each thread takes the name of the MA-Box object where it stops.

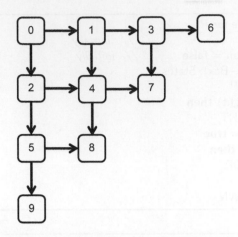

FIGURE 12.9

Array layout for MA-Box objects.

Prove that MA-Boxes can be used to solve $(2n + 1)$-renaming.

- Prove that each thread eventually stops at some MA-Box object.
- How many MA-Box objects do you need in the array if you know in advance a bound $n + 1$ on the number of threads?

CHAPTER

Task Solvability in Different Models

13

We have seen that one way to analyze complex models of computation is to break protocol executions into *layers*. In this chapter, we show how to exploit layering to compute inductively the connectivity of protocol complexes in various models of computation. The advantage of layering is that we can treat connectivity as an *invariant*, established in the first layer and preserved in later layers. Understanding the computational power of a model splits into two tasks: We use model-specific reasoning to analyze connectivity for a single-layer complex, and we use model-independent reasoning to analyze how individual layers compose.

The principal technical idea introduced in this chapter is a proof that if the single-layer complex is *shellable*, a combinatorial property defined later, then multilayer compositions preserve connectivity under certain easilycheckable conditions. These are theorems of combinatorial topology, independent of any model of computation.

We then show that for several models of computation, each single-layer complex is indeed shellable, so it becomes a straightforward exercise to derive tight (or nearly tight) bounds on when and if one can solve k-set agreement.

Distributed Computing Through Combinatorial Topology. http://dx.doi.org/10.1016/B978-0-12-404578-1.00013-9

13.1 Shellability

Roughly speaking, a pure n-dimensional abstract simplicial complex is *shellable* if it can be decomposed into components whose unions and intersections have a nice regular structure amenable to the use of the Nerve lemma. Not all complexes have such a nice structure, but the complexes that arise in several important models of concurrent computation do. Specifically, exploiting shellability allows us to break computations of arbitrary length into simpler, well-structured *layers* and to argue that each layer preserves connectivity. For some (asynchronous) models, the protocol complexes remain connected forever, yielding impossibility results. For other (synchronous and semisynchronous) models, the protocol complexes remain connected for only a bounded number of layers, yielding communication complexity lower bounds.

Recall that a *facet* of a complex is a maximal simplex. A simplicial complex C is *shellable* if its facets can be arranged in a linear order $\phi_0, \ldots, \phi_t$, called a *shelling order*, in such a way that the subcomplex $(\cup_{i=0}^{k-1} \phi_i) \cap \phi_k$ is the union of $(\dim \phi_k - 1)$-faces of ϕ_k for $0 < k \leq t$. All the shellable complexes considered here are *pure*; their facets have the same dimension.

Figure 13.1 shows the construction of an octahedron, a simple shellable 2-complex that represents all the possible ways to assign binary values to three processes. The construction starts with a single 2-simplex (triangle) in the upper left and adjoins new 2-simplices along either one, two, or (at lower right) three edges. Figure 13.2 shows a simple complex that is *not* shellable; it consists of two 2-simplices that are joined at a single vertex.

Mathematical Note 13.1.1. A facet (maximal simplex) that is added by attaching all of its proper faces is called a *spanning simplex*. It can be shown that any shellable complex is topologically equivalent (more precisely, homotopy equivalent) to a set of spheres attached at a single point (called a *wedge of spheres*). Each such sphere corresponds to a spanning simplex in the shellable complex, and that sphere has the same dimension as the corresponding spanning simplex.

Fact 13.1.2. If C is a pure, shellable k-complex, then C is $(k-1)$-connected.

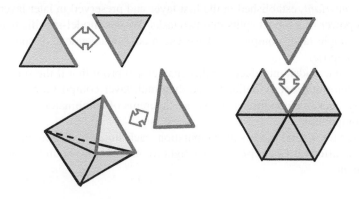

FIGURE 13.1

Constructing a shellable complex.

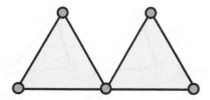

FIGURE 13.2

Example of a nonshellable complex.

The following alternative formulation for shellability, though less concise, is easier to use in proofs.

Fact 13.1.3.   Let $\mathcal{K}$ be a simplicial complex with facets $\phi_0, \ldots, \phi_t$. This sequence is a shelling order if and only if, for every i, j such that $0 \leq i < j \leq t$, there exists ϕ_k satisfying

1. $0 \leq k < j$,
2. $\phi_i \cap \phi_j \subseteq \phi_k \cap \phi_j$, and
3. $|\phi_j \setminus \phi_k| = 1$.

Informally, these conditions capture what happens as we glue each successive facet ϕ_j onto the complex. The first condition simply says we are considering how a newly appended facet, ϕ_j, intersects an earlier facet, ϕ_k. The second condition says that we can ignore "small" intersections between ϕ_j and any preceding ϕ_i because any such intersection is contained within a larger intersection with ϕ_k. The third condition says ϕ_j is glued onto ϕ_k along an $(n-1)$-face. To prove that a complex $\mathcal{K}$ is shellable, we will display a total order on its facets satisfying these conditions.

13.2 Examples

Let $\sigma = \{s_0, \ldots, s_n\}$ be an n-simplex, together with a total ordering on its vertices, indicated here by index order. The *face ordering* $<_f$ on faces of σ is defined as follows.

Definition 13.2.1.   Each face τ of σ has an associated *signature*, denoted $\tau[\cdot]$, an $(n+1)$-element Boolean string whose i^{th} value, denoted $\tau[i]$, is $\perp$ if $s_i \in \tau$, and $\top$ otherwise. (Note that $\perp$ denotes presence and $\top$ denotes absence!) Faces are ordered by their signatures: $\tau_0 <_f \tau_1$ if $\tau_0[\cdot]$ is lexicographically less than $\tau_1[\cdot]$, where $\perp < \top$.

For this ordering and for others, we use the obvious notational extensions: $\tau_0 \leq_f \tau_1$ if $\tau_0 <_f \tau_1$ or $\tau_0 = \tau_1$, and $\tau_1 >_f \tau_0$ if $\tau_0 <_f \tau_1$.

The simplex whose signature bits are all $\perp$ is equal to σ and is ordered before any of its proper faces. The simplex whose signature bits are all $\top$ is equal to $\emptyset$ and is ordered after any of the faces.

The following lemma, illustrated by Figure 13.3, shows the techniques which we will later use to prove more complex results.

Lemma 13.2.2.   *For all $d, 0 \leq d \leq n$, the simplicial complex $skel^d(\sigma)$ is shellable. (see Figure 13.3).*

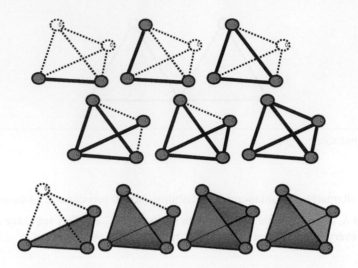

FIGURE 13.3

Illustration for Lemma 13.2.2: shelling orders for the 1- and 2-skeletons of a tetrahedron Δ^2.

Proof. Let $\text{Faces}_d(\sigma)$ be the set of d-faces of σ, which is also the set of facets of the complex $\text{skel}^d \sigma$. We claim that the ordering $<_f$ is a shelling order for $\text{Faces}_d(\sigma)$. Let $\phi_i <_f \phi_j$ be two d-faces in $\text{Faces}_d(\sigma)$.

Consider their signatures. Let ℓ be the least index such that $\phi_i[\ell] \neq \phi_j[\ell]$. Because $\phi_i <_f \phi_j$, $\phi_i[\ell] < \phi_j[\ell]$, meaning that $\phi_i[\ell] = \bot$ and $\phi_j[\ell] = \top$. Equivalently, $s_\ell \in \phi_i$, but $s_\ell \notin \phi_j$.

Because ϕ_i and ϕ_j both have dimension d, each contains exactly $d + 1$ vertices, implying there is an index m, $\ell < m \leq n + 1$ such that $s_m \in \phi_j$ but $s_m \notin \phi_i$, implying that $\phi_j[m] < \phi_j[m]$.

Construct ϕ_k by replacing s_m in ϕ_j with s_ℓ:

$$\phi_k = (\phi_j \setminus \{s_m\}) \cup \{s_\ell\}.$$

This simplex has the signature

$$\phi_k[q] = \begin{cases} \phi_j[q] & \text{if } q \notin \{\ell, m\} \\ \bot & \text{if } q = \ell \\ \top & \text{if } q = m. \end{cases}$$

We must check the three conditions of Fact 13.1.3.

By construction, ℓ is the first index at which the signatures of ϕ_k and ϕ_j differ. Because $\phi_k[\ell] = \bot$ and $\phi_j[\ell] = \top$, it follows that $\phi_k <_f \phi_j$, satisfying the first condition.

Also by construction, $\phi_j \cap \phi_k = \phi_j \setminus \{s_m\}$. Because $s_m \notin \phi_i$, $\phi_i \cap \phi_j \subseteq \phi_k \cap \phi_j$, satisfying the second condition.

Finally, s_m is the only vertex in ϕ_j not in ϕ_k, so $|\phi_j \setminus \phi_k| = |\{s_m\}| = 1$, satisfying the third condition. $\square$

13.3 Pseudospheres

Consider the input complexes for the 2- and 3-process *binary consensus* task, in which each process starts with a binary value, and all must agree on some process's input. Each process in the set $\{P, Q\}$ or $\{P, Q, R\}$ is independently assigned a value from $\{0, 1\}$. As shown in Figure 13.4, the resulting complex for two processes is a rectangle and for three processes is an octahedron. Each process is assigned 0 in the upper face and 1 in the lower face. The processes are assigned mixed values in the simplices joining these faces. These complexes are homeomorphic to the 1- and 2-spheres, and we leave it as an exercise to show that the complex constructed by independently assigning binary values to a set of $n + 1$ processes is homeomorphic to a combinatorial n-sphere.

What if we independently assign values from a larger set? Figure 13.5 shows a schematic picture of the complex constructed by independently assigning each process a value from $\{0, 1, 2\}$. It consists of

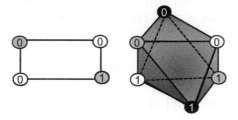

FIGURE 13.4

The pseudosphere complex $\Psi(\{P, Q\}, \{0, 1\})$ (left) is a rectangle, and $\Psi(\{P, Q, R\}, \{0, 1\})$ (right) is an octahedron.

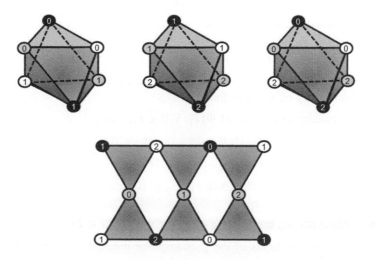

FIGURE 13.5

An exploded view of the pseudosphere complex $\Psi(\{P, Q, R\}, \{0, 1, 2\})$. Colors indicate process names, and vertices with the same color and value are the same. The simplices labeled with one or two values form three octahedrons, shown at the top. They are linked by six triangles labeled with three values, shown at the bottom.

three octahedrons, one for each pair of values, where the octahedron generated by $\{0, 1\}$ is "glued" to the octahedron generated by $\{0, 2\}$ at the simplex in which all processes are assigned 0, and similarly for the other values. These octahedrons are further linked by six triangles labeled with all three values. When we move from assigning two values to three, the resulting complex is no longer homeomorphic to a sphere, but it shares many of the combinatorial properties of spheres. For this reason, we call such a complex a *pseudosphere*. We are now ready to define pseudospheres formally.

Definition 13.3.1.   Let I be a finite index set. For each $i \in I$, let P_i be a process name, indexed so that if $i \neq j$, then $P_i \neq P_j$, and let V_i be a set. The *pseudosphere* complex $\Psi(P_i, V_i | i \in I)$ is defined as follows:

- Every pair (P_i, v) is a vertex, where $v \in V_i$.
- For any index set $J \subseteq I$, the set $\{(P_j, v_j) | j \in J, v_j \in V_j\}$ is a simplex if the P_j are distinct.

Note that this definition ensures that each facet of $\Psi(P_i, V_i | i \in I)$ is properly colored with all process names, so a pseudosphere is pure. For the index set $[n]$, we sometimes use the notation $\Psi(P_0, V_0; \ldots; P_n, V_n)$. Often, but not always, the V_i are the same, so we write $\Psi(U, V)$ as shorthand for $\Psi(P_i, V | P_i \in U)$.

The following simple properties follow directly from the definitions and are left as exercises. (Recall that we do not always distinguish between a simplex and the power set complex of all its faces.)

Fact 13.3.2.   If each V_i is a singleton set $\{v_i\}$, then the pseudosphere is isomorphic to a single simplex:

$$\Psi(P_i, V_i | i \in I) \cong \{v_i | i \in I\}.$$

Fact 13.3.3.   If there is a $j \in I$ such that $V_j = \emptyset$, then

$$\Psi(P_i, V_i | i \in I) \cong \Psi(P_i, V_i | i \in I \setminus \{j\}).$$

Pseudospheres are closed under intersection.

Fact 13.3.4.
$$\Psi(P_i, U_i | i \in I) \cap \Psi(P_i, V_i | i \in I) = \Psi(P_i, U_i \cap V_i | i \in I).$$

Next, we show that pseudospheres are shellable. Assume we have a total order on each V_i. Define the *pseudosphere ordering* $<_p$ on facets as follows.

Definition 13.3.5.   Consider facets ϕ, ϕ' of $\Psi(P_i, V_i | i \in I)$, where

$$\phi = \{(P_0, u_0), \ldots, (P_n, u_n)\}$$
$$\phi' = \{(P_0, v_0), \ldots, (P_n, v_n)\}.$$

We order facets lexicographically by value: define $\phi <_p \phi'$ if there is an $\ell, 0 \leq \ell \leq n$, such that $u_i = v_i$ for $0 \leq i < \ell$, and $u_\ell < v_\ell$.

Theorem 13.3.6.   *The order $<_p$ is a shelling order for* $\Psi(P_i, V_i | i \in I)$.

Proof.   Let $\phi_0, \ldots, \phi_t$ be the facets of the pseudosphere indexed according to the pseudosphere ordering. Consider ϕ_i and ϕ_j, where $i < j$.

$$\phi_i = \{(P_0, u_0), \ldots, (P_n, u_n)\}$$
$$\phi_j = \{(P_0, v_0), \ldots, (P_n, v_n)\}$$

Let ℓ be the least index such that $u_\ell < v_\ell$. Replacing (P_ℓ, v_ℓ) in ϕ_j with (P_ℓ, u_ℓ) yields a facet ϕ_k. We must check the three conditions of Fact 13.1.3. By construction, $\phi_k <_p \phi_j$, so $k < j$, $\phi_j \cap \phi_i \subseteq \phi_j \cap \phi_k$, and $|\phi_j \setminus \phi_k| = 1$. $\square$

Corollary 13.3.7.   *If the V_i are non-empty, then $\Psi(P_i, V_i | i \in I)$ is $(|I| - 2)$-connected.*

For $n + 1$ processes, $|I| = n + 1$, then $\Psi(P_i, V_i | i \in I)$ is $(n - 1)$-connected.

13.4 **Carrier maps and shellable complexes**

Recall from Chapter 3 that for simplicial complexes $\mathcal{C}$ and $\mathcal{D}$, a *carrier map* $\Phi : \mathcal{C} \to 2^{\mathcal{D}}$ takes each simplex of $\mathcal{C}$ to a subcomplex of $\mathcal{D}$. A carrier map is *rigid* if $\Phi(\sigma)$ is pure of dimension $\dim \sigma$, for all $\sigma \in \mathcal{C}$, and we say that Φ *preserves intersections* if $\Phi(\sigma \cap \tau) = \Phi(\sigma) \cap \Phi(\tau)$; such a carrier map is called *strict*. We also use the shorthand notation $\Phi(\mathcal{K})$ to denote $\cup_{\sigma \in \mathcal{K}} \Phi(\sigma)$.

We decompose computations into a sequence of *layers*. Each layer is defined by a carrier map that carries the complex representing the possible configurations before the layer to the complex representing the possible configurations after the layer. In synchronous models, we can think of layers as happening one at a time: First, layer one executes, then layer two, and so on. In asynchronous models, however, like the layered asynchronous snapshot model used in earlier chapters, the layered structure represents information flow, not temporal order. Steps of different layers can take place concurrently, but information flows only from earlier layers to later ones.

We use the following invariance argument to analyze when k-set agreement is impossible in such a model. The input complex, where each process independently chooses an input value, is shellable and therefore $(k - 1)$-connected. It is enough to show that each layer's carrier map preserves $(k - 1)$-connectivity. When we define these carrier maps, we do not need to consider all executions permitted by the model, only the "worst-case" ones that preserve connectivity. For impossibility results, if a protocol cannot solve a task on a subset of executions of a model, it can certainly not solve it on all executions.

We will make use of the *Nerve lemma* (Lemma 10.4.2), repeated here:

> Let $\{\mathcal{K}_i | i \in I\}$ be a cover for a complex $\mathcal{K}$. For any index set $J \subset I$, define $\mathcal{K}_J = \cap_{j \in J} \mathcal{K}_j$. If each $\mathcal{K}_J$ is either $(k - |J| + 1)$-connected or empty, then $\mathcal{K}$ is k-connected if and only if the nerve complex $\mathcal{N}(\mathcal{K}_i | i \in I)$ is also k-connected.

We will also use the following *special case* of the Nerve lemma (Lemma 10.4.3).

> If $\mathcal{K}$ and $\mathcal{L}$ are complexes such that $\mathcal{K}$ and $\mathcal{L}$ are k-connected and $\mathcal{K} \cap \mathcal{L}$ is $(k - 1)$-connected, then $\mathcal{K} \cup \mathcal{L}$ is also k-connected.

Recall that the *codimension* of an m-simplex in a pure n-complex $\mathcal{C}$, written $\mathrm{codim}\, \sigma$, is $n - m$. We distinguish yet another class of carrier maps.

Definition 13.4.1.   Assume $\mathcal{K}$ and $\mathcal{L}$ are simplicial complexes, such that $\mathcal{K}$ is pure. A q-*connected carrier map* $\Phi : \mathcal{K} \to 2^{\mathcal{L}}$ is a rigid strict carrier map such that for each $\sigma \in \mathcal{K}$, the simplicial complex $\Phi(\sigma)$ is $(q - \mathrm{codim}\, \sigma)$-connected.

For a n-complex $\mathcal{K}$, these conditions can be restated as follows: For n-simplices σ the subcomplexes $\Phi(\sigma)$ are q-connected, for $(n - 1)$-simplices σ the subcomplexes $\Phi(\sigma)$ are $(q - 1)$-connected, $\ldots$, for $(n - q)$-simplices σ the subcomplexes $\Phi(\sigma)$ are 0-connected (that is, *path-connected*), and for $(n - q - 1)$-simplices σ the subcomplexes $\Phi(\sigma)$ are (-1)-connected (that is, *non-empty*). These conditions place no restrictions on simplices of dimension $n - q - 2$ and lower.

Let us now state the following obvious property: if $\mathcal{K}$ is an n-dimensional pure simplicial complex, $\tilde{\mathcal{K}}$ is an m-dimensional pure simplicial subcomplex of $\mathcal{K}$, and $\Phi : \mathcal{K} \to 2^{\mathcal{L}}$, is a q-connected carrier map, then the restriction of Φ to $\tilde{\mathcal{K}}$ is a $(q - n + m)$-connected carrier map.

The next two lemmas illustrate why shellability is such a powerful notion. Informally, if a carrier map, operating on a shellable complex, preserves shellability locally, then it preserves connectivity globally, giving us a tool for *local* reasoning about connectivity.

Lemma 13.4.2. *If $\mathcal{K}$ is a pure shellable simplicial complex, and $\Phi : \mathcal{K} \to 2^{\mathcal{L}}$ a q-connected carrier map, then the simplicial complex $\Phi(\mathcal{K})$ is q-connected.*

Proof. Let $\mathcal{K} = \bigcup_{i=0}^{t} \phi_i$, where the ϕ_i are the facets of $\mathcal{K}$ indexed in a shelling order. We argue by induction on t.

For the base case, $\mathcal{K} = \phi_0$. Because ϕ_0 is a facet of $\mathcal{K}$ of codimension 0, $\Phi(\phi_0)$ is q-connected by hypothesis.

For the induction hypothesis, assume

$$\mathcal{L} = \bigcup_{i=0}^{t-1} \Phi(\phi_i) = \Phi\left(\bigcup_{i=0}^{t-1} \phi_i\right)$$

is q-connected. Because ϕ_t is a facet of $\mathcal{K}$, $\mathcal{M} = \Phi(\phi_t)$ is q-connected by hypothesis. If we show that $\mathcal{L} \cap \mathcal{M}$ is $(q - 1)$-connected, then the claim follows from Lemma 10.4.3, the special case of the Nerve Lemma. Now, since Φ is strict, we have,

$$\mathcal{L} \cap \mathcal{M} = \Phi\left(\bigcup_{i=0}^{t-1} \phi_i\right) \cap \Phi(\phi_t) = \Phi\left(\phi_t \cap \bigcup_{i=0}^{t-1} \phi_i\right).$$

We know that the simplicial complex $\phi_t \cap \bigcup_{i=0}^{t-1} \phi_i$ is pure of dimension $n - 1$. It is shellable (see Exercise 13.1), and Φ restricted to it is $(q - 1)$-connected. By induction on the dimension of $\mathcal{K}$ we know that $\Phi\left(\phi_t \cap \bigcup_{i=0}^{t-1} \phi_i\right)$ is $(q - 1)$-connected; which is exactly what we need.

Because strict carrier maps preserve intersections,

$$\bigcup_{i=0}^{t-1} \left(\Phi(\phi_i) \cap \Phi(\phi_t)\right) = \bigcup_{i=0}^{t-1} \Phi(\phi_i \cap \phi_t).$$

Because $\mathcal{K}$ is shellable, for every $\phi_i, 0 \leq i < t$, there is a $k < t$ such that $\phi_i \cap \phi_t \subseteq \phi_k \cap \phi_t$, hence $\Phi(\phi_i \cap \phi_t) \subseteq \Phi(\phi_k \cap \phi_t)$. Moreover, $\psi_i = \phi_k \cap \phi_t$ is a face of ϕ_t of codimension 1, and because Φ is a strict carrier map,

$$\Phi(\phi_i \cap \phi_t) \subseteq \Phi(\phi_k \cap \phi_t) = \Phi(\psi_i).$$

It follows that for some index set I,

$$\bigcup_{i=0}^{t-1} \Phi(\phi_i \cap \phi_t) = \bigcup_{i \in I} \Phi(\psi_i).$$

We now use the Nerve lemma to compute the connectivity of $\mathcal{L} \cap \mathcal{M}$. For any $J \subseteq I$, let $\psi_J = \cap_{i \in J} \psi_i$. Then

$$\bigcap_{j \in J} \Phi(\psi_j) = \Phi(\bigcap_{j \in J} \psi_j) = \Phi(\psi_J).$$

Because the ψ_i are faces of ϕ_t of codimension 1, ψ_J is a $(\dim \phi_t - |J|)$-dimensional face of ϕ_t. It has codimension $|J|$ in $\mathcal{K}$, so $\Phi(\psi_J)$ is $(q - |J|)$-connected by hypothesis. For later use in computing the nerve, note that if $|J| \leq q + 1$, $\Phi(\psi_J)$ is non-empty.

To apply the Nerve lemma, note that each $\Phi(\psi_J)$ is $((q-1) - |J| + 1)$-connected. The $\Phi(\psi_j)$ subcomplexes cover $\mathcal{L} \cap \mathcal{M}$, and as noted, the intersection of any $q+1$ or fewer is non-empty. It follows that the nerve of this covering has $|I|$ vertices, of which any $q + 1$ form a simplex, so the nerve contains the q-skeleton of an $|I|$-simplex, which is $(q-1)$-connected, by Fact 13.2.2. The Nerve lemma implies that $\mathcal{L} \cap \mathcal{M}$ is $(q-1)$-connected.

Since $\mathcal{L}$ and $\mathcal{M}$ are q-connected, and their intersection is $(q-1)$-connected, their union $\mathcal{L} \cup \mathcal{M} = \Phi(\mathcal{K})$ is q-connected by Lemma 10.4.3, the special-case Nerve lemma.   □

Definition 13.4.3. A *shellable carrier map* $\Phi : \mathcal{K} \to 2^{\mathcal{L}}$ is a rigid strict carrier map such that for each $\sigma \in \mathcal{K}$, the simplicial complex $\Phi(\sigma)$ is shellable.

Assume the simplicial complex $\mathcal{K}$ is pure of dimension n. A shellable carrier map $\Phi : \mathcal{K} \to 2^{\mathcal{L}}$ is an $(n-1)$-connected carrier map, but not necessarily vice-versa.

Lemma 13.4.4. *Consider a sequence of pure complexes and carrier maps*

$$\mathcal{K}_0 \xrightarrow{\Phi_0} \mathcal{K}_1 \xrightarrow{\Phi_1} \mathcal{K}_2,$$

where Φ_0 is a shellable carrier map, and Φ_1 is a q-connected carrier map. Then the composition $\Phi_1 \circ \Phi_0$ is a q-connected carrier map.

Proof. Note that $\Phi_1 \circ \Phi_0$ is a rigid strict carrier map because it is a composition of two rigid strict carrier maps. It remains to check that $(\Phi_1 \circ \Phi_0)(\sigma)$ is $(q - \operatorname{codim} \sigma)$-connected for all σ.

For each $\sigma \in \mathcal{K}_0$, apply Lemma 13.4.2, substituting $\Phi_0(\sigma)$ for $\mathcal{K}$, $q - \operatorname{codim} \sigma$ for q, $\mathcal{K}_2$ for $\mathcal{L}$, and Φ_1 for Φ. Because Φ_0 is a shellable carrier map, $\Phi_0(\sigma)$ is pure and shellable of dimension $\dim \sigma$. Because Φ_1 is a q-connected carrier map, Because Φ_1 is a q-connected carrier map, its restriction to the simplicial complex $\Phi_0(\sigma)$ is $(q - \operatorname{codim} \sigma)$-connected. It follows from Lemma 13.4.2 that $\Phi_1(\Phi_0(\sigma)) = (\Phi_1 \circ \Phi_0)(\sigma)$ is $(q - \operatorname{codim} \sigma)$-connected.   □

Lemma 13.4.5. *Consider a sequence of complexes and carrier maps*

$$K_0 \xrightarrow{\Phi_0} K_1 \xrightarrow{\Phi_1} \cdots \xrightarrow{\Phi_\ell} K_{\ell+1},$$

such that the carrier maps $\Phi_0, \ldots, \Phi_{\ell-1}$ are shellable, and the carrier map Φ_ℓ is q-connected. Then the composition $\Phi_\ell \circ \cdots \circ \Phi_1 \circ \Phi_0$ is a q-connected carrier map.

Proof. By induction on ℓ. The claim is immediate when $\ell = 0$, and the induction step follows directly from Lemma 13.4.4. $\qquad\square$

Theorem 13.4.6. *Let $(\mathcal{I}, \mathcal{P}, \Xi)$ be an $(n+1)$-process protocol, such that $\mathcal{I}$ is shellable. If the rigid strict carrier map Ξ can be decomposed into a sequence of layers, $\Phi_0, \ldots, \Phi_N$, where each Φ_i is a shellable carrier map, then the protocol complex $\mathcal{P}$ is $(n-1)$-connected.*

Proof. The dimension of $\mathcal{I}$ is n, hence all the protocol complexes between the layers are pure of dimension n. In particular, the source simplicial complex of Φ_N is pure of dimension n, so it follows that Φ_N is an $(n-1)$-connected carrier map. Lemma 13.4.5 implies that the composition $\Phi_N \circ \cdots \circ \Phi_1 \circ \Phi_0$ is an $(n-1)$-connected carrier map. Furthermore, since $\mathcal{I}$ is shellable, Lemma 13.4.2 implies that the protocol complex $\Xi(\mathcal{I}) = (\Phi_N \circ \cdots \circ \Phi_1 \circ \Phi_0)(\mathcal{I})$ is $(n-1)$-connected. $\qquad\square$

By Theorem 10.3.1,

Corollary 13.4.7. *If $(\mathcal{I}, \mathcal{P}, \Xi)$ satisfies the conditions of Theorem 13.4.6, then it cannot solve n-set agreement.*

13.5 Applications

In this section, we show how to apply shellability to several layered models of computation, along with Corollary 13.4.7. For each model, we consider an adversary A with minimal core size c. We use two strategies for applying layered decompositions. For asynchronous models, we construct per-layer carrier maps that are shellable and that do not fail any processes, implying that no protocol can solve k-set agreement in a finite number of layers. For synchronous and semisynchronous models, we construct per-layer carrier maps that are shellable and that fail a certain number of processes in each layer, implying a lower bound on the number of layers needed to solve k-set agreement.

For generality, we will use the adversary model introduced in Section 5.4. Recall that an adversary is defined by its cores, where a *core* is a minimal set of processes that will not all fail in any execution. An alternative definition is in terms of its survivor sets, where a *survivor set* is a minimal set of processes that intersects every core.

We will need the following lemma about adversaries.

Lemma 13.5.1. *Assume that we have $n+1$ processes and an adversary A with faulty set complex $\mathcal{F}$. Let c be A's minimal core size, and let s be A's maximal survivor set size. Then we have $n+2 \geq c+s$.*

Proof. Since c is the minimal core size, for all $\sigma \subseteq [n]$, such that $|\sigma| \leq c-1$, we have $\sigma \in \mathcal{F}$. Hence all the facets of $\mathcal{F}$ have size at least $c-1$, implying that $s \leq n+1-(c-1)$, which is the same as to say $n+2 \geq c+s$. $\qquad\square$

13.5.1 Asynchronous message passing

In this model, a set of $n+1$ asynchronous, crash-prone processes communicate by sending messages. As illustrated in Figure 13.6, in each layer, each process sends its state to every process, including itself, and waits until it receives a message for that layer from every process in a survivor set. This model generalizes the message-passing model considered in Section 5.5 to encompass adversaries.

```
// N is number of layers, n + 1 the number of processes
AsynchronousMP(v_i: Value): Value
    view: set of Value := {v_i}          // initial  view is input value
    for ℓ := 0 to N − 1 do
        send(P, view, ℓ) to all
        do  // collect  values  from a  survivor  set
            upon receive(Q, u, ℓ) do
                view : = view ∪ {(Q, u)}
            until view contains a survivor set
    return δ(view)          // apply decision map to final view
```

FIGURE 13.6

Asynchronous message passing: *N*-layer full-information protocol.

Because the model is asynchronous, a message sent from one process to another in a particular layer might not be delivered in that layer. Suppose P sends a message to Q in layer ℓ. If Q receives that message in an earlier layer, then Q buffers the message and delivers it in layer ℓ, whereas if Q receives that message in a later layer, it discards it.

By the model-independent Theorem 13.4.6, it is enough to focus on the single-layer carrier map for this model:

$$\Phi_a(\sigma) = \Psi(\sigma, \{\tau \subseteq \sigma \,|\, \tau \text{ contains a survivor set for A}\}),$$

where $\Psi(\cdot)$ is the pseudosphere operator. Here is how to interpret this map: The input simplex σ is a set of nonfaulty processes, together with their local states. This set includes a survivor set, although it may be larger. Each participating process in σ broadcasts its vertex and waits until it receives a message for that layer from every process in a survivor set, discarding any messages arriving late from earlier layers. Each message received is a vertex of σ, and the set of messages received in that layer form a face τ of σ, where τ contains a survivor set. Because each process can be assigned any such face independently, the resulting complex is a pseudosphere. This pseudosphere is constructed over σ itself because no process in σ fails in this execution. The pseudosphere is generated by considering all possible ways in which messages can be delayed. Later, we will consider single-layer carrier maps in which processes do fail.

Note that if σ does not contain a survivor set, then we define $\Phi_a(\sigma)$ to be empty. If the codimension of σ is less than c, the minimum core size, then σ contains a survivor set, so $\Phi_a(\sigma)$ is nonempty.

Lemma 13.5.2.   *The single-layer carrier map Φ_a is a shellable carrier map.*

Proof.   The complex $\Phi_a(\sigma)$ is non-empty only when $\text{codim}\,\sigma < c$, in which case the complex is a non-empty pseudosphere, which is pure and shellable by Theorem 13.3.6.   □

From Theorem 13.4.6,

Theorem 13.5.3.   *If $(\mathcal{I}, \mathcal{P}, \Xi)$ is an asynchronous message-passing protocol against an adversary with minimum core size c, then for all $\sigma \in \mathcal{I}$, $\Xi(\sigma)$ is $(c - 2 - \text{codim}\,\sigma)$-connected, and the protocol complex $\mathcal{P}$ is $(c - 2)$-connected.*

From Theorem 10.3.1,

Corollary 13.5.4.   *There is no asynchronous message-passing protocol for $(c - 1)$-set agreement against an adversary with minimum core size c.*

This bound is tight. Here is an asynchronous message-passing protocol for c-set agreement. Pick a core C of size c, and have each member broadcast its input value. Each process waits until it hears a value from a process in C and decides that value. The protocol must terminate because the processes in C cannot all fail, and no more than c distinct values may be chosen.

13.5.2 Synchronous message passing

In the synchronous message-passing model with crash failures, there are $n + 1$ processes, where each process's initial state consists of its input value. Computation proceeds in a sequence of layers. In each layer, each process sends its state to all the other processes and receives states from the others. A process's new state is the set of messages it received in that layer. A process can fail by crashing during a layer, in which case it may send its state to an arbitrary subset of the processes in that layer. In later layers, the crashed process never sends another message. Figure 13.7 shows pseudo-code for a protocol in this model.

We want to derive a lower bound on the number of layers needed to solve k-set agreement against an adversary **A** in this model. We will show that there is no N-layer protocol for k-set agreement if $n \geq (N + 1)k$ and $Nk < c$. By Theorem 13.4.6, the lower bound question can be rephrased: For how many layers can **A** maintain a shellable carrier map?

Because we are proving a lower bound, we can restrict our attention to any subset of the executions permitted by **A**. In particular, we consider the executions in which exactly k processes fail in each layer. We can now motivate our constraints on n, k, and N. The constraint $n \geq (N + 1)k$ ensures that at least $k + 1$ nonfaulty processes survive to the end of N layers, which is necessary because otherwise k-set agreement would be possible in these restricted executions. (The matching lower bound shows that this

```
// N is number of layers, n + 1 the number of processes
SynchronousMP(v_i: Value): Value
    view: set of Value := {v_i}          // initial view is input value
    for ℓ := 0 to N − 1 do
        send(P, view, ℓ) to all
        do // collect values from a survivor set
            upon receive(Q, u, ℓ) do
                view := view ∪ {(Q, u)}
            until all messages for layer ℓ received
    return δ(view)          // apply decision map to final view
```

FIGURE 13.7

Synchronous message passing: *N*-layer full-information protocol.

assumption does not sacrifice generality.) The constraint $Nk < c$ ensures that the cumulative number of failures does not exceed c, ensuring that all such executions are permitted by **A**.

Unlike the asynchronous model, where each layer uses the same carrier map, we will define a distinct carrier map for each layer:

$$\mathcal{K}_0 \xrightarrow{\Phi_0} \mathcal{K}_1 \xrightarrow{\Phi_1} \cdots \xrightarrow{\Phi_{N-1}} \mathcal{K}_N,$$

where $\mathcal{K}_0$ is the input complex, each $\mathcal{K}_i$ is the image of $\mathcal{K}_{i-1}$ under Φ_{i-1}, and

$$\Phi_i(\sigma) = \bigcup_{\tau \in \mathrm{Faces}_{n-(i+1)k}\sigma} \Psi(\tau, [\tau, \sigma]). \tag{13.5.1}$$

For each $\mathcal{K}_i$, and each σ in $\mathcal{K}_i$, $\tau \subseteq \sigma$ is the subset consisting of processes that do not fail. Each non-faulty process receives a message from all the nonfaulty processes in τ and from an arbitrary subset of the faulty processes in $\sigma \setminus \tau$, yielding a face ϕ of σ *between* τ and σ: $\tau \subseteq \phi \subseteq \sigma$. Recall that $[\tau, \sigma]$ denotes the set of simplices between τ and σ. For a given τ, these faces are assigned independently; hence the resulting complex is a pseudosphere. Because the cumulative number of failures is $i \cdot k < c$, each facet of $\Phi_i(\sigma)$ corresponds to an execution of **A**.

Figure 13.8 shows the single-layer protocol complex for processes P_0, P_1, and P_2, where exactly one failure occurs in each execution, operating on a single input complex. The complex is the union of three pseudospheres, one for each possible failure. Each vertex is colored black, gray, or white for P_0, P_1, or P_2 and is labeled with the process names from which messages were received (for example, "01" means that messages were received from P_0 and P_1 but not P_2).

We now construct a shelling order for $\Phi_i(\sigma)$. We assign each facet ϕ a *signature*, which is a vector of faces of σ whose i^{th} entry, written $\phi[i]$, is defined as follows:

$$\phi[q] = \begin{cases} \tau & \text{if } (P_q, \tau) \in \phi, \\ \top & \text{otherwise.} \end{cases}$$

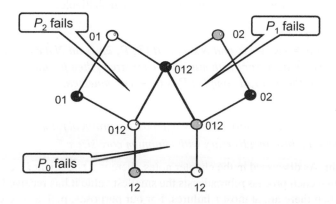

FIGURE 13.8

Synchronous message-passing protocol complex for three processes and one failure. Each vertex is colored to indicate its process name and is labeled with indexes of processes from which messages were received.

Each facet has a unique signature, and vice versa. Facets are ordered lexicographically by their signatures: Faces of σ are compared by the face ordering, so for any face $\tau \subset \sigma, \sigma < \tau < \top$.

Lemma 13.5.5. *For each facet σ of $\mathcal{K}_{i-1}$, the signatures define a shelling order on the facets of $\Phi_i(\sigma)$.*

Proof. Let $\phi_0, \ldots, \phi_t$ be the induced order on the facets of $\Phi_s(\sigma)$.

Let ϕ_i and ϕ_j be facets where $i < j$. We need to find $\phi_k, k < j$, such that $\phi_i \cap \phi_j \subseteq \phi_k \cap \phi_j$, and $|\phi_j \setminus \phi_k| = 1$. Let ℓ be the least index such that $\phi_i[\ell] < \phi_j[\ell]$. Because $\top$ is maximal in this ordering, $\phi_i[\ell] \neq \top$, and because σ is minimal, $\phi_i[\ell] \neq \sigma$.

There are two cases. First, suppose $\phi_j[\ell] \neq \top$. Construct ϕ_k by replacing P_ℓ's label with σ:

$$\phi_k[q] = \begin{cases} \sigma & \text{if } q = \ell, \text{ and} \\ \phi_j[q] & \text{otherwise.} \end{cases}$$

We now check the shellability conditions. Because ϕ_j and ϕ_k differ only at element ℓ, and $\phi_k[\ell] < \phi_j[\ell]$, $\phi_k < \phi_j$ and $\phi_k[\ell] < \phi_j[\ell]$. Finally, $\phi_j \cap \phi_i \subseteq \phi_j \cap \phi_k$ because $v_\ell = \phi_k \setminus \phi_j$ is not in ϕ_i.

Second, suppose $\phi_j[\ell] = \top$. Because ϕ_i and ϕ_j have the same dimension, there must be an index $m > \ell$ such that $\phi_i[m] = \top$ and $\phi_j[m] \neq \top$. Construct ϕ_k from ϕ_j by replacing the entry ℓ with σ and entry m with $\top$:

$$\phi_k[q] = \begin{cases} \sigma & \text{if } q = \ell \\ \top & \text{if } q = m, \text{ and} \\ \phi_j[q] & \text{otherwise.} \end{cases}$$

The first element at which ϕ_j and ϕ_k disagree is ℓ, and $\phi_k[\ell] < \phi_j[\ell]$, so $\phi_k < \phi_j$. Because m is the only entry in ϕ_j but not in ϕ_i, $|\phi_j \setminus \phi_k| = 1$. Because that entry is also not in ϕ_i, $\phi_j \cap \phi_i \subseteq \phi_j \cap \phi_k$. $\square$

Lemma 13.5.6. *For $0 \leq i \leq N$, Φ_i is a shellable carrier map.*

Proof. By Lemma 13.5.5, for all $\sigma \in \mathcal{K}_i$, $\Phi_i(\sigma)$ is pure and shellable. $\square$

From Theorem 13.4.6,

Theorem 13.5.7. *Let $n \geq (N + 1)k$ and $Nk < c$. If $(\mathcal{I}, \mathcal{P}, \Xi)$ is an N-layer synchronous message-passing protocol against an adversary with minimum core size c, then for all $\sigma \in \mathcal{I}$, $\Xi(\sigma)$ is $(k - 1 - \text{codim } \sigma)$-connected, and the protocol complex $\mathcal{P}$ is $(k - 1)$-connected.*

From Theorem 10.3.1,

Corollary 13.5.8. *There is no synchronous message-passing protocol for k-set agreement that decides in $\lfloor \frac{c-1}{k} \rfloor$ layers or less against an adversary with minimum core size c.*

This bound is tight. As discussed in the chapter notes, there is a t-resilient k-set agreement protocol for this model in which each process rebroadcasts the smallest value it has received. This protocol runs in $\lfloor \frac{t}{k} \rfloor + 1$ layers when there are at most t failures. For our purposes, pick a core C of minimal size c and have the processes in C run a $(c - 1)$-resilient k-set agreement protocol. The remaining processes "eavesdrop" on the protocol's messages and make the same final decisions as the participants. The result is a k-set agreement protocol that runs in $\lfloor \frac{c-1}{k} \rfloor + 1$ layers.

13.5.3 Asynchronous snapshot memory

We now turn our attention from message-passing to shared memory. This model is similar to the one considered in Section 5.4, except that we will use separate write and snapshot operations instead of immediate snapshots. As shown in Figure 13.9, the processes share a two-dimensional array mem[·][·], indexed by layer number and by process name. Each memory location is initialized to $\perp$. In layer ℓ, each P_i writes its state to mem[ℓ][i] and takes repeated snapshots of row ℓ until the set of processes that have written at that layer includes a survivor set.

Let σ be an n-simplex where each vertex is labeled with a distinct process name.

Definition 13.5.9. A *survivor chain* for σ is a sequence of faces of σ: $\sigma_0, \ldots, \sigma_k$ such that names(σ_0) contains a survivor set for A, and $\sigma_0 \subset \cdots \subset \sigma_k = \sigma$.

We denote the set of survivor chains for σ by Chains(σ).

Figure 13.10 shows an example of a 3-process survivor chain $\overline{\sigma} = (\sigma_0, \sigma_1, \sigma_2)$, where σ_0 is a vertex labeled with process P (shown as black), σ_1 is an edge labeled with P and Q (black and gray), and σ_2 is a solid triangle labeled with P, Q, and R (black, gray, and white). In this example, we assume $\{P\}$ is a survivor set.

For a survivor chain $\overline{\sigma} = \sigma_0 \subset \cdots \subset \sigma_k$, and $P_i \in \Pi$, let $\overline{\sigma}_i$ be the suffix of $\overline{\sigma}$ of sets containing P_i. The single-layer carrier map is defined by:

$$\Phi_m(\sigma) = \bigcup_{\overline{\sigma} \in \text{Chains}(\sigma)} \Psi(\sigma; \overline{\sigma}_0, \ldots, \overline{\sigma}_{\dim \sigma}).$$

Here is how to interpret this carrier map: Each process in σ writes its vertex to the memory, waits until a survivor set has written, and then takes a snapshot, yielding a face of σ. The order in which vertices are written defines a sequence of faces of σ ordered by inclusion, starting with a single vertex and ending with σ itself. The suffix of this sequence consisting of faces that contain a survivor set defines a survivor chain $\overline{\sigma}$. The snapshot by P_i returns a simplex from this chain. Because P_i must read its own vertex, the simplex it chooses must be an element of $\overline{\sigma}_i$. Except for that constraint, however, each P_i independently chooses a simplex from $\overline{\sigma}_i$, so the complex is a pseudosphere.

```
// N is number of layers, n + 1 the number of processes
AsynchronousRW(v_i: Value): Value
  view: set of Value := {v_i}          // initial view is input value
  for ℓ := 0 to N − 1 do
    mem[ℓ][i] := view
    do  // collect values from a survivor set
      view := snapshot(mem[ℓ][*])
    until view contains a survivor set
  return δ(view)       // apply decision map to final view
```

FIGURE 13.9

Asynchronous snapshot memory: *N*-layer full-information protocol for P_i.

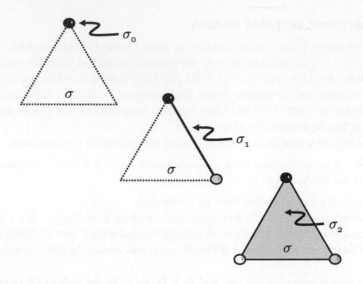

FIGURE 13.10

Example survivor chain $\bar{\sigma} = (\sigma_0, \sigma_1, \sigma_2)$.

Following Definition 13.2.1, each facet of $\Phi_m(\sigma)$ has a signature, and facets are ordered by their signatures.

Lemma 13.5.10.   *For any input simplex σ containing a survivor set, Φ_m is a shellable carrier map.*

Proof.   Facets are ordered by their signatures. Index the facets $\phi_0, \ldots, \phi_t$ in signature order. As usual, if ϕ_i and ϕ_j are facets where $i < j$, we need to find $\phi_k, k < j$, such that $\phi_i \cap \phi_j \subseteq \phi_k \cap \phi_j$, and $|\phi_j \setminus \phi_k| = 1$. Let ℓ be the least index such that $\phi_i[\ell] < \phi_j[\ell]$ in the face ordering. Because σ is minimal in the face ordering, ϕ_i is a proper face of σ. Construct ϕ_k to have the following signature:

$$\phi_k[q] = \begin{cases} \sigma & \text{if } q = \ell, \\ \phi_j[q] & \text{otherwise.} \end{cases}$$

Note that (1) $\phi_k \in \Phi_m(\sigma)$ because replacing any element of a survivor chain with σ is still a survivor chain; (2) $\phi_k < \phi_j$ because ℓ is the least index where their signatures differ, and $\sigma < \phi_j[\ell]$; (3) $|\phi_j \setminus \phi_k| = 1$ because they differ in only one vertex, and (4) $\phi_i \cap \phi_j \subseteq \phi_k \cap \phi_j$, because that vertex is not in ϕ_i.   □

From Theorem 13.4.6,

Theorem 13.5.11.   *If $(\mathcal{I}, \mathcal{P}, \Xi)$ is an asynchronous snapshot memory protocol against an adversary with minimum core size c, then for all $\sigma \in \mathcal{I}$, $\Xi(\sigma)$ is $(c - 2 - \operatorname{codim} \sigma)$-connected, and the protocol complex $\mathcal{P}$ is $(c - 2)$-connected.*

From Theorem 10.3.1,

Corollary 13.5.12.   *There is no asynchronous snapshot memory protocol for $(c - 1)$-set agreement against any adversary with minimum core size c.*

This bound is tight. Here is a simple c-set agreement protocol against this adversary. Pick a core C of size c. Each process in C writes its input to a distinguished shared memory location, initialized to hold a null value. Each process chooses the first nonnull value it reads from a process in C. This protocol must terminate because C is a core, and the adversary cannot fail every process in C.

13.5.4 Semisynchronous message passing

In this model, a set Π of $n + 1$ processes exchange messages, but the time between two consecutive process steps is at least d_0 and at most d_1, and the time to deliver a message is at most d_m, where $d_m \gg d_0$. Because d_m is much larger than d_0, we will assume d_m is an exact multiple of d_0 to avoid tedious round-off calculations. The values d_0, d_1, and d_m are known constants. Let $d_s = d_1/d_0$. Failures and timing are controlled by an adversary A with minimal core size c. We will derive a lower bound on the time needed to solve k-set agreement against A in this model.

We first use topological arguments to prove a lower bound for a wait-free adversary. We then use an argument by reduction to extend the bound to general adversaries.

A *fast* execution is defined as follows. Each layer takes exactly time d_m. Each process sends a message in each layer, and the messages sent during a layer are delivered at the very end of that layer (at multiples of time d_m). All processes take steps in lock-step as quickly as possible (at multiples of time d_0). The interval between process steps is called a *microlayer*, and there are $\mu = d_m/d_0$ microlayers per layer. Within a layer, microlayers are indexed from 0 to $\mu - 1$, and each message is labeled with the index of the microlayer in which it was sent. This model is illustrated in Figure 13.14.

A *failure pattern* is a sequence of $n + 1$ integers, each between 0 and μ. The ith element of the sequence is the microlayer in which P_i fails, or μ if it does not fail. The *view* of a process P_i at the end of a layer is the failure pattern it observed, identifying each P_j to the first microlayer in which P_i fails to receive a message from P_j, or to μ if no messages are missing. Note that a process's view may not be the same as the actual failure pattern, because a process that fails in microlayer m may still send messages during that microlayer.

Failure patterns are ordered lexicographically. If ℓ is the first index where failure patterns F and G disagree, then $F < G$ if and only if $F(\ell) > G(\ell)$ (note the reversed comparison). The minimal failure pattern, denoted $\tilde{F}$, assigns μ to each process and corresponds to an execution in which no failures were observed.

To ensure that all complexes are pure, we consider only a subset of the possible executions. (Recall that any impossibility result that holds for a restricted adversary also holds for a more powerful adversary.) First, we consider executions in which exactly k processes fail in each layer and all faulty processes fail in the same microlayer. Let $F(\tau, m)$ be the failure pattern in which all processes not in τ fail at microlayer m. Finally, we require that if the faulty processes fail in microlayer m, then for every non-faulty process $P_i \in \text{names}(\tau)$, either P_i observes all failures to have occurred in microlayer m (that is, failure pattern $F(\tau, m)$) or it observes them in microlayer $m + 1$ (failure pattern $F(\tau, m + 1)$).

Although these restrictions are not necessary to prove shellability, they substantially reduce the size of the complex and simplify our analysis.

As in the synchronous case, these executions define a sequence of complexes and carrier maps

$$\mathcal{K}_0 \xrightarrow{\Phi_0} \mathcal{K}_1 \xrightarrow{\Phi_1} \cdots \xrightarrow{\Phi_{N-1}} \mathcal{K}_N,$$

where $\mathcal{K}_0$ is the input complex, and each $\mathcal{K}_i$ is the image of $\mathcal{K}_{i-1}$ under Φ_{i-1}. As in the synchronous case, we assume $n > (N+1)k$ and $Nk < c$. We now construct $\Phi_i(\cdot)$.

As illustrated in Figure 13.11, the protocol complex for the single-layer execution under failure pattern $F(\tau, m)$ is just the pseudosphere

$$\Phi(\tau, m) = \Psi(\tau, \{F(\tau, m), F(\tau, m+1)\}) \simeq \Psi(\tau, \{0, 1\}).$$

(Topologically, this complex is a combinatorial sphere of the same dimension as τ.) Moreover, it is a shellable complex, whose minimal facet in the lexicographic order associates each vertex of τ with $F(\tau, m+1)$ and whose maximal facet associates each vertex with $F(\tau, m)$.

The following complex encompasses all the executions where the processes in τ do not fail:

$$\Phi(\tau) = \bigcup_{m=0}^{\mu} \Phi(\tau, m).$$

As illustrated in Figure 13.12, for a fixed τ, as m ranges from 0 to μ, the complexes $\Phi(\tau, m)$ are linked in a chain, where the maximal facet of each $\Phi(\tau, m)$ is identified with the minimal facet of $\Phi(\tau, m+1)$.

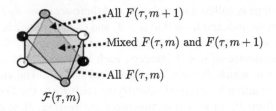

FIGURE 13.11

The octahedron $\Phi(\tau, m)$ represents the executions where the faulty processes all fail at microlayer m. In the top face, the three nonfaulty processes observe the failures at microlayer $m+1$ and in the bottom face at microlayer m. In the other faces, the observations are mixed.

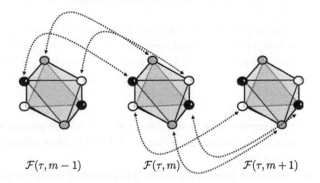

FIGURE 13.12

An exploded view of $\Phi(\tau)$. Vertices linked by dashed arrows are the same: The octahedrons $\Phi(\tau, m)$ and $\Phi(\tau, m+1)$ intersect at the top face of $\Phi(\tau, m)$, which is also the bottom face of $\Phi(\tau, m+1)$.

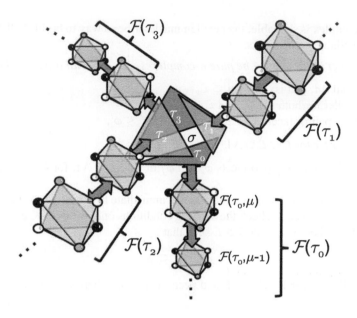

FIGURE 13.13

A schematic view of $\Phi(\sigma)$. Each τ_i is an $(n-k)$-face of σ. The chain of octahedron extending from each τ_i represents $\Phi(\tau_i)$, the executions in which the processes in τ_i are nonfaulty. Each octahedron represents $\Phi(\tau_i, m)$, the executions in which the faulty processes all fail at microlayer m. The chains come together in the k-skeleton of σ, representing the executions where each nonfaulty process observes the failure pattern $\tilde{F}$, where no failures are detected.

Finally, in the full single-layer protocol complex, illustrated schematically in Figure 13.13, the components range over all $(n-k)$-faces of σ:

$$\Phi_i(\sigma) = \bigcup_{\tau \in \mathrm{Faces}^{n-(i+1)\cdot k}(\sigma)} \Phi(\tau).$$

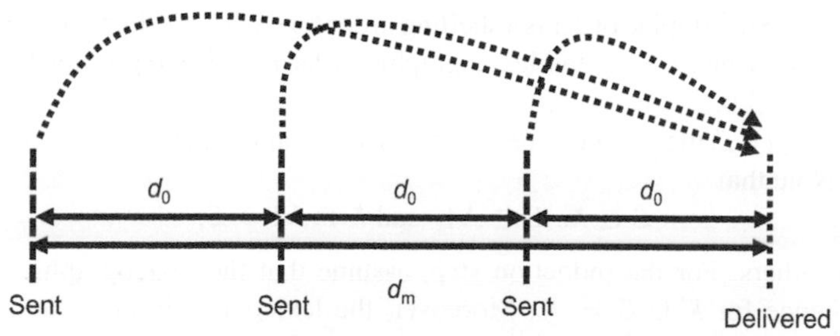

FIGURE 13.14

Each layer takes time exactly d_m, and all messages sent during a layer are delivered at the end of that layer. Processes run in lock-step at maximum speed, taking a step at intervals of duration d_0.

To show that this complex is shellable, the next lemma describes a way to build shellable complexes by "concatenating" shellable components.

Lemma 13.5.13. *Let $\mathcal{K}$, $\mathcal{L}$, and $\mathcal{M}$ be pure n-complexes and $<$ a total order on their facets. If*

- $\mathcal{L}$ *separates* $\mathcal{K}$ *and* $\mathcal{M}$, *that is:* $\mathcal{K} \cap \mathcal{M} \subseteq \mathcal{L}$,
- $<$ *is a shelling order for both* $\mathcal{K} \cup \mathcal{L}$ *and* $\mathcal{L} \cup \mathcal{M}$, *and*
- $\mathcal{L} \leq \mathcal{M}$: *for every pair of facets* ϕ *of* $\mathcal{L}$ *and* ϕ' *of* $\mathcal{M}$, $\phi \leq \phi'$,

then $<$ is a shelling order for $\mathcal{K} \cup \mathcal{L} \cup \mathcal{M}$.

Proof. We must show that any two facets ϕ_i and ϕ_j of $\mathcal{K} \cup \mathcal{L} \cup \mathcal{M}$, for $i < j$, satisfy the shelling conditions.

If $\phi_i, \phi_j \in \mathcal{K} \cup \mathcal{L}$ or $\phi_i, \phi_j \in \mathcal{L} \cup \mathcal{M}$, the claim is immediate because $<$ is a shelling order for both $\mathcal{K} \cup \mathcal{L}$ and $\mathcal{L} \cup \mathcal{M}$, so we need to check the shelling conditions only for $\phi_i \in \mathcal{K}$ and $\phi_j \in \mathcal{M}$. Because $\mathcal{L}$ separates $\mathcal{K}$ and $\mathcal{M}$, there is a facet $\phi \in \mathcal{L}$ such that

$$\phi_j \cap \phi_i \subseteq \phi$$

Because $<$ is a shelling order for $\mathcal{L} \cup \mathcal{M}$, and because $\phi < \phi_i$, there is a facet $\phi_k \in \mathcal{L} \cup \mathcal{M}$, where $k < j$, such that

$$\phi \cap \phi_i \subseteq \phi_k \cap \phi_i$$
$$|\phi_j \setminus \phi_k| = 1.$$

Putting these observations together,

$$\phi_j \cap \phi_i \subseteq \phi$$
$$\phi_j \cap \phi_i \subseteq \phi \cap \phi_i$$
$$\phi_j \cap \phi_i \subseteq \phi_k \cap \phi_i,$$

satisfying the shelling conditions. □

We now use several steps to construct a shelling order for $\Phi_i(\sigma)$.

Lemma 13.5.14. *The lexicographic order $<$ is a shelling order for each $\Phi(\tau)$.*

Proof. We show the lexicographic order is a shelling order for each $\cup_{i=0}^{\ell} \Phi(\tau, i)$. We argue by induction on ℓ. For the base case, when $\ell = 0$, the lexicographic order is a shelling order for the pseudosphere $\Phi(\tau, 0)$ by Theorem 13.3.6.

Let $\mathcal{K}$ be $\cup_{i=0}^{\ell-1} \Phi(\tau, i)$, $\mathcal{L}$ the simplex $\{(P, F(\tau, \ell)) | P \in \text{names}(\tau)\}$ and its faces, and $\mathcal{M}$ the pseudosphere $\Phi(\tau, \ell)$. Note that

$$\mathcal{L} \subset \mathcal{K}, \mathcal{L} \subset \mathcal{M}, \text{ and } \mathcal{K} \cap \mathcal{M} = \mathcal{L},$$

so $\mathcal{L}$ separates the others. For the induction step, assume that the lexicographic order is a shelling order for $\mathcal{K}$, and hence for $\mathcal{K} \cup \mathcal{L} = \mathcal{K}$. Moreover, the lexicographic order is a shelling order for $\mathcal{L} \cup \mathcal{M} = \mathcal{M}$ because $\mathcal{M}$ is a pseudosphere. Because $\mathcal{L}$ is the minimal facet of $\mathcal{M}$, $\mathcal{L} \leq \mathcal{M}$. It follows from Lemma 13.5.13 that the lexicographic order is a shelling order for $\mathcal{K} \cup \mathcal{L} \cup \mathcal{M} = \cup_{i=0}^{\ell} \Phi(\tau, i)$. □

Let $\sigma = \{s_0, \ldots, s_n\}$. Earlier, we introduced the following lexicographical order on $(n-k)$-faces of σ. If τ_0 and τ_1 are $(n-k)$-faces of σ, and ℓ is the least index at which their signatures disagree, then $\tau_0 < \tau_1$ if and only if $\tau_0[\ell] = \bot$ (that is, $s_\ell \in \tau_0$) and $\tau_1[\ell] = \top$ ($s_\ell \notin \tau_1$). Let $<$ be the following total order on facets of $\Phi_i(\sigma)$: Order every facet of $\Phi(\tau_0)$ before any facet of $\Phi(\tau_1)$, and within each $\Phi(\tau)$, order the facets lexicographically.

Lemma 13.5.15.   *The order $<$ is a shelling order for $\Phi_i(\sigma)$.*

Proof.   Let $\tau_0, \ldots, \tau_t$ be the $(n-k)$-faces of σ indexed in lexicographic order. We argue by induction on ℓ that the order $<$ is a shelling order for $\cup_{m=0}^{\ell}\Phi(\tau_m)$.

For the base case, when $\ell = 0$, the lexicographic order is a shelling order for the complex $\Phi(\tau_0)$ by Lemma 13.5.14.

Let $\mathcal{K}$ be $\cup_{m=0}^{\ell-1}\Phi(\tau_m)$, $\mathcal{L}$ the simplex $\left\{(P, \tilde{F}) | P \in \text{names}(\tau_\ell)\right\}$ and its faces, and $\mathcal{M}$ the complex $\Phi(\tau_\ell)$. Note that the unique facet of $\mathcal{L}$ is the lexicographically minimal facet of $\mathcal{M} = \Phi(\tau_\ell)$.

For the induction step, assume that $<$ is a shelling order for $\cup_{m=0}^{\ell-1}\Phi(\tau_m)$. Let ϕ_i and ϕ_j be facets of $\cup_{m=0}^{\ell}\Phi(\tau, m)$, where $i < j$. The shelling properties are immediate if $\phi_i, \phi_j \in \mathcal{K}\cup\mathcal{L}$ or $\phi_i, \phi_j \in \mathcal{L}\cup\mathcal{M}$, so consider the case where $\phi_i \in \mathcal{K}$ and $\phi_j \in \mathcal{M}$.

Suppose $\phi_i \in \Phi(\tau_i, m_i) \subset \mathcal{K}$, $\phi_j \in \Phi(\tau_j, m_j) \subset \mathcal{M}$, and $\phi_i \cap \phi_j \neq \emptyset$. Because τ_i and τ_ℓ are distinct n-simplices, and $\tau_i < \tau_\ell$ in the lexicographic order, there is an index p such that $\tau_i[p] \neq \top$ and $\tau_j[p] = \top$ and an index $q > p$ such that $\tau_i[q] = \top$ and $\tau_j[q] \neq \top$. Because $\phi_i \cap \phi_j \neq \emptyset$, there is an index r such that $\phi_i[r] = \phi_j[r] \neq \bot$.

We claim that $\mathcal{K} \cap \mathcal{M} \subseteq \mathcal{L}$. Every failure pattern labeling a vertex in ϕ_i assigns microlayer $\mu + 1$ to every process in $\text{names}(\tau_i)$ and the same microlayer m_i to every other process. Similarly, every failure pattern labeling a vertex in ϕ_j assigns microlayer $\mu + 1$ to every process in $\text{names}(\tau_j)$ and the same microlayer m_j to every other process. In the non-empty intersection, $\phi_i[r](p) = \mu + 1 = \phi_j[r](p) = m_j$, and $\phi_j[r](q) = \mu+1 = \phi_i[q] = m_i$. As a result, every failure pattern in $\phi_j \cap \phi_i$ assigns microlayer $\mu + 1$ to every process, implying that $\mathcal{K} \cap \mathcal{M} \subseteq \mathcal{L}$.

Next, we claim that $<$ is a shelling order for both $\mathcal{K} \cup \mathcal{L}$ and $\mathcal{L} \cup \mathcal{M}$. By the induction hypothesis, $<$ is a shelling order for $\mathcal{K}$, so we need to check the shelling conditions only for $\phi_i \in \mathcal{K}$ and $\phi_j \in \mathcal{L}$. Define ϕ_k by the following signature:

$$\phi_k[m] = \begin{cases} \tilde{F} & \text{If } m = p \\ \top & \text{If } m = q \\ \phi_j[m] & \text{otherwise.} \end{cases}$$

It is easy to check the shelling conditions: $\phi_k < \phi_j$, $\phi_i \cap \phi_j \subseteq \phi_k \cap \phi_j$, and $|\phi_j \setminus \phi_k| = 1$. It follows that $<$ is a shelling order for $\mathcal{K} \cup \mathcal{L}$. Because $\mathcal{L} \subset \mathcal{M}$ and $<$ is a shelling order for $\mathcal{M}$, it follows that $<$ is also a shelling order for $\mathcal{L} \cup \mathcal{M}$.

Finally, we claim that $\mathcal{L} \leq \mathcal{M}$. $\mathcal{L}$ has a single facet, in which the vertices of τ_ℓ are labeled with the minimal failure pattern $\tilde{F}$, which is lexicographically less than any other facet of $\Phi(\tau_\ell) = \mathcal{M}$.

By Lemma 13.5.13, $<$ is a shelling order for $\mathcal{K} \cup \mathcal{L} \cup \mathcal{M} = \cup_{m=0}^{\ell}\Phi(\tau_m)$.   □

Lemma 13.5.16.   *For $0 \leq i \leq N$, Φ_i is a shellable carrier map.*

Proof.   We note that $\mathcal{K}_i = \Phi_{i-1}(\mathcal{K}_{i-1})$ is the union of pseudospheres, and is therefore pure and shellable by Lemma 13.5.15 .   □

From Theorem 13.4.6,

Theorem 13.5.17. *Let $N < \frac{n}{k}$. If $(\mathcal{I}, \mathcal{P}, \Xi)$ is an N-layer semisynchronous message-passing protocol against a wait-free adversary, then for all $\sigma \in \mathcal{I}$, $\Xi(\sigma)$ is $(k-1-\text{codim } \sigma)$-connected, and the protocol complex $\mathcal{P}$ is $(k-1)$-connected.*

From Theorem 10.3.1,

Corollary 13.5.18. *If $N < \frac{n}{k}$, then no N-layer protocol can solve k-set agreement in time $N \cdot d_m$ against a wait-free adversary.*

This lower bound of Nd_m is considered short. We now show how to add a "stretched" layer at the end of length $d_s d_m$, which is much longer. Assume for now that $n + 1 = (N + 1)k + 1$. We will relax this assumption later.

Lemma 13.5.19. *There is no semisynchronous message-passing protocol for k-set agreement against the wait-free adversary for $(N + 1)k + 1$ processes that runs in time less than $Nd_m + d_s d_m$.*

Proof. Let $\Phi_\epsilon^{N+1}(\mathcal{I})$ denote the protocol complex at time $(N + 1)d_m - \epsilon$ for $d_m > \epsilon > 0$ at time ϵ before the start of layer $N + 1$. (See Figure 13.15.) No process can decide at this time because no process has received a message since time Nd_m, so any decision it could make after waiting without a message could have been made when it received its last message at time Nd_m, contradicting Corollary 13.5.18.

Let $\Phi_\infty^{N+1}(\mathcal{I})$ denote the protocol complex corresponding to the following executions: Run a fast execution for N layers as before, but at the start of layer $N + 1$, fail all processes but P, and run P as slowly as possible, taking steps at multiples of time d_1. At time $Nd_m + d_s d_m$, P will time out, but at time $Nd_m + d_s d_m - \epsilon$, this execution is indistinguishable to P from the corresponding execution in $\Phi_\epsilon^{N+1}(\mathcal{I})$, and therefore $\Phi_\infty^{N+1}(\mathcal{I})$ cannot solve k-set agreement. (See Figure 13.16.) $\square$

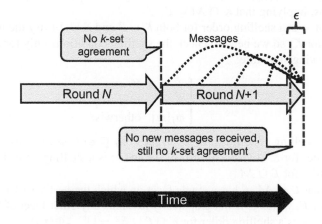

FIGURE 13.15

After N fast layers, at time Nd_m, no k-set agreement protocol is possible. At time $(N+1)d_m - \epsilon$, the adversary can ensure that no messages sent in layer $N + 1$ have been delivered, hence no k-set agreement protocol is possible.

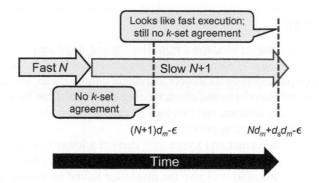

FIGURE 13.16

Take the execution shown in Figure 13.15 and "stretch" the last layer, failing all processes but one. That process will time out after time $Nd + Cd$, but at time ϵ it times out. The execution is indistinguishable from the fast execution where that process is unable to decide.

So far, our analysis applies only to the wait-free adversary. To extend this result to a general adversary A, pick a core C of minimal size c and a set S that intersects C in a single process such that $S \cup C = \Pi$, the complete set of processes.

Theorem 13.5.20.   *There is no semisynchronous message-passing protocol for k-set agreement against an adversary with minimum core size $c = (N + 1)k + 1$ that runs in time less than $Nd_m + d_s d_m$.*

Proof.   Assume, by way of contradiction, that such a protocol exists. We will show that any such protocol can be transformed into a wait-free $(c + 1)$-process protocol that completes in that time, contradicting Lemma 13.5.19.

Pick a core C of minimal size c, and a set S that intersects C in a single process P, such that $S \cup C = \Pi$, the complete set of processes. Consider the following restricted set of executions:

- The processes in S take steps simultaneously, including send and receive steps.
- If any process in S crashes, they all crash simultaneously.
- When a message is delivered to one process in S, it is simultaneously delivered to all processes in S.
- When a message is received from one process in S, it is simultaneously received from all processes in S.
- Messages between processes in S are delivered quickly, at multiples of d_m/d_s steps.

Because the processes in S start with the same inputs and receive the same full-information messages at the same time, they send the same messages each time, and if one can decide an output value, so can the rest.

Because the processes in S behave identically in these executions, they can be simulated by a single process P in a wait-free $(c + 1)$-process protocol. When a process Q receives a message from P, it acts as if it had received simultaneous messages from all processes in S. If the simulated $(n + 1)$-process protocol against the adversary A can decide in time less than $Nd_m + d_s d_m$, so can the simulated $(c + 1)$-process protocol against the wait-free adversary, contradicting Lemma 13.5.19. $\square$

13.6 Chapter notes

Much of the material in this chapter is adapted from Herlihy and Rajsbaum [86].

Fischer, Lynch, and Paterson [55] used operational arguments to show that asynchronous consensus is impossible in message-passing systems subject to a single failure. Biran, Moran, and Zaks [19] later recast this argument in combinatorial terms, characterizing which tasks can be solved in terms of graph connectivity. The later use of combinatorial topology to analyze a broader class of problems [91] can be viewed as a generalization of their approach.

Dolev and Strong [48] and Fischer and Lynch [57] derived a lower bound of $t + 1$ layers to solve consensus in a synchronous message-passing system subject to t failures, either crash failures or Byzantine failures. Junqueira and Marzullo [98] give the first lower bound on consensus expressed in terms of cores and survivor sets, and Wang and Song [144] give a simple $(c - 1)$-layer consensus protocol running against and adversary with minimum core size c. The first lower bound for k-set agreement in the synchronous message-passing model is due to Chaudhuri, Herlihy, Lynch, and Tuttle [39], who used simplicial complexes and Sperner's lemma to show a tight lower bound of $\frac{f}{k} + 1$ layers in systems subject to f failures.

Loui and Abu-Amara [110] were the first to show that consensus is impossible in asynchronous-write memory. Herlihy [78] defined the notion of *consensus number* to characterize the synchronization power of other shared-memory primitives. The first proof that k-set agreement is impossible in asynchronous read-write memory is due to three simultaneous papers by Borowsky and Gafni [23], who used a graph-theoretic model; Saks and Zaharoglou [135], who used a model based on point-set topology; and Herlihy and Shavit [91].

Attiya, Dwork, Lynch, and Stockmeyer [13], as discussed earlier, derived a lower bound of time $(f - 1)d_m + d_s d_m$ for consensus and an algorithm that takes time $2f d_m + d_s d_m$. Their result is tight, like ours, to within a factor of 2. Their lower-bound proof introduced the "stretching" construction that we use here.

Michailidis [116] and Attiya *et al.* [10] improve this earlier result by giving semisynchronous protocols for k-set agreement that take time $\left\lfloor \frac{f}{k} \right\rfloor d_m + 2d_s d_m$ against the symmetric adversary.

The layered (or round-by-round) approach has been used before many times, in particular to unify and simplify the analysis of different models of computation [26,120]. Herlihy, Rajsbaum, and Tuttle [88,89] showed that connectivity arguments for message-passing models could be cast into a common framework. They introduced the notion of pseudospheres and later proposed the notion of an *absorbing sequence*, a notion similar to but not identical with shellability.

Junqueira and Marzullo [99] introduced the core/survivor-set formalism for characterizing general adversaries used here and derived the first lower bounds for synchronous consensus against such an adversary. Delporte-Gallet et al. [45] were the first to prove several important lower bounds on k-set agreement in asynchronous read-write memory against an adversary. Herlihy and Rajsbaum [84] gave the first direct application of combinatorial topology to the asynchronous read-write memory model against an adversary.

The semisynchronous consensus protocol used to prove that Theorem 13.5.20 is tight is due to Attiya, Lynch, Dolev, and Stockmeyer [13].

13.7 Exercises

Exercise 13.1. Show that any n-dimensional pure simplicial complex with $n + 2$ vertices is shellable. List all shelling orders of its facets.

Exercise 13.2. In the n-dimensional *crosspolytope*, the vertices are all the permutations of $(\pm 1, 0, 0, \ldots, 0)$, and a set of vertices $v_0, \ldots, v_n$ defines a simplex if, for $0 \leq i \leq n$, at most one vertex has a nonzero i^{th} coordinate. Show that $\Psi(P_i, \{0, 1\} \mid i \in [0 : n])$ is isomorphic to the n-dimensional crosspolytope.

Exercise 13.3. Prove Fact 13.3.2: If each V_i is a singleton set $\{v_i\}$, then the pseudosphere is isomorphic to a single simplex:

$$\Psi(P_i, V_i \mid i \in I) \cong \{v_i \mid i \in I\}.$$

Exercise 13.4. Prove Fact 13.3.3: if there is a $j \in I$ such that $V_j = \emptyset$, then

$$\Psi(P_i, V_i \mid i \in I) \cong \Psi(P_i, V_i \mid i \in I \setminus \{j\}).$$

Exercise 13.5. Prove Fact 13.3.4:

$$\Psi(P_i, U_i \mid i \in I) \cap \Psi(P_i, V_i \mid i \in I) = \Psi(P_i, U_i \cap V_i \mid i \in I).$$

Exercise 13.6. Let $\mathcal{K}$ be a complex constructed by taking the union of pseudospheres:

$$\mathcal{K} = \bigcup_{i \in I} \Psi(U, V_i).$$

For $k < n$, use the Nerve lemma show that $\mathcal{K}$ is k-connected if and only if the nerve complex $\mathcal{N}(V_i \mid i \in I)$ is k-connected.

Exercise 13.7. Prove that each of the carrier maps defined in Section 13.5 is indeed a carrier map.

This page is intentionally left blank

Simulations and Reductions for Colored Tasks

CHAPTER OUTLINE HEAD

In this chapter we show how to use the results of Chapter 11 to study simulations and reductions for colored tasks. Earlier, in Chapter 7, we defined the notion of a simulation for colorless tasks and showed how to construct reductions from one failure model to another. Because the tasks under consideration were colorless, we had considerable flexibility to move between models that encompass different numbers of processes.

In the simulations considered in this chapter, we consider models with the same processes but different communication models. In particular, we will (at last) prove that the shared-memory models we have used interchangeably in this book really are equivalent.

To illustrate the duality of operational and combinatorial arguments, some of our proofs will be operational (we construct an explicit simulation) and some combinatorial (we use topological properties to show that a simulation exists, without constructing it).

14.1 Model

As usual, a model of computation is given by a set of process names, a communications medium such as shared memory or message passing, a timing model such as synchronous or asynchronous, and a failure model given by an adversary. In this chapter, we restrict our attention to asynchronous shared-memory models with the wait-free adversary (any subset of processes can fail).

Distributed Computing Through Combinatorial Topology. http://dx.doi.org/10.1016/B978-0-12-404578-1.00014-0
© 2014 Elsevier Inc. All rights reserved.

Recall that a protocol $(\mathcal{I}, \mathcal{P}, \Xi)$ *solves* a (colored) task $(\mathcal{I}, \mathcal{O}, \Delta)$ if there is a color-preserving simplicial map $\delta : \mathcal{P} \to \mathcal{O}$ *carried by* Δ: for every simplex σ in $\mathcal{I}$, $\delta(\Xi(\sigma)) \subseteq \Delta(\sigma)$. Operationally, each process finishes the protocol in a local state that is a vertex of $\mathcal{P}$ of matching color and then applies δ to choose an output value at a vertex of matching color in $\mathcal{O}$.

As before, a reduction is defined in terms of two models of computation, a model $\mathbb{R}$ (called the *real model*) and a model $\mathbb{V}$ (called the *virtual model*). They have the same set of process names and the same colored input complex $\mathcal{I}$, but their protocol complexes may differ. For example, the real model might be snapshot memory, whereas the simulated model might be layered immediate snapshot memory.

The definitions of simulation and reduction are similar to those used for colorless tasks in Chapter 7, with the important distinction that all maps must be color-preserving.

Definition 14.1.1.   A (real) model $\mathbb{R}$ *reduces to* a (virtual) model $\mathbb{V}$ if, for any (colored) task, given a protocol for that task in $\mathbb{V}$, one can construct a protocol in $\mathbb{R}$.

A systematic way to construct a reduction is to use simulation.

Definition 14.1.2.   Let $(\mathcal{I}, \mathcal{P}, \Xi)$ be a protocol in $\mathbb{R}$ and $(\mathcal{I}, \mathcal{P}', \Xi')$ a protocol in $\mathbb{V}$. A *simulation* is a color-preserving simplicial map

$$\phi : \mathcal{P} \to \mathcal{P}'$$

such that, for each simplex σ in $\mathcal{I}$, ϕ maps $\Xi(\sigma)$ to $\Xi'(\sigma)$.

Informally, each process executing the real protocol chooses a simulated execution for itself in the virtual protocol, where each virtual process has the same input as the real process that simulates it. Unlike for colorless tasks, ϕ must be color-preserving (hence noncollapsing): Each real process may choose a simulated execution and output value for itself only.

The semi-commuting diagram of simulations and reductions shown in Figure 14.1 is essentially the same as the earlier one in Chapter 7, with the important distinction that we now require all complexes to be properly colored by process names, and all maps and carrier maps must be color-preserving.

The proof of the following theorem is nearly identical to the proof of Theorem 7.2.5 and is omitted.

Theorem 14.1.3.   *If every protocol in $\mathbb{V}$ can be simulated by a protocol in $\mathbb{R}$, then $\mathbb{R}$ reduces to $\mathbb{V}$.*

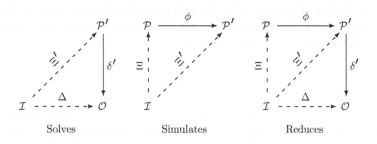

Solves   Simulates   Reduces

FIGURE 14.1

Carrier maps are shown as dashed arrows, simplicial maps as solid arrows. On the left, $\mathcal{P}'$ via δ' solves the colorless task $(\mathcal{I}, \mathcal{O}, \Delta)$. In the middle, $\mathcal{P}$ simulates $\mathcal{P}'$ via ϕ. On the right, $\mathcal{P}$ via the composition of ϕ and δ' solves $(\mathcal{I}, \mathcal{O}, \Delta)$.

14.2 Shared-memory models

As an application for colored simulations, we prove that five natural models of asynchronous shared memory are equivalent: If a task has a protocol in one model, then it has a protocol in them all. This equivalence frees us to use whichever model is most convenient for expressing particular algorithms or impossibility results. (see Figure 14.2).

Perhaps the most natural model of shared memory is the *read-write* model (Figure 14.2). The processes share a one-dimensional array of words, mem[·]. In one atomic step, a process can read or write any individual word. Reading or writing multiple words requires multiple steps, which may be interleaved with steps of other processes. This model is natural in the sense that it is close to the memory model provided by most real computers Without loss of generality (see the chapter notes), we may restrict our attention to read-write protocols where P_i reads any word but writes only mem[i].

Initially, each process's view is its input value. For N layers, P_i writes its view to mem[i] and then copies the memory, one word at a time, into its view. A process's read and subsequent writes are distinct events and can be interleaved with reads and writes of other processes. After the last layer, P_i returns an output value by applying a task-specific decision map δ_i to its final view.

In the (simple) *snapshot* model (Figure 14.3), $n + 1$ processes share a memory array mem[0..n], where P_i can write to mem[i] and take a snapshot of the entire memory. Initially, each process's view is its input. For N layers, P_i writes its view to mem[i] and then takes a snapshot of the entire memory, which becomes its new view. A process's read and subsequent snapshot are distinct events and can be separated by reads and snapshots of other processes. After the last layer, P_i returns an output value by applying a task-specific decision map δ_i to its final view.

In the *layered snapshot* model (Figure 14.4), the processes share a *two-dimensional* memory array mem[0..N − 1][0..n]. As usual, P_i's initial view is its input value. At layer ℓ, P_i writes its state to mem[ℓ][i] and takes a snapshot of that layer's row, mem[ℓ][*], which becomes its new view. After N layers, P_i returns an output value by by applying a task-specific decision map δ_i to its final view.

In the (simple) *immediate snapshot* model (Figure 14.5), $n + 1$ processes share a memory array mem[0..n], where P_i can write to mem[i] and take an immediate snapshot of the entire memory. Initially, each process's view is its input. For N layers, P_i writes its view to mem[i] and, in the very next

```
// N is number of layers, n + 1 the number of processes
shared mem: array[0..n] of Value
ReadWriteProtocol(v_i: value): Value
  view: set of Value := {v_i}   // initial view is input value
  for ℓ := 0 to N − 1 do
    mem[i] := view
    for j := 0 to n do
      view[j] := mem[j]
  return δ(view)        // apply decision map to final view
```

FIGURE 14.2

Full-information read-write protocol for P_i.

```
// N is number of layers, n + 1 the number of processes
shared mem: array[0..n] of Value
SnapshotProtocol(v_i: value): Value
  view: set of Value := {v_i}   // initial view is input value
  for ℓ := 0 to N − 1 do
    mem[i] := view
    view := snapshot(mem[*])
  return δ(view)      // apply decision map to final view
```

FIGURE 14.3

Full-information snapshot protocol for P_i.

```
shared mem: array[0..N−1][0..n] of Value
LayeredSnapshotProtocol(v_i: value): Value
  view: set of Value := {v_i}   // initial view is input value
  for ℓ := 0 to N − 1 do
    mem[ℓ][i] := view
    view := snapshot(mem[ℓ][*])
  return δ(view)      // apply decision map to final view
```

FIGURE 14.4

Full-information layered snapshot protocol for P_i.

```
shared mem: array[0..n] of Value
ImmediateSnapshotProtocol(v_i: value): Value
  view: set of Value := {v_i}   // initial view is input value
  for ℓ := 0 to N − 1 do
    immediate
      mem[i] := view
      view := snapshot(mem[*])
  return δ(view)      // apply decision map to final view
```

FIGURE 14.5

Full-information immediate snapshot protocol for P_i.

step, takes a snapshot of the entire memory, which becomes its new view. Processes can write concurrently and take snapshots concurrently, but each snapshot happens immediately after its matching write. After the last layer, P_i returns an output value by applying a task-specific decision map δ_i to its final view.

In the *layered immediate snapshot* model (Figure 14.6), the processes share a two-dimensional memory array $\mathsf{mem}[0..N − 1][0..n]$. As usual, P_i's initial view is its input value. At layer ℓ, P_i writes its state to $\mathsf{mem}[\ell][i]$ and, in the very next step, takes a snapshot of that layer's row, $\mathsf{mem}[\ell][*]$, which

```
shared mem: array[0..N−1][0..n] of Value
LayeredImmediateSnapshotProtocol(v_i: value): Value
  view: set of Value := {v_i}   // initial view is input value
  for ℓ := 0 to N − 1 do
    immediate
      mem[ℓ][i] := view
      view := snapshot(mem[ℓ][*])
  return δ(view)        // apply decision map to final view
```

FIGURE 14.6

Full-information layered immediate snapshot protocol for P_i.

becomes its new view. Processes can write concurrently and take snapshots concurrently, but each process's snapshot happens immediately after its matching write. After N layers, P_i returns an output value by by applying a task-specific decision map $δ_i$ to its final view. Just as for the snapshot models, this model differs from the simple immediate snapshot model by using a *clean memory* for each layer, which might, for all we know, sacrifice computational power.

14.3 Trivial reductions

Some reductions between these models are trivial. If a task has a protocol in the read-write model, then it has a protocol in all the other models (see Exercise 14.1). The other direction is nontrivial: For all we know, the ability to take a consistent memory snapshot may add computational power.

If a task has a protocol in either of the layered models, then it has a protocol in the corresponding simple model (see Exercise 14.2). Each of the layered models uses "clean" memory for each layer, which makes analysis easier. In principle, however, there may be (immediate) snapshot protocols that cannot be expressed as layered protocols.

Finally, if a task has a protocol in either of the snapshot models, then it has a protocol in the corresponding immediate snapshot model (see Exercise 14.3). For the other direction, it will require a nontrivial argument to show that the ability to take an immediate snapshot is not more computationally powerful than the ability to take a snapshot.

Figure 14.7 shows the five shared-memory models. An arrow from one model to another indicates a reduction: If a protocol exists in the target model, then one exists in the source model. Solid arrows represent trivial implications, whereas dashed arrows represent nontrivial implications. Each dashed arrow is labeled with the chapter section number where we prove that implication.

14.4 Layered snapshot from read-write

For our first nontrivial reduction, we will show how to simulate any layered snapshot protocol by a read-write protocol. Since each layer of a layered snapshot protocol uses a disjoint region of memory, it is enough to show how to simulate a single layer's snapshot.

The simulation is shown in Figure 14.8. It consists of two protocols: a simulated write and a simulated snapshot. Each memory word has two components: its *version number*, which is incremented each time

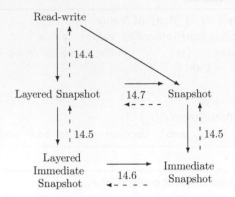

FIGURE 14.7

Reductions between various shared-memory models. An arrow from one model to another indicates that the existence of a protocol in the first model implies the existence of a protocol in the second. Trivial reductions are shown as solid arrows, and nontrivial reductions are shown as dashed arrows. Each dashed arrow is labeled with the chapter section number where we prove that implication.

```
1   mem: array[0..n ][0.. n] of Value // initially ⊥
2
3   WriteImpl(v_i: value)
4     int version := mem[i].version // read old version number
5     mem[i] := ⟨version + 1, v_i⟩ //write new version number and data together
6
7   SnapshotImpl(): array [0.. n] of Value
8     collect0 : array [0.. n] of value
9     collect1 : array [0.. n] of value
10    repeat
11      for j := 0 to n do
12        collect0 [j] := mem[j]
13      for j := 0 to n do
14        collect1 [j] := mem[j]
15      if version numbers in collect0 and collect1 agree then
16        return collect0
```

FIGURE 14.8

Simulating a snapshot with a read-write protocol.

the word is written, and its *data*, which is the value of interest. The simulated write (Lines 3-5) reads the version number, increments it, and atomically writes the new version number along with the data. Because each word is written by only one process, there is no danger that concurrent writes will interfere.

A *collect* is the nonatomic act of copying the register values one by one into an array. If we perform two collects, one after the other, and both collects read the same version numbers for each word, then no process wrote to the memory between the start of the first collect and the start of the second, so the result is a snapshot of the memory during that interval. We call such a pair of collects a *clean double collect*.

The simulated snapshot repeatedly executes pairs of collects (Lines 11 and 13) and returns after completing a clean double collect (Line 16).

Lemma 14.4.1. *The protocol of Figure 14.8 is a wait-free simulation of a single-round layered snapshot protocol.*

Proof. As noted, the result of a clean double collect is a snapshot.

The simulated write is clearly wait-free because it consists of a single read followed by a write. Assume, by way of contradiction, that there is an execution in which the simulated snapshot runs forever, without completing a clean double collect. A pair of collects can disagree only if some process writes between the first and second collect. But there are $n + 1$ processes, and each writes only once in a layer, so a pair of collects can fail to be clean at most $n + 1$ times. □

Given a layered snapshot protocol, we can substitute this simulation in each layer to get a read-write protocol. If follows that each of the two models can simulate the other, so they have equivalent computational power.

14.5 Immediate snapshot from snapshot

Next we show how to implement an immediate snapshot using snapshots. As a result, the immediate snapshot model reduces to the snapshot model: Given a protocol in the immediate snapshot model, one can replace each immediate snapshot with its snapshot implementation, yielding a snapshot protocol for the same task. Because our construction actually uses layered snapshots, the layered immediate snapshot model reduces to the layered snapshot model.

The protocol appears in Figure 14.9. Each process executes $n + 1$ layers. At layer ℓ, each P_i writes its input to mem$[\ell][i]$ and takes a snapshot of mem$[\ell][*]$. If it observes that at least $n + 1 - \ell$ processes have written values in the row for layer ℓ, then it returns that snapshot.

```
1   mem: array[0..n ][0.. n] of Value // initially ⊥
2   ImmediateSnapshotImpl(vi: value): array [0.. n] of Value
3      for ℓ := 0 to n do
4         mem[ℓ][i] := vi
5         view: array of Value := snapshot(mem[ℓ][*])
6         if view has at least n + 1 − ℓ values written then
7            return view
```

FIGURE 14.9

Simulating an immediate snapshot with snapshots.

This protocol satisfies the following invariant: At most $n + 1 - q$ processes have written at layer q or higher. This property clearly holds at the start, when no processes have written at any layers. We say that P_i at layer q creates a *pending* write at Line 5 when it observes that fewer than $n + 1 - q$ processes have written at layer q. It completes the pending write in the next loop iteration at Line 4 when it writes to $\mathsf{mem}[q + 1][i]$.

Consider a protocol state where, at each layer q, for $0 \le q \le n$, k_q processes have written at layer q but no higher, and there would be ℓ_q processes if all pending writes were to complete. Following a state change, we denote the new values of these quantities by k'_q and ℓ'_q. We will show that the following property is invariant: If all pending writes are completed, then there are $n + 1 - q$ or fewer processes at round q or higher:

$$\sum_{i=q}^{n} \ell_i \le n + 1 - q.$$

If this property holds in a state, it cannot be violated by completing a pending write, because the values of ℓ_i are unchanged. What if a process at round p creates a new pending write? The only value of ℓ'_q that could change is $q = p + 1$. If a process that wrote at round p is about to write at $p + 1$, it must have observed that

$$\sum_{i=p}^{n} k_i < n + 1 - p.$$

Completing pending writes at round p and higher can add at most k_p processes to round p and higher:

$$\sum_{i=p+1}^{n} \ell'_i \le \sum_{i=p}^{n} k_i.$$

Combining these inequalities yields

$$\sum_{i=0}^{p+1} \ell'_i \le n - p,$$

satisfying the invariant.

Here is why the results satisfy the immediate snapshot properties: Each process that halts at layer p returns the same $n + 1 - p$ values, and the set of values returned by a process at layer p contains the set of values returned from level $p + 1$.

Naturally, this protocol can be repeated to implement layered immediate snapshots, showing that the layered immediate snapshot model reduces to the layered snapshot model.

14.6 Immediate snapshot from layered immediate snapshot

We now present a combinatorial argument that the immediate snapshot model can be simulated by the layered immediate snapshot model. Since simulation in the other direction is trivial, this result implies that the two models are equivalent.

Our strategy is simple: We will show that the protocol complex for every layered immediate snapshot protocol is a manifold. It follows that the protocol complex is link-connected, so Corollary 11.5.8 implies that the immediate snapshot protocol can be simulated by a layered immediate snapshot protocol.

To describe layered immediate snapshot protocols, we require some additional notation. An $(n+1)$-element *layer vector* $\vec{R}$ is a map from process names to nonnegative integers. The ith entry of the layer vector $\vec{R}$ indicates that process P_i executes $\vec{R}[i]$ immediate snapshot phases. For any set of processes $U \subseteq [n]$, define $\vec{R} \setminus U$ to be the layer vector where each active process in U participates in one phase less:

$$(\vec{R} \setminus U)[i] = \begin{cases} \max(0, \vec{R}[i] - 1), & \text{if } i \in U; \\ \vec{R}[i], & \text{otherwise.} \end{cases}$$

We use $\text{Ch}^{\vec{R}}(\sigma)$ to denote the immediate snapshot protocol complex starting from σ, where each P_i runs for $\vec{R}[i]$ layers, independently of the information it receives.

The decomposition of $\text{Ch}^{\vec{R}}(\sigma)$ corresponding to various choices of U is illustrated in Figure 14.10 for three processes. In this figure, we use the notation $\text{Ch}^{p,q,r}(\sigma)$ for the complex where P_0 executes p layers, P_1 executes q layers, and P_2 executes r layers.

If $\vec{R} = \langle 1, \ldots, 1 \rangle$, the vector consisting of all 1's, then $\text{Ch}^{\vec{R}}(\sigma) = \text{Ch}(\sigma)$. Note that if $\vec{R} = \langle r, \ldots, r \rangle$, the vector whose entries are all $r > 1$, then $\text{Ch}^r(\sigma)$ and $\text{Ch}^{\vec{R}}(\sigma)$ are *not* isomorphic!

Recall from Section 10.2.3 that for any configuration C and any $U \subseteq [n]$, we write $C \uparrow U$ to denote the configuration reached from C by running the processes in U in the next layer. When the configuration C is an initial configuration identified with input simplex σ, we use the notation $\sigma \uparrow U$. Also recall that for a protocol $(\mathcal{I}, \mathcal{P}, \Xi)$, $(\Xi \downarrow U)(C)$ denotes the complex of executions where, starting from C, the processes in U halt without taking further steps, and the rest finish the protocol. For example, $(\text{Ch}^{\vec{R}} \downarrow V)(\sigma \uparrow U)$ is the complex of multilayer immediate snapshot executions in which, starting

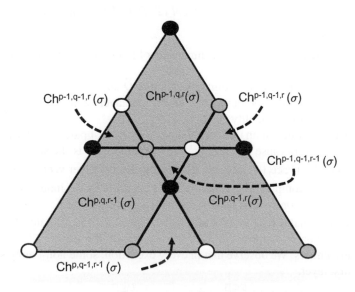

FIGURE 14.10

How a three-process immediate snapshot complex is subdivided into smaller complexes. Here, $\text{Ch}^{p,q,r}(\sigma)$ is the complex where P_0 executes p layers, P_1 executes q layers, and P_2 executes r layers. In each "corner" subcomplex, one process executes one fewer layer; in each "side" complex, two processes execute one fewer layer; and in the "central" complex, all three processes execute one fewer layer.

from input simplex σ, the processes in U simultaneously write and read, the processes in V then halt, and the remaining processes run to completion. Proposition 10.2.9 states that

$$\Xi(C \uparrow V) \cap \Xi(C \uparrow U) = (\Xi \downarrow W)(C \uparrow U \cup V)$$

where W, the set of processes that take no further steps, satisfies

$$W = \begin{cases} V, & \text{if } V \subseteq U; \\ U, & \text{if } U \subseteq V; \\ U \cup V, & \text{otherwise.} \end{cases}$$

The next lemma states that the complex reached by executions where the processes in U form the first layer is isomorphic to the complex reached by executions where the processes in U each execute one fewer layer.

Lemma 14.6.1.   *Assume as above that we are given a protocol* $(\mathcal{I}, \mathcal{P}, \mathrm{Ch}^{\vec{R}})$, $\sigma \in \mathcal{I}$, *and* U *is a set of processes; then we have*

$$\mathrm{Ch}^{\vec{R} \backslash U}(\sigma) \cong \mathrm{Ch}^{\vec{R}}(\sigma \uparrow \tilde{U}) \tag{14.6.1}$$

where $\tilde{U} = \{i \in U \mid \vec{R}[i] \geq 1\}$.

Proof.   Let σ_U be the face of σ labeled with names in U. After the first phase, each process in U has state σ_U, the set of values it read, whereas the others, which did not take a step, still have their original inputs. From that point on, the execution is the same as if each process in U had had σ_U as input but executed one fewer layers.   □

Theorem 14.6.2.   *Assume we are given a protocol* $(\mathcal{I}, \mathcal{P}, \mathrm{Ch}^{\vec{R}})$. *Then for any* $\sigma \in \mathcal{I}$, *the simplicial complex* $\mathrm{Ch}^{\vec{R}}(\sigma)$ *is a manifold.*

Proof.   First we argue by induction on n, the dimension of σ. For the base case, when $n = 0$, the claim is immediate.

   Assume that $\mathrm{Ch}^{\vec{Q}}(\tau)$ is a manifold for dim $(\tau) < n$, and $\vec{Q}$ any (dim $\tau + 1$)-element layer vector.

   We impose the following partial order on $(n + 1)$-element layer vectors: $\vec{Q} \prec \vec{R}$ if, for all i, $\vec{Q}[i] \leq \vec{R}[i]$, and for at least one i, the inequality is strict.

   We argue by structural induction on this partial order. For the first part of the base case, suppose that $\vec{R}[i] = 0$ for some $i \in [n]$. Because P_i takes no steps, its final state is the same as its initial state, so $\mathrm{Ch}^{\vec{R}}(\sigma)$ is a cone over $\mathrm{Ch}^{\vec{R}'}(\mathrm{Face}_i(\sigma))$, where $\vec{R}'$ is the n-element layer vector constructed from $\vec{R}$ by omitting the i-th entry. The complex $\mathrm{Ch}^{\vec{R}'}(\mathrm{Face}_i(\sigma))$ is a manifold by the induction hypothesis for n, so the claim follows because a cone over a manifold is also a manifold (see Exercise 9.8). For the second part of the base case, note that if $\vec{R}[i] = 1$ for all $i \in [n]$, then $\mathrm{Ch}^{\vec{R}}(\sigma)$ is a manifold by Theorem 9.2.8.

   For the induction step on $\vec{R}$, suppose that $\mathrm{Ch}^{\vec{Q}}(\sigma)$ is a manifold for every $\vec{Q} \prec \vec{R}$. Let $\vec{R}$ be a layer vector with all nonero entries. We observe that in every execution, some non-empty set of processes U participates in the first layer.

$$\mathrm{Ch}^{\vec{R}}(\sigma) = \bigcup_{\substack{U \subseteq \Pi \\ U \neq \emptyset}} \mathrm{Ch}^{\vec{R}}(\sigma \uparrow U) \cong \bigcup_{\substack{U \subseteq \Pi \\ U \neq \emptyset}} \mathrm{Ch}^{\mathcal{R} \backslash U}(\sigma)$$

Because $\vec{R}$ has no zero entries, $\vec{R} \setminus U \prec \vec{R}$, so by the induction hypothesis for layer vectors, each $\mathrm{Ch}^{\mathcal{R} \backslash U}(\sigma)$ is a manifold; thus every $(n - 1)$-simplex in each such subcomplex is a face of one or two

n-simplices in that subcomplex. We must check that no $(n-1)$-simplex that lies in an intersection of these subcomplexes is a face of more than two n-simplices.

By Proposition 10.2.9,

$$\mathrm{Ch}^{\vec{R}}(\sigma \uparrow U) \cap \mathrm{Ch}^{\vec{R}}(\sigma \uparrow V) = (\mathrm{Ch}^{\vec{R}} \downarrow W)(\sigma \uparrow U \cup V)$$

for a non-empty set of processes W. If $|W|$, the number of processes that fail, exceeds one, then this complex has dimension less than $n-1$ and cannot contain any $n-1$ simplices. In particular, any intersection of more than two subcomplexes has too small a dimension to contain any $(n-1)$-simplices. It follows that there is an $(n-1)$ simplex τ on the boundary of two subcomplexes $\mathrm{Ch}^{\vec{R}}(\sigma \uparrow U)$ and $\mathrm{Ch}^{\vec{R}}(\sigma \uparrow V)$ exactly when $U = \{P_i\}$, where P_i is the unique process name *not* in $names(\tau)$, and $P_i \in V$:

$$\tau \in \mathrm{Ch}^{\vec{R}}(\sigma \uparrow \{P_i\}) \cap \mathrm{Ch}^{\vec{R}}(\sigma \uparrow V) = (\mathrm{Ch}^{\vec{R}} \downarrow \{P_i\})(\sigma \uparrow V).$$

First we claim that τ is a face of exactly one n-simplex in $\mathrm{Ch}^{\vec{R}}(\sigma \uparrow \{P_i\})$. Operationally, the argument states that τ is generated by an execution in which P_i is the only process to execute in the first layer, but does not take any more steps until all the remaining processes have halted at vertices of τ. After that, P_i runs deterministically by itself until it halts.

More precisely, by Lemma 14.6.1,

$$\mathrm{Ch}^{\vec{R}}(\sigma \uparrow \{P_i\}) \cong \mathrm{Ch}^{\vec{R} \setminus \{P_i\}}(\sigma).$$

If we restrict our attention to executions where P_i does not participate,

$$(\mathrm{Ch}^{\vec{R}} \downarrow \{P_i\}(\sigma \uparrow \{P_i\}) \cong \mathrm{Ch}^{\vec{R} \setminus \{P_i\}}(\mathrm{Face}_i\, \sigma).$$

Any $(n-1)$-simplex $\tau \in (\mathrm{Ch}^{\vec{R}} \downarrow \{P_i\}(\sigma \uparrow \{P_i\})$ is isomorphic to an $(n-1)$-simplex $\tau' \in \mathrm{Ch}^{\vec{R} \setminus \{P_i\}}(\mathrm{Face}_i\, \sigma)$ where τ' is a face of a unique n-simplex in $\mathrm{Ch}^{\vec{R} \setminus \{P_i\}}(\mathrm{Face}_i\, \sigma)$. It follows that τ is a face of a unique n-simplex in $(\mathrm{Ch}^{\vec{R}} \downarrow \{P_i\}(\sigma \uparrow \{P_i\})$.

The same argument, with minor changes, shows that τ is a face of a unique n-simplex of $\mathrm{Ch}^{\vec{R}}(\sigma \uparrow V)$. It follows that τ is a face of exactly two n-simplices in $\mathrm{Ch}^{\vec{R}}(\sigma)$, as desired.

Finally, we need to check that $\mathrm{Ch}^{\vec{R}}(\sigma)$ is strongly connected. It is enough to show that we can link any simplex $\tau_0 \in \mathrm{Ch}^{\vec{R}}(\sigma \uparrow U)$ to any $\tau_1 \in \mathrm{Ch}^{\vec{R}}(\sigma \uparrow \{P_i\})$ for any $P_i \in U$. Pick an $(n-1)$-simplex $\tau \; \tau_0 \in \mathrm{Ch}^{\vec{R}}(\sigma \uparrow U) \cap \tau_1 \in \mathrm{Ch}^{\vec{R}}(\sigma \uparrow \{P_i\})$. Because $\mathrm{Ch}^{\vec{R}}(\sigma \uparrow U)$ is a manifold, it contains an n-simplex τ_{01} such that $\tau \subset \tau_{01}$ and a chain linking τ_0 to τ_{01}. In the same way, $\mathrm{Ch}^{\vec{R}}(\sigma \uparrow \{P_i\})$ contains an n-simplex τ_{10} such that $\tau \subset \tau_{10}$ and a chain linking τ_1 to τ_{10}. Concatenating these chains gives a chain linking τ_0 and τ_1.

Given any $\tau_i \in \mathrm{Ch}^{\vec{R}}(\sigma \uparrow U)$ and $\tau_j \in \mathrm{Ch}^{\vec{R}}(\sigma \uparrow V)$, where $P_i \in U$ and $P_j \in V$, we can use the previous construction to link any simplex in $\mathrm{Ch}^{\vec{R}}(\sigma \uparrow U)$ to $\mathrm{Ch}^{\vec{R}}(\sigma \uparrow \{P_i\})$, to $\mathrm{Ch}^{\vec{R}}(\sigma \uparrow \{P_i, P_j\})$, to $\mathrm{Ch}^{\vec{R}}(\sigma \uparrow \{P_j\})$, to any simplex in $\mathrm{Ch}^{\vec{R}}(\sigma \uparrow V)$. $\square$

The protocol complex for an immediate snapshot protocol where each process executes ℓ layers is just $\mathrm{Ch}^{\ell}\mathcal{I}$, the ℓ-fold subdivision of the input complex. In Chapter 16 we show that for every $\sigma \in \mathcal{I}$, the simplicial complex $\mathrm{Ch}^{\ell}(\sigma)$ is a subdivision of σ such that for every $\tau \subseteq \sigma$, the simplicial complex $\mathrm{Ch}^{\ell}(\tau)$ is the corresponding subdivision of τ.

We know from Theorem 10.4.7 that $\mathrm{Ch}^{\vec{R}}(\mathcal{I})$ is n-connected and from Theorem 9.2.8 that it is a manifold. Because the complex is a manifold, each link is a combinatorial sphere, so the complex is link-connected. It follows from Corollary 11.5.8 that there is a simulation

$$\phi : \mathrm{Ch}^N \mathcal{I} \to \mathrm{Ch}^{\vec{R}} \mathcal{I}$$

for some $N > 0$.

14.7 Snapshot from layered snapshot

We now show how a layered snapshot protocol can simulate a snapshot protocol. Starting from input simplex σ, as the protocol unfolds each process incrementally constructs the simulated state of the snapshot protocol. For clarity, we refer to writes and snapshots that occur in the simulated protocol $\mathcal{P}(\cdot)$ as *simulated writes* and *simulated snapshots*.

Without loss of generality, we will simulate a snapshot protocol where every process runs for $N > 0$ layers. Protocols where different processes run for different numbers of layers can be simulated simply by having processes run for redundant layers. Each process repeatedly writes an approximation of the snapshot memory it is simulating. While the protocol is in progress, this simulated memory will be incomplete or inconsistent, but by the protocol's end, the simulated memory will become complete and consistent.

Figure 14.11 shows the simulation protocol. A simulated memory is an $(n + 1)$-element array of memory values (Line 5), where each memory entry can be either $\bot$, a single input value, or an array of $n + 1$ memory values, depending on the layer being simulated. For process P_i, each memory value has an implicit *clock* value:

$$\mathrm{clock}_i(v) = \begin{cases} 0 & \text{if } v = \bot \\ 1 & \text{if } v \text{ is an input value, and} \\ 1 + \mathrm{clock}_i(v[i]) & \text{otherwise.} \end{cases}$$

The clock value counts the number of layers P_i must have taken to have produced such a value. Given a two-dimensional array of simulated memories $a[\cdot][\cdot]$, the function $\mathrm{latest}(a)$ returns a single simulated memory constructed by taking their entry-wise maximums:

$$\mathrm{latest}(a)[i] = a[j][i] \text{ such that } j = \mathrm{argmax}_j(\mathrm{clock}_i(a[j][i])).$$

That is, for each i, j is chosen to maximize $\mathrm{clock}_i(a[j][i])$. Given a simulated memory m, the function $\mathrm{vectorClock}(m)$ produces an integer vector whose i^{th} value is $\mathrm{clock}_i(m[i])$, which we call a *vector clock*. If c is a vector clock, define

$$|c| = \sum_{i=0}^{n} c[i].$$

The processes share an $N \times n$ array $\mathsf{mem}[\cdot][\cdot]$ of simulated memories. Each P_i initializes its simulated memory to hold its own input at its own position (Line 11) and $\bot$ elsewhere. At each layer ℓ, P_i writes its simulated memory to $\mathsf{mem}[\ell][i]$ and then takes a snapshot of all the simulated memories for layer ℓ. It then constructs a single memory by merging the latest entries for each process (Line 15) and constructs that simulated memory's vector clock (Line 16). If P_i's own clock value in the simulated memory is less than its value for ℓ, then P_i has not yet simulated all its writes. The value $|c|$ computed on Line 16 is the total number of writes reflected in the simulated memory. If P_i observes that this total equals the layer number (Line 18), then the simulated memory it has assembled is a consistent simulated snapshot, so it writes that snapshot to its own entry in the simulated shared memory (Line 19) and resumes the next layer. After N layers, it decides its own entry in the simulated memory.

Lemma 14.7.1.   *For each process, the total number of simulated writes that P_i observes is less than or equal to its layer number:* $\ell \le |c|$.

Proof.   By induction. For the base case, at layer 0, P_i has written its input to $s[i]$, so $c \ge 1$.

For the induction step, suppose P_i computes vector clock c at layer $\ell - 1$ and vector clock c' at layer ℓ. By the induction hypothesis, $\ell - 1 \le |c|$. If $\ell - 1 < |c|$, then $\ell \le |c| \le |c'|$. If, instead, $\ell - 1 = |c|$, then P_i completes a simulated snapshot at layer $\ell - 1$ (Line 18), so it writes a new snapshot to $s[i]$, ensuring that at the following layer, $|c'|$ is at least one larger than $|c|$:

$$\ell - 1 = |c|$$
$$\ell = |c| + 1$$
$$\ell \le |c'|.$$

$\square$

```
1   protocol SimpleRWSim
2     // shared memory being simulated
3     mem: array [0..N][0..n] of Value
4
5     LayeredSnapshotSim($v_i$: Value): array of Value
6       // my copy of the simulated memory, initially all ⊥
7       s: array[0..n] of Value := {⊥,...,⊥}
8       // my vector clock, initially all 0
9       c: array [0..n] of int := {0,...,0}
10
11      s[i] := v_i                    // initial state is input
12      for ℓ := 0 to N do
13        mem[ℓ][i] := s[i]
14        a := snapshot(mem[ℓ][*])  // snapshot layer ℓ memories
15        s := latest(a)              // merge latest entries
16        c := vectorClock(s)        // reconstruct vector clock
17        if (c[i] < N) then         // more snapshots to simulate?
18          if |c| = N then          // current simulated snapshot complete
19            s[i] := s               // record simulated snapshot
20      return s[i]]                   // decide after last layer
```

FIGURE 14.11

Simulating a snapshot protocol with layered snapshots.

Lemma 14.7.2.   *All processes that complete a simulated snapshot at layer ℓ do so with the same vector clock.*

Proof.   Consider a layer ℓ. By construction, all processes in S_i take the same snapshot and construct the same vector clock c_i. By Lemma 14.7.1, $\ell \leq c_i$ for processes that have not completed all their writes. If the processes in S_i and S_j both complete a simulated snapshot, then $|c_i| = |c_j| = \ell$. If $i < j$, then for $0 \leq k \leq n$, $c_i[k] \leq c_j[k]$, implying that if $|c_i| = |c_j| = \ell$, then for $0 \leq k \leq n$, $c_i[k] = c_j[k]$.   □

Lemma 14.7.3.   *If N is sufficiently large, then in any state where not every process has completed, some process will eventually complete a simulated snapshot.*

Proof.   By Lemma 14.7.1, the total number of completed snapshots and writes is at least ℓ. The claim follows because ℓ is continually increasing.   □

Lemma 14.7.4.   *Every non-faulty process completes all its simulated snapshots and simulated writes at or before layer $(n + 1)N$.*

Proof.   A process P_i is *incomplete* if it has not completed N simulated writes and snapshots. We first claim that if N is sufficiently large and there are incomplete processes, then some incomplete process will eventually complete a snapshot. If not, then every incomplete process's vector clock $|c|$ remains constant while ℓ continues to advance. By Lemma 14.7.1, $\ell \leq |c|$, so eventually ℓ must catch up, and the snapshot will complete.

By Lemma 14.7.3, some process continually completes and writes a simulated snapshot. Some process P_i eventually completes ℓ_i such snapshots and drops out, leaving the rest to continue. Eventually, they all complete.

Once P_i has completed, its value of $c[i]$ is N. The last process to complete sees $c[i] = N$ for every i, so $|c| = (n + 1)N = \ell$.   □

Theorem 14.7.5.   *The protocol of Figure 14.11 simulates a snapshot memory.*

Proof.   Each simulated snapshot is ordered when it completes at Line 18, and each simulated write is ordered just before the first snapshot that includes it.

More precisely, consider the vector clocks $c_0, c_1, \ldots, c_N$, where c_0 is the all-zero vector and c_i is the vector clock corresponding to the ith completed simulated snapshot, completed in layer $|c_i|$. We have seen that $c_i < c_{i+1}$. Let R_i be the set of processes that completed a simulated snapshot in layer $|c_i|$, and let W_j be the set of processes P_j such that $c_{i-1}[j] < c_i[j]$. The snapshots in R_i can be ordered arbitrarily, and the value s of the snapshot satisfies vectorClock$(s) = c_i$. Each write in W_j is ordered just before these snapshots, in arbitrary order, yielding the schedule

$$W_1, R_1, \ldots, W_{(n+1)N}, R_{(n+1)N}.$$

    □

This completes the reductions shown in Figure 14.7. Since each model can be reduced to any other, they are all equivalent.

14.8 **Chapter Notes**

Lamport [105] shows the equivalence of various read-write memory models, including the equivalence of the single and multiple-writer models. Afek et al. [2] and Anderson [6] proposed the first read-write protocol for wait-free atomic snapshot. These algorithms and proofs are covered in textbooks by Attiya and Welch [17] and by Herlihy and Shavit [92].

The algorithm of Figure 14.9 (recall Exercise 4.12) is the iterative version of the recursive algorithm by Gafni and Rajsbaum [67]. It is adapted from Borowsky and Gafni [24]. See also the surveys by Herlihy, Rajsbaum and Raynal [87,129].

The algorithm of Figure 14.11 is adapted from Gafni and Rajsbaum [67]. The first simulation algorithm for the layered model is by Borowsky and Gafni [26]. Extensions of this algorithm to failure detector models were studied by Rajsbaum, Raynal, and Travers [131].

14.9 **Exercises**

Exercise 14.1.   Explain why the existence of a read-write protocol for a task implies the existence of a protocol in all of the other shared-memory models.

Exercise 14.2.   Explain why the existence of a layered snapshot protocol for a task implies the existence of a snapshot protocol. Do the same for layered immediate snapshot and snapshot protocols.

Exercise 14.3.   Explain why the existence of a snapshot protocol for a task implies the existence of an immediate snapshot protocol. Do the same for layered snapshot and layered immediate snapshot protocols.

Exercise 14.4.   In the worst case, the protocol in Figure 14.8 can make $2n$ collects before returning. Modify this protocol so that it makes $n + 1$ collects in the worst case.

Exercise 14.5.   Give a direct proof that if we replace snapshots with immediate snapshots, the protocol of Figure 14.11 allows the layered immediate snapshot model to simulate the snapshot model.

Exercise 14.6.   In the algorithm of Figure 14.9, show that one could replace snapshots with non-atomic collects (sequentially reading the array).

This page is intentionally left blank

Classifying Loop Agreement Tasks

15

A task T *implements* task T' if one can construct a wait-free protocol for T' by calling any protocol for T, possibly followed by access to a shared read-write or snapshot memory. This notion of implementation induces a partial order on tasks and allows us to *classify* tasks by partitioning them into disjoint *classes* such that tasks in the same class implement one another. In this sense, the tasks in the same class are computationally equivalent. One class is *more powerful* than another if any task in the first class implements any class in the second, but not vice versa.

Recall from Section 5.6.2 that *loop agreement* is a family of colorless tasks for which the existence of a wait-free read-write protocol is undecidable. Here we give a complete classification of loop agreement tasks. Each loop agreement task can be assigned an *algebraic signature* consisting of a group G and a distinguished element g in G. Remarkably, this signature completely characterizes the task's computational power. If T and T' are loop agreement tasks with respective signatures $\langle G, g \rangle$ and $\langle G', g' \rangle$, then T implements T' if and only if there exists a group homomorphism $h : G \to G'$ carrying g to g'. In short, the algorithmic problem of determining how to implement one loop agreement task in terms of another reduces to a problem in group theory.

We will see that the loop agreement task corresponding to 3-process, 2-set agreement belongs to the most powerful class in the classification (the maximal element in the classification order), whereas

Distributed Computing Through Combinatorial Topology. http://dx.doi.org/10.1016/B978-0-12-404578-1.00015-2

(various forms of) approximate agreement belong to the weakest class (the minimal element). In between is an infinite number of inequivalent classes.

The material in this chapter assumes that the reader has some familiarity with elementary abstract algebra, including finitely generated groups, group homomorphisms, and presentations of finite groups.

15.1 The fundamental group

15.1.1 Basic definitions

Let $\mathcal{K}$ be a 2-dimensional simplicial complex, and let I be the unit interval $[0, 1]$. Recall from Section 5.6.1 that a *path* in a complex $\mathcal{K}$ is a continuous map $\alpha : I \rightarrow |\mathcal{K}|$. If $\alpha(0) = \alpha(1) = x$, the path is a *loop* with base point x. Two loops with the same base point are *homotopic* if one can be continuously deformed to the other while leaving the base point fixed. A loop α is *simple* if its restriction to $[0, 1)$ is injective.

Two paths α and β can be *concatenated* to form another path $\alpha * \beta$ if $\alpha(1) = \beta(0)$:

$$\alpha * \beta(s) = \begin{cases} \alpha(2s) & \text{for } 0 \leq s \leq \frac{1}{2} \\ \beta(2s - 1) & \text{for } \frac{1}{2} \leq s \leq 1 \end{cases}.$$

In particular, loops with the same base point can be concatenated in two different ways.

The *fundamental group* $\pi_1(\mathcal{K}, x_0)$ of $\mathcal{K}$ is defined as follows: The elements of the group are equivalence classes under homotopy of loops with base point x_0. The group operation "·" on these equivalence classes is concatenation of their representatives, i.e., $[\alpha] \cdot [\beta] := [\alpha * \beta]$. It is a standard exercise to check that this operation defines a group whose identity element is the equivalence class $[x_0]$ of the constant loop $\alpha(s) = x_0$ and where the inverse of $[\alpha]$ is obtained by traversing α in the opposite direction,

$$\alpha^{-1}(t) := \alpha(1 - t), \quad \text{for } t \in [0, 1],$$

and $[\alpha]^{-1} := [\alpha^{-1}]$. The identity element is the equivalence class of contractible loops, those that can be continuously deformed to x_0.

For example, the fundamental group of the circle is isomorphic to the group of integers under addition; this group is also called the *infinite cyclic* group. The constant loop corresponds to the *identity element* 0; concatenating the constant loop to any other loop does not change its homotopy type. By convention, the loop that wraps around the circle once in the clockwise direction corresponds to the generator 1, and counter-clockwise corresponds to the generator -1. Any loop is homotopic to one that "wraps" around the circle k times, where positive k is clockwise and negative k is counter-clockwise. Furthermore, any two loops that wrap around the circle a different number of times are not homotopic.

If the complex $\mathcal{K}$ is path-connected, its fundamental group is independent of the base point, up to isomorphism. For simplicial complexes, the notions of connected and path-connected coincide, and all the complexes we consider are connected, so we often write $\pi_1(\mathcal{K})$ in place of $\pi_1(\mathcal{K}, x_0)$.

Now let $\mathcal{L}$ be another connected simplicial complex, let $f : |\mathcal{K}| \rightarrow |\mathcal{L}|$ be a continuous map, and let α be a loop in $|\mathcal{K}|$. The composition $f \circ \alpha$ is a loop in $|\mathcal{L}|$. Define the map *induced* by f, $f_* : \pi_1(\mathcal{K}) \rightarrow \pi_1(\mathcal{L})$ to be $f_*([\alpha]) = [f \circ \alpha]$. It is a standard fact that f_* is a group homomorphism.

Recall that an *edge loop* is a loop whose base point is a vertex of $\mathcal{K}$ and whose path is a sequence of oriented edges in $\mathcal{K}$. Since in a simplicial complex every loop with a base at a vertex is homotopic to a standard loop associated with an edge loop, every element of the fundamental group based at a vertex has a representative that is associated with an edge loop.

15.1.2 A representation of the fundamental group associated with a spanning tree

Assume that $\mathcal{K}$ is a finite connected 2-dimensional simplicial complex. Let $\mathcal{T}$ be a rooted spanning tree of $\mathrm{skel}^1(\mathcal{K})$. There is a standard representation of the fundamental group $\pi_1(\mathcal{K})$ in terms of generators and relations, which we now proceed to describe. Let r denote the root of $\mathcal{T}$, and let $v_0, v_1, \ldots, v_s$ denote all the vertices of $\mathcal{K}$.

- **Generators of $\pi_1(\mathcal{K})$.** For each $0 \leq i < j \leq s$ such that (v_i, v_j) is an edge of $\mathcal{K}$ that does not belong to $\mathcal{T}$, we take a generator g_{ij}. For convenience, for all pairs (i, j) such that (v_i, v_j) is an edge of $\mathcal{T}$, we shall set by convention $g_{ij} := 1$. We shall also set $g_{ji} := g_{ij}^{-1}$ for all $i < j$, and $g_{ii} := 1$ for all i.
- **Relations of $\pi_1(\mathcal{K})$.** Whenever the vertices v_i, v_j, v_k form a 2-simplex of $\mathcal{K}$, we have a relation

$$g_{ij} \circ g_{jk} = g_{ik}. \tag{15.1.1}$$

Note the following special cases of the relation (15.1.1). If (v_i, v_j) and (v_j, v_k) are edges of $\mathcal{T}$, then automatically $g_{ik} = 1$. If (v_i, v_j) is an edge of $\mathcal{T}$ and k is arbitrary, then $g_{ik} = g_{jk}$.

To see that the group defined by these generators and relations is isomorphic to the fundamental group $\pi_1(\mathcal{K})$, simply send each generator g_{ij} to the standard loop associated with the edge loop constructed by concatenating the following three paths:

(1) From r to v_i along the tree
(2) The edge $\{v_i, v_j\}$
(3) The edge path back from v_j to r along the tree

In particular, we see that the fundamental group of a finite complex is finitely generated.

An alternative way to think of this representation is as follows: Consider the space X obtained from $|\mathcal{K}|$ by shrinking the tree $\mathcal{T}$ to a point. This new space has one vertex; all the edges are now loops and all the triangles are now 2-cells, which may have one, two, or three boundary edges. The space X is topologically the same as $|\mathcal{K}|$; speaking formally, it is homotopy equivalent to $|\mathcal{K}|$. In particular, the fundamental groups are the same. The above representation can now be viewed as the representation of the fundamental group of X, with edges of $\mathcal{K}$ giving generators and 2-cells giving relations. As mentioned, there are then three different kinds of relations. The 2-cell, having only one boundary edge, gives relation $g_{ij} = 1$. The 2-cell with two boundary edges will give a relation $g_{ij} = g_{ik}$ or $g_{ij} = g_{kj}$. Finally, the 2-cell with three boundary edges will give a relation of the type (15.1.1).

15.2 Algebraic signatures

Let $\mathcal{K}$ again be a 2-dimensional connected simplicial complex, and consider a loop agreement task Loop $(\mathcal{K}, \lambda)$. The triangle loop $\lambda = (v_0, v_1, v_2, p_{01}, p_{12}, p_{20})$ represents an element $[c(\lambda)]$ of the fundamental group $\pi_1(\mathcal{K}, v_0)$.

Definition 15.2.1.   We call the pair $(\pi_1(\mathcal{K}, x_0), [c(\lambda)])$ the *algebraic signature* of the loop agreement task Loop $(\mathcal{K}, \lambda)$.

Let $\mathcal{L}$ be another 2-dimensional connected simplicial complex, and consider a loop agreement task Loop $(\mathcal{L}, \mu)$, where $\mu = (w_0, w_1, w_2, q_{01}, q_{12}, q_{20})$ is some triangle loop. We use the notation

$$h : (\pi_1(\mathcal{K}), [c(\lambda)]) \to (\pi_1(\mathcal{L}), [c(\mu)])$$

to denote any group homomorphism h from $\pi_1(\mathcal{K})$ to $\pi_1(\mathcal{L})$ that maps $[c(\lambda)]$ to $[c(\mu)]$, where, on the left side, v_0 is taken to be the base point and, on the right side, w_0 is taken to be the base point.

Fact 15.2.2.   Assume that x and y are two distinct points of a topological space X, and assume that $\rho, \rho' : [0, 1] \to X$ are two simple paths from x to y such that $|\rho| = |\rho'|$. Then the paths ρ and ρ' are homotopic.

Furthermore,

$$f : (\mathcal{K}, \lambda) \to (\mathcal{L}, \mu)$$

denotes a continuous map from $|\mathcal{K}|$ to $|\mathcal{L}|$ such that $f(v_i) = w_i$ and $f(|p_{ij}|) \subseteq |q_{ij}|$ for all i, j. Let f_* denote the homomorphism on underlying fundamental groups induced by f.

Lemma 15.2.3.   *If $f : (\mathcal{K}, \lambda) \to (\mathcal{L}, \mu)$ is a continuous map as stated, then $f_*([c(\lambda)]) = [c(\mu)]$.*

Proof.   We can use Fact 15.2.2 to reparameterize $c(\lambda)$ and obtain a loop $\ell : [0, 1] \to |\mathcal{K}|$ such that $f \circ l = c(\mu)$. This of course means $f_*([l]) = [c(\mu)]$. Since $c(\lambda)$ and ℓ are homotopic, we conclude that $f_*([c(\lambda)]) = [c(\mu)]$.   □

We can show that the converse also holds. To start with, we have the following general result.

Lemma 15.2.4.   *Assume that $\mathcal{K}$ and $\mathcal{L}$ are finite simplicial complexes, the complex $\mathcal{K}$ is 2-dimensional, x is a vertex of X, and y is a vertex of Y. Assume $\varphi : \pi_1(\mathcal{K}, x) \to \pi_1(\mathcal{L}, y)$ is a group homomorphism. Then there exists, a continuous map $f : |\mathcal{K}| \to |\mathcal{L}|$ such that $f(x) = y$ and $f_* = \varphi$.*

Proof.   Let $\mathcal{T}$ be a spanning tree of $\text{skel}^1 \mathcal{K}$, rooted at x. Set as before $X := |\mathcal{K}|/|\mathcal{T}|$, which is the space obtained from $|\mathcal{K}|$ by shrinking $|\mathcal{T}|$ to a point, which we call t. Let $q : |\mathcal{K}| \to X$ denote the quotient map, which takes $|\mathcal{T}|$ to t. Since $\mathcal{T}$ is a tree, $q_* : \pi_1(\mathcal{K}, x) \to \pi_1(X, t)$ is an isomorphism, which we use to identify these two groups.

We now proceed to define a continuous map $g : X \to |\mathcal{L}|$. First we set $g(t) := y$. Next, take any directed edge e of X. Identify it with a characteristic loop $e : [0, 1] \to X$, $e(0) = e(1) = t$. It corresponds to an element $[e] \in \pi_1(X, t) \simeq \pi_1(\mathcal{K}, x)$. Let ℓ_e be a representative loop of $\varphi([e]) \in \pi_1(\mathcal{L}, y)$, $\ell_e : [0, 1] \to \mathcal{L}$, $\ell_e(0) = l_e(1) = y$. Define $g(e(q)) := l_e(q)$ for all $q \in [0, 1]$. This defines a continuous function g on the 1-skeleton of X.

As a last step, we can extend g to the rest of X, going cell by cell. For each 2-cell σ, g can be extended to σ if and only if its boundary loop is contractible in $|\mathcal{L}|$. Clearly, the boundary loop of σ represents the trivial element of the fundamental group $\pi_1(X, t)$. Since the map φ is a group homomorphism, the image of the boundary map under g represents the trivial element of $\pi_1(\mathcal{L}, y)$. Thus g can be extended to σ, and since this can be done for every 2-cell, we are done.

Finally, we construct the continuous map $f : |\mathcal{K}| \to |\mathcal{L}|$ as a composition of g and q.   □

We now apply this result to the special situation of triangle loops.

Lemma 15.2.5. *Assume $\mathcal{K}$ and $\mathcal{L}$ are 2-dimensional connected simplicial complexes, λ is a triangle loop in $\mathcal{K}$, and μ is a triangle loop in $\mathcal{L}$. If there exists a group homomorphism $h : (\pi_1(\mathcal{K}), [c(\lambda)]) \rightarrow (\pi_1(\mathcal{L}), [c(\mu)])$, then there exists a continuous map $f : (\mathcal{K}, \lambda) \rightarrow (\mathcal{L}, \mu)$.*

Proof. By Lemma 15.2.4, there exists a continuous map $g : |\mathcal{K}| \rightarrow |\mathcal{L}|$, which takes $[c(\lambda)]$ to $[c(\mu)]$. This means that g takes the loop $c(\lambda)$ to a loop ℓ, which is homotopic to $c(\mu)$. Let H be the loop homotopy from ℓ to $c(\mu)$. It is a standard topological fact, which we use here without proof, that the homotopy H can be extended to deform the whole map g to a map $f : |\mathcal{K}| \rightarrow |\mathcal{L}|$. This map f is continuous and takes $c(\lambda)$ to $c(\mu)$.

We finish the proof by remarking that the claim that H can be extended to all of $|\mathcal{K}|$ follows from the fact that $\lambda : S^1 \rightarrow |\mathcal{K}|$ is a *cofibration* (see the chapter notes). $\qquad \square$

Combining Lemmas 15.2.3 and 15.2.5 yields

Theorem 15.2.6. *There exists a continuous map $f : (\mathcal{K}, \lambda) \rightarrow (\mathcal{L}, \mu)$ if and only if there exists a homomorphism $h : (\pi_1(\mathcal{K}), [c(\lambda)]) \rightarrow (\pi_1(\mathcal{L}), [c(\mu)])$.*

15.3 Main theorem

In this section, we demonstrate the equivalence of the existence of a continuous map $f : (\mathcal{K}, \lambda) \rightarrow (\mathcal{L}, \mu)$ and the existence of an implementation of Loop $(\mathcal{L}, \mu)$ by an instance of Loop $(\mathcal{K}, \lambda)$. This will imply our main result, Theorem 15.3.8.

15.3.1 Map implies protocol

Recall that a simplicial map $\phi : \mathcal{K} \rightarrow \mathcal{L}$ is a *simplicial approximation* of a continuous map $f : |\mathcal{K}| \rightarrow |\mathcal{L}|$ if, for every point x in $|\mathcal{K}|$, $|\phi|(x)$ lies in the carrier of $f(x)$. The Simplicial Approximation Theorem 3.7.5 guarantees that every such f has a simplicial approximation $\phi : \text{Bary}^N \mathcal{K} \rightarrow \mathcal{L}$ for large enough N.

Lemma 15.3.1. *Let $\mathcal{K}$ and $\mathcal{L}$ be simplicial complexes with respective simplicial subcomplexes $\mathcal{K}_0$ and $\mathcal{L}_0$, and let $f : |\mathcal{K}| \rightarrow |\mathcal{L}|$ be a continuous map such that $f(|\mathcal{K}_0|) \subseteq |\mathcal{L}_0|$. If $\phi : \text{Bary}^N(\mathcal{K}) \rightarrow \mathcal{L}$ is a simplicial approximation to f, then $|\phi|(|\mathcal{K}_0|) \subseteq |\mathcal{L}_0|$.*

Proof. Let x be a point of $|\mathcal{K}_0|$. The carrier of $f(x)$ is in $|\mathcal{L}_0|$, since $f(|\mathcal{K}_0|) \subseteq |\mathcal{L}_0|$. Because ϕ is a simplicial approximation to f, $|\phi|(x) \in |\mathcal{L}_0|$. $\qquad \square$

In the *barycentric agreement* task, processes start with vertices in a simplex σ in a complex $\mathcal{L}$ (of arbitrary dimension), and they must converge to vertices of a simplex in $\text{Bary}^N(\sigma)$, the N-th barycentric subdivision of the input simplex.

We can now settle one direction of our main theorem.

Lemma 15.3.2. *If there exists a continuous map $f : (\mathcal{K}, \lambda) \rightarrow (\mathcal{L}, \mu)$, then an instance of Loop $(\mathcal{K}, \lambda)$ implements Loop $(\mathcal{L}, \mu)$.*

Proof. By the Simplicial Approximation Theorem, f has a simplicial approximation

$$\phi : \text{Bary}^N(\mathcal{K}) \rightarrow \mathcal{L}$$

for some $N \geq 0$. Assume that we have $\lambda = (v_0, v_1, v_2, p_{01}, p_{12}, p_{20})$ and that $\mu = (w_0, w_1, w_2, q_{01}, q_{12}, q_{20})$. Assume that a process has input w_i. We proceed as follows.

- Let this process run the wait-free protocol for Loop $(\mathcal{K}, \lambda)$ with input v_i, and let o_i be its output.
- Run the wait-free read-write N-th barycentric agreement protocol for $\mathcal{K}$ with o_i as input, and let z_i be its output.
- Choose $\phi(z_i)$ and halt.

The entire protocol is wait-free because all its parts are wait-free.

The outputs of the Loop $(\mathcal{K}, \lambda)$ protocol lie on a single simplex of $\mathcal{K}$, and the outputs of the N-th barycentric agreement protocol lie on a single simplex of $\text{Bary}^N(\mathcal{K})$. Because ϕ is a simplicial map, the decision values lie on a single simplex of $\mathcal{L}$.

Suppose the processes have two distinct inputs w_i and w_{i+1} (where we identify w_0 with w_3). The outputs of the Loop $(\mathcal{K}, \lambda)$ protocol lie on a single simplex of $p_{i,i+1}$, and the outputs of the N-th barycentric agreement protocol, lie on a single simplex of $\text{Bary}^N(p_{i,i+1})$. Because λ and μ are edge loops, $p_{i,i+1}$ and $q_{i,i+1}$ can be considered subcomplexes of $\mathcal{K}$ and $\mathcal{L}$ respectively. Because $f(|p_{i,i+1}|) \subseteq |q_{i,i+1}|$, Lemma 15.3.1 states that the simplicial approximation ϕ carries $\text{Bary}^N(p_{i,i+1})$ to $q_{i,i+1}$. All processes thus choose vertices in a simplex of $q_{i,i+1}$.

Suppose all processes have the same input w_i. The outputs of the Loop $(\mathcal{K}, \lambda)$ protocol are all v_i, and the outputs of the N-th barycentric agreement protocol are all v_i. Since ϕ carries v_i to w_i, all processes thus choose w_i. $\square$

15.3.2 Protocol implies map

We now turn our attention to the other direction: If a protocol exists by which an instance of Loop $(\mathcal{K}, \lambda)$ implements Loop $(\mathcal{L}, \mu)$, then so does a continuous map $f : (\mathcal{K}, \lambda) \rightarrow (\mathcal{L}, \mu)$. Assume that we have $\lambda = (v_0, v_1, v_2, p_{01}, p_{12}, p_{20})$ and that $\mu = (w_0, w_1, w_2, q_{01}, q_{12}, q_{20})$. Our basic strategy is the following: We may assume without loss of generality that the protocol has two phases. In the first phase, it calls a "subroutine" to solve Loop $(\mathcal{K}, \lambda)$, and in the second phase, it uses the result as input to a pure read-write phase. We treat the read-write phase as a protocol in its own right, where each process has a vertex of $\mathcal{K}$ as input and chooses a vertex of $\mathcal{L}$ as output.

Formally, there is a simple way to transform $\mathcal{K}$ into an output complex.

Definition 15.3.3. Let $\mathcal{K}$ be a complex. The *colorized* complex $\widetilde{\mathcal{K}}$ is defined as follows:

- The vertices of $\widetilde{\mathcal{K}}$ are all combinations of the form $\langle P, k \rangle$, where P is a process name and k a vertex of $\mathcal{K}$.
- Vertices $\langle P_0, k_0 \rangle, \ldots, \langle P_m, k_m \rangle$ span an m-simplex in $\widetilde{\mathcal{K}}$ if and only if the P_i are distinct, and $k_0, \ldots, k_m$ (not necessarily distinct) span a simplex in $\mathcal{K}$.

Let $\pi : \widetilde{\mathcal{K}} \rightarrow \mathcal{K}$ be the projection simplicial map that discards colors.

The read-write phase can now be recast as the decision task $\langle \widetilde{\mathcal{K}}, \widetilde{\mathcal{L}}, \Delta \rangle$, where Δ is the "colorized" version of the loop agreement relation. Let $\tilde{v}_i$ denote the maximal simplex of $\widetilde{\mathcal{K}}$ such that $\pi(\tilde{v}_i) = \{v_i\}$. Let $\tilde{p}_{ij}$ be the subcomplex of $\widetilde{\mathcal{K}}$ such that, for all $S \in \tilde{p}_{ij}$, $\pi(S) \subseteq p_{ij}$, and similarly for $\tilde{w}_i$ and $\tilde{q}_{ij}$. The relation Δ carries each $\tilde{v}_i$ to $\tilde{w}_i$ and each simplex of $\tilde{p}_{ij}$ to $\tilde{q}_{ij}$.

The circumstances under which a decision task has a wait-free read-write implementation are given by the following *asynchronous computability theorem*:

Theorem 15.3.4. *A decision task $\langle \mathcal{I}, \mathcal{O}, \Delta \rangle$ has a wait-free protocol using read-write memory if and only if there exists a chromatic subdivision σ of $\mathcal{I}$ and a color-preserving simplicial map*

$$\mu : \sigma(\mathcal{I}) \to \mathcal{O}$$

such that, for each vertex s in $\sigma(\mathcal{I})$, $\mu(s) \in \Delta(\mathrm{Car}(s, \mathcal{I}))$.

Applying this theorem to the read-write phase yields a color-preserving simplicial map

$$\mu : \sigma(\widetilde{\mathcal{K}}) \to \widetilde{\mathcal{L}}$$

that carries each $\tilde{v}_i$ to $\tilde{w}_i$ and carries each $\tilde{p}_{ij}$ to $\tilde{q}_{ij}$.

Composing μ with the color-discarding projection map π yields a simplicial map

$$\mu_0 : \sigma(\widetilde{\mathcal{K}}) \to \mathcal{L}.$$

The map μ_0 carries each $\tilde{v}_i$ to w_i and carries each $\tilde{p}_{ij}$ to q_{ij}.

We now claim that we can assume without loss of generality that $\mathcal{K}$ has a 3-coloring.

Lemma 15.3.5. *If $\mathcal{K}$ is a 2-dimensional complex, then $\mathrm{Bary}(\mathcal{K})$ has a 3-coloring.*

Proof.   Assign each x in $\mathrm{Bary}(\mathcal{K})$ the label $\dim(\mathrm{Car}(x, \mathcal{K}))$. The result is a 3-coloring.   □

Lemma 15.3.6. *The tasks $\mathrm{Loop}(\mathcal{K}, \lambda)$ and $\mathrm{Loop}(\mathrm{Bary}(\mathcal{K}), \mathrm{Bary}(\lambda))$ are equivalent; an instance of one implements the other.*

Proof.   To implement $\mathrm{Loop}(\mathrm{Bary}(\mathcal{K}), \mathrm{Bary}(\lambda))$, each process runs the protocol for $\mathrm{Loop}(\mathcal{K}, \lambda)$ and feeds the output to a round of barycentric agreement.

To implement $\mathrm{Loop}(\mathcal{K}, \lambda)$, each process runs the protocol for $\mathrm{Loop}(\mathrm{Bary}(\mathcal{K}), \mathrm{Bary}(\lambda))$, yielding output k. Each process then chooses any vertex in $\mathrm{Car}(k, \mathcal{K})$.   □

Assume that $\mathcal{K}$ is 3-colorable. Pick a 3-coloring for $\mathcal{K}$ with the first three process names, and let $\mathcal{K}^*$ be the resulting colored complex. Clearly, $\mathcal{K}$ and $\mathcal{K}^*$ are isomorphic (the only difference is the labeling of vertices). Let v_i^* and p_{ij}^* denote the images of v_i and p_{ij} in $\mathcal{K}^*$.

Note that $\mathcal{K}^*$ is a subcomplex of $\widetilde{\mathcal{K}}$, so $\sigma(\mathcal{K}^*)$ is a subcomplex of $\sigma(\widetilde{\mathcal{K}})$. We now have a simplicial map

$$\mu_1 : \sigma(\mathcal{K}^*) \to \mathcal{L},$$

the restriction of μ_0. Thus, the map μ_1 carries each v_i^* to w_i, and each p_{ij}^* to q_{ij}.

Lemma 15.3.7. *If an instance of $\mathrm{Loop}(\mathcal{K}, \lambda)$ implements $\mathrm{Loop}(\mathcal{L}, \mu)$, then there exists a continuous map $f : (\mathcal{K}, \lambda) \to (\mathcal{L}, \mu)$.*

Proof.   The map μ_1 constructed above induces a continuous map

$$|\mu_1| : |\sigma(\mathcal{K}^*)| \to |\mathcal{L}|$$

carrying each v_i^* to w_i and each $|p_{ij}^*|$ to q_{ij}. Since $|\sigma(\mathcal{K}^*)| = |\mathcal{K}|$, $|\mu_1|$ is the desired continuous map $|\mathcal{K}| \to |\mathcal{L}|$ carrying each v_i to w_i and each p_{ij} to q_{ij}.   □

Theorem 15.3.8. *An instance of* $(\mathcal{K}, \lambda)$ *implements* $(\mathcal{L}, \mu)$ *if and only if there exists a group homomorphism* $h : (\pi_1(\mathcal{K}), [c(\lambda)]) \rightarrow (\pi_1(\mathcal{L}), [c(\mu)])$.

Proof. From Lemmas 15.3.2, 15.3.7, and Theorem 15.3.8. □

15.4 Applications

We start with a known result, but one that illustrates the power of the theorem.

Proposition 15.4.1. *(3,2)-set agreement has no wait-free implementation using read-write registers.*

Proof. Recall that (3,2)-set agreement can be viewed as Loop $(\mathrm{skel}^1 \Delta^2, \zeta)$, where ζ is the triangle loop $(0, 1, 2, ((0, 1)), ((1, 2)), ((2, 0)))$. It is a standard result that $\pi_1(\mathrm{skel}^1 \Delta^2)$ is infinite cyclic with generator $[\zeta]$. Implementing $(3, 2)$-set agreement wait-free is the same as implementing it with 0-th barycentric agreement (Δ^2, ζ). Because $\pi_1(\Delta^2)$ is trivial, the only homomorphism $h : (\pi_1(\Delta^2), [\zeta]) \rightarrow (\pi_1(\mathrm{skel}^1 \Delta^2), [\zeta])$ is the trivial one. It carries $[\zeta]$ to the identity element of the group and not to $[\zeta]$. □

It is now easy to identify the most powerful and least powerful loop agreement tasks. We say that a task is *universal* if it implements any loop agreement task whatsoever.

Proposition 15.4.2. *(3,2)-set agreement is universal.*

Proof. As was just mentioned, (3,2)-set agreement is Loop $(\mathrm{skel}^1 \Delta^2, \zeta)$, where ζ is as above. It is a standard result that $\pi_1(\mathrm{skel}^1 \Delta^2)$ is the infinite cyclic group $\mathbf{Z}$ with generator $[\zeta]$. To implement any Loop $(\mathcal{L}, \mu)$, let $h([\zeta]) = [c(\mu)]$. □

Proposition 15.4.3. *Uncolored simplex agreement is implemented by any loop agreement task.*

Proof. The complex for uncolored simplex agreement has a trivial fundamental group because it is a subdivided simplex, and hence its polyhedron is a convex subset of Euclidean space. To implement this task with any Loop $(\mathcal{L}, \mu)$, let h of every element be the identity. □

As another example that illustrates the power of the theorem, we show that the loop agreement tasks classification is undecidable. In fact, we prove a more specific result: It is undecidable to compute if a loop agreement task belongs to the weakest class (of loop agreement tasks that are equivalent to uncolored simplex agreement, by the previous proposition).

The following gives a different proof of the result that wait-free task solvability is undecidable.

Proposition 15.4.4. *It is undecidable whether an instance of uncolored simplex agreement implements* Loop $(\mathcal{L}, \mu)$.

Proof. By Theorem 15.3.8, an instance of uncolored simplex agreement (Δ^2, ζ) implements Loop $(\mathcal{L}, \lambda)$ if and only if there exists a group homomorphism $h : (\pi_1(\Delta^2), [\zeta]) \rightarrow (\pi_1(\mathcal{L}), [c(\mu)])$. Since $\pi_1(\Delta^2)$ is the trivial group, h exists if and only if $[c(\mu)]$ is the identity of $\pi_1(\mathcal{L})$, that is, if and only if μ is contractible in $\mathcal{L}$. But it is a classic result that loop contractibility is undecidable in 2-dimensional complexes. □

15.5 **Torsion classes**

The *torsion number* of $(\mathcal{K}, \lambda)$ is the least positive integer k such that $[c(\lambda)]^k$ is the identity element of $\pi_1(\mathcal{K})$; in group theory this is called the *order* of the element $[c(\lambda)]$ of $\pi_1(\mathcal{K})$. If no such k exists, then the order is infinite. Every loop agreement task has a well-defined torsion number. Define *torsion class* k to be the tasks with torsion number k.

As an example of a loop agreement task with a nontrivial (i.e., not 1 and not ∞) torsion number, let $\mathcal{K}$ be complex whose polyhedron is isomorphic to a Moebius strip, and choose a triangle loop λ that geometrically corresponds to the "equator" of the strip. Then $c(\lambda)$ is the generator of $\pi_1(\mathcal{K}, k_0)$ and has torsion class 2.

How much information does a task's torsion number convey? The following properties follow directly from Theorem 15.3.8:

- If a task in class k (finite) implements a task in class ℓ, then $\ell | k$ (ℓ divides k).
- Each torsion class includes a *universal* task that solves any loop agreement task in that class.
- A universal task for class k (finite) is also universal for any class ℓ where $\ell | k$.
- A universal task for class k (finite) does not implement any task in class ℓ if ℓ does not divide k.
- A universal task for class ∞ is universal for any class k.

Torsion classes form a coarser partitioning than our complete classification, but they are defined in terms of simpler algebraic properties, and they too have an interesting combinatorial structure.

15.6 **Conclusions**

We have established a connection between the computational power of a class of distributed tasks and the structure of the fundamental groups of topological spaces related to the tasks. This connection raises a number of open problems. Can we use a similar approach to characterize the computational power of tasks other than loop agreement tasks? A potential obstacle here is that Theorem 15.3.8 is known to be false above dimension two, so it may be necessary to settle for weaker characterizations in higher dimensions. Can we characterize the computational power of *compositions* of tasks? In the implementations considered here, we compose one copy of a protocol for $(\mathcal{K}, \lambda)$ with an arbitrary read-write protocol to construct a protocol for $(\mathcal{L}, \mu)$. Can we give a similar characterization for multiple tasks in terms of the fundamental groups of their components? There is a need for further investigation into these questions.

15.7 **Chapter notes**

Most of the results in this chapter originally appeared in Herlihy and Rajsbaum [83].

The notion of loop agreement task was extended to *degenerate loop agreement* tasks by Liu et al. [108,109], A degenerate loop agreement task is defined by a graph (1-complex) $\mathcal{K}$ with two distinguished vertices, a and b. Each process starts with a binary input (0 or 1). Each process halts with a vertex from $\mathcal{K}$. If the inputs of all participating processes are 0, the processes halt with output value a, and if all values are 1, they halt with b. If the inputs are mixed, each process processes halts with a vertex from $\mathcal{K}$, and all vertices chosen must lie on an edge or vertex (0- or 1-simplex) of $\mathcal{K}$. The degenerate loop agreement tasks fall into two equivalence classes: those that are equivalent to consensus and those that are equivalent to read-write memory. The "consensus" class is universal in the sense that any task in that class implements any degenerate loop agreement task. The "read-write" class is the weakest class in the sense that any degenerate loop agreement task implements any task from the read-write class.

Fraignaud, Rajsbaum, and Travers [59] use techniques similar to the ones in this chapter to obtain a classification of locality-preserving tasks in terms of their computational power, considering a relationship with covering spaces (the classic algebraic topology notion).

More on cofibrations can be found in Kozlov [100, Chapter 7]. The representation of the fundamental group of a simplicial complex from subSection 15.1.2 is taken from Armstrong [7, p. 135]. Notions related to the torsion number appear in Munkres [124, p. 22] and in Armstrong [7, p. 178].

15.8 Exercises

Exercise 15.1. Find a loop agreement task Loop $(\mathcal{K}, \lambda)$ that is equivalent to the 2-set agreement but such that the fundamental group of $\mathcal{K}$ is not $\mathbb{Z}$.

Exercise 15.2. Consider the loop agreement task Loop $^*(\mathcal{K}, \lambda)$, where we no longer assume that $\mathcal{K}$ is 2-dimensional. What can be said about the solvability of Loop $^*(\mathcal{K}, \lambda)$?

Exercise 15.3. Construct an infinite sequence of loop agreement tasks Loop $(\mathcal{K}_1, \lambda_1)$, Loop $(\mathcal{K}_2, \lambda_2), \ldots$, such that for all $i < j$ the task Loop $(\mathcal{K}_i, \lambda_i)$ is implemented by the task Loop $(\mathcal{K}_j, \lambda_j)$, but not vice versa.

Exercise 15.4. Give an example of a loop agreement task that does not solve the universal task in its torsion class.

Immediate Snapshot Subdivisions

16

CHAPTER OUTLINE HEAD

Throughout this book, we have relied on the fact that Ch Δ^n, the standard chromatic subdivision of the n-simplex Δ^n defined in Chapter 3, is indeed a subdivision of Δ^n. In this chapter, we give a rigorous proof of this claim.

The complex Ch Δ^n captures all single-layer immediate snapshot executions for $n + 1$ processes. Recall that a single-layer immediate snapshot execution is given by a schedule $S_0, \ldots, S_d$, where each S_i is the set of processes that participate in step i, the S_i are disjoint, and the complete set of processes $[n]$ is their union. We call $S_0, \ldots, S_d$ an *ordered partition* of $[n]$.

16.1 A glimpse of discrete geometry

16.1.1 Polytopes

Recall from basic linear algebra that a *hyperplane* in $\mathbb{R}^{d+1}$ is a solution of a single linear equation $\sum_{i=0}^{d+1} c_i x_i = 0$. Given such an equation, we also have two *open half-spaces*: the set of points for which $\sum_{i=0}^{d+1} c_i x_i < 0$ and the set for which $\sum_{i=0}^{d+1} c_i x_i > 0$. It also defines two *closed half-spaces*: the unions of the hyperplane with the open half-spaces.

A $(d + 1)$-dimensional convex *polytope*[1] P is the convex hull of finitely many points in $\mathbb{R}^{d+1}$, where we assume that not all points lie on the same hyperplane. A *face* of a polytope P is the intersection of P with a hyperplane that does not intersect the interior of P. An $(d + 1)$-dimensional polytope

---

[1] This is a special case of what is called polytope in some literature, which will be sufficient for our purposes.

Distributed Computing Through Combinatorial Topology. http://dx.doi.org/10.1016/B978-0-12-404578-1.00016-4

P is bounded by a number of d-dimensional faces, which are themselves polytopes. In fact, P is an intersection of the closed half-spaces associated to its d-faces.

16.1.2 Schlegel diagrams

In general, it can be complicated to prove that one simplicial complex is a subdivision of another, even for a subdivision of a simplex. A straightforward argument would need to go into the technical details of topology of a geometric simplicial complex, possibly having to deal with explicit point descriptions as convex combinations of the vertices and so on.

Fortunately, in this case there is a short-cut: *Schlegel diagrams*. Informally, a Schlegel diagram is constructed by taking a "photograph" (perspective projection) of the polytope from a vantage point just outside of it, centered over a chosen d-face F. Since the polytope is convex, it is possible to choose the vantage point so that all the faces project onto F and the projections of disjoint faces are themselves disjoint.

Let us make this more specific. Pick a d-dimensional face F of the polytope P. As noted, F itself is a d-dimensional polytope obtained as an intersection of P with some hyperplane H so that the rest of the polytope lies entirely on one side of this hyperplane. Let H_- denote the open half-space bordered by H, which does not intersect the polytope P. Choose a point x in H_- very close to the barycenter of F (in fact, any point in the interior of F will do). Now project the boundary of the polytope P along the rays connecting it to x into the hyperplane H. If x is sufficiently close to the barycenter of F, the image of that projection will be contained in F. (In fact, topologically it will be precisely F.) Furthermore, by linearity, the images of the faces on the boundary of P, excluding F itself, will constitute a polyhedral subdivision of F, which we denote $\mathrm{Sch}_F(P)$. See Figure 16.1 for examples of Schlegel diagrams of a tetrahedron, cube, and dodecahedron.

16.1.3 Schlegel diagrams of cross-polytopes

A polytope whose d-faces are simplices is said to be *simplicial*. If the polytope P is simplicial, then the Schlegel diagram is a simplicial subdivision of the d-simplex F. We now consider a specific simplicial polytope in $\mathbb{R}^{d+1}$. For $i = 0, \ldots, d$, let e_i denote the point whose i^{th} coordinate is 1, and all other coordinates are 0. Let P_d be the convex hull of the point set $\{e_0, \ldots, e_d, -e_0, \ldots, -e_d\}$. It is easy to see that these $2d + 2$ points are in convex position and that the obtained polytope is simplicial. This is the so-called *cross-polytope*. Since for any pair of d-faces of P_d there exists a symmetry of P_d moving one of the faces to the other one, the Schlegel diagram will not depend on which face of P_d we choose, so we just write $\mathrm{Sch}P_d$. Examples of Schlegel diagrams for crosspolytopes of dimensions 1 and 2 are shown in Figure 16.2. Note that the boundary complex of P_d is precisely the simplicial join of $d + 1$ copies of the simplicial complex consisting of two points with no edge between them.

We shall need the following combinatorial description of the simplicial complex $\mathrm{Sch}P_d$. The set of vertices is indexed by $V = \{(i, s) \mid 0 \le i \le d, s \in \{+, -\}\}$, where for every $i = 0, \ldots, d$, the pair $(i, +)$ denotes the inner point of $\mathrm{Sch}P_d$ corresponding to the ith axis, whereas the pair $(i, -)$ denotes the vertex of the d-simplex used as the initial face for constructing the Schlegel diagram, corresponding to the ith axis. The d-dimensional simplices of $\mathrm{Sch}P_d$ are all tuples $((0, s_0), \ldots, (d, s_d))$ such that $(s_0, \ldots, s_d) \ne (-, \ldots, -)$.

Clearly, the advantage of using Schlegel diagrams is that one gets the fact that the diagram is a subdivision of the face for free. Figure 16.2 shows that $\mathrm{Sch}P_1$ is isomorphic to $\mathrm{Ch}\,\Delta^1$, whereas $\mathrm{Sch}P_2$

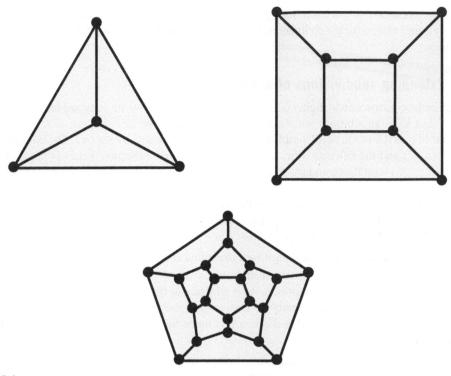

FIGURE 16.1

Schlegel diagrams of tetrahedron, cube, and dodecahedron.

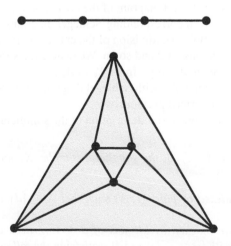

FIGURE 16.2

Schlegel diagrams of 1 and 2-dimensional cross-polytopes.

is different from Ch Δ^2. To get from SchP_2 to Ch Δ^2 one needs to further subdivide each of the edges of the triangle and extend these subdivisions to the subdivision of S^2. For higher d, one needs to do this several times.

16.1.4 Extending subdivisions of simplices

To generalize the construction to higher dimensions, we need the following standard fact about simplicial complexes. Let $\mathcal{K}$ be an arbitrary simplicial complex, and let σ be any simplex of $\mathcal{K}$. Recall that the (closed) star of σ is the union of all simplices that contain σ, denoted by $\text{st}_K(\sigma)$. The complex $\mathcal{K}$ is the union of $\text{St}(\sigma, \mathcal{K})$ and the deletion $\text{dl}(\sigma, \mathcal{K})$. The intersection of these two pieces is precisely the join of the link $\text{Lk}(\sigma, \mathcal{K})$ with the boundary $\partial \sigma$. The closed star itself is the simplicial join of σ with its link (see Section 3.3).

Now assume that $\mathcal{X}$ is a subdivision of σ that only subdivides the interior of σ while leaving the boundary of σ unchanged. A Schlegel diagram is an example of such a subdivision. In this case, the join of $\mathcal{X}$ with $\text{Lk}(\sigma, \mathcal{K})$ is a subdivision of $\text{St}(\sigma, \mathcal{K})$.

Fact 16.1.1. Whenever we have a join of two simplicial complexes $\mathcal{X} * \mathcal{Y}$, we can replace $\mathcal{X}$ with any simplicial subdivision $\widetilde{\mathcal{X}}$ and the obtained complex $\widetilde{\mathcal{X}} * \mathcal{Y}$ will be a subdivision of $\mathcal{X} * \mathcal{Y}$. One can see this geometrically if one remembers the realization of the join from Section 3.3. Indeed, when $\mathcal{X}$ and $\mathcal{Y}$ are embedded in complementary dimensions and the join is obtained by drawing all the line segments connecting points in $\mathcal{X}$ and in $\mathcal{Y}$, then it is immediate that replacing $\mathcal{X}$ with its subdivision will subdivide the join.

Back to our subdivision: We notice that since it does not change the link of σ, and it does not change the boundary $\partial \sigma$, it will also not change their join. Since this is precisely the space along which we attach $\sigma * \text{Lk}(\sigma, \mathcal{K})$, we can extend our local subdivision to a global subdivision of the entire $\mathcal{K}$.

We now have all the tools at hand to describe how to obtain the chromatic subdivision Ch Δ^n. Start with $\mathcal{X}_0 = \Delta^d$. Subdivide it as a Schlegel diagram of the cross-polytope. Proceed with the faces of Δ^d of codimension 1; replace them with corresponding Schlegel diagrams and extend these subdivisions using the argument above to the global subdivision of the entire complex. After this, proceed to do the same for the faces of Δ^d of codimension 2 and so on. We denote the simplicial complexes constructed in this way by $\mathcal{X}_0, \ldots, \mathcal{X}_{d-1}$. In short: To go from $\mathcal{X}_k$ to $\mathcal{X}_{k+1}$, we replace all boundary simplices of codimension k of the original simplex Δ^d with Schlegel diagrams SchP_{d-k} and then extend this to the subdivision of the entire $\mathcal{X}_k$ as described previously.

We are now ready to give a combinatorial description of the simplicial structure that we get at every step of the process.

Proposition 16.1.2. *For $k = 0, \ldots, d$, the simplicial complex X_k has the following combinatorial description:*

- *The vertices of X_k are indexed by pairs (i, A) such that $A \subseteq [d], i \in A$, and either $|A| = 1$ or $|A| \geq d - k + 2$.*
- *The d-simplices of X_k are indexed by all sets of $d + 1$ vertices corresponding to tuples $\sigma = ((i_0, A_0), \ldots, (i_d, A_d))$, with $\{i_0, \ldots, i_d\} = [d]$, satisfying the following conditions:*

  (1) $|A_0| \leq \cdots \leq |A_d|$

(2) $A_p \subseteq A_q$ *whenever* $p < q$ *and* $|A_q| \neq 1$

(3) $|A_{d-k+2}| \geq 2$

Proof. First we note that for $k = 0$, we get only vertices (i, A), with $|A| = 1$, i.e., the vertices $(0, \{0\}), \ldots, (d, \{d\})$. These form a single d-simplex; so our combinatorial description is correct for $X_0 = \Delta^d$.

Let us now show that for $k = 0, \ldots, d - 1$, the transformation from X_k to X_{k+1} produces the combinatorial simplicial structure described in the proposition. For convenience, we define $r(\sigma)$ to be the minimal index such that $A_{r(\sigma)} = 1$, and set $R(\sigma) = \{i_0, \ldots, i_{r(\sigma)}\}$.

First we consider what happens to vertices. For every simplex of $\partial \Delta^d$ of codimension k, i.e., for every subset $S \subseteq [d]$ such that $|S| = d + 1 - k$, we add new vertices $(s_0, S), \ldots, (s_{d-k}, S)$, where $S = \{s_0, \ldots, s_{d-k}\}$. This is consistent with the Schlegel construction and with the description of the set of vertices of X_k.

Next we analyze what happens with the d-simplices. Many d-simplices stay intact. Those that get subdivided are of the form $\sigma = ((i_0, A_0), \ldots, (i_d, A_d))$ such that $r(\sigma) = d - k + 1$. Each such σ gets replaced by new d-simplices, which are obtained as follows: Choose a non-empty subset $S \subseteq R(\sigma)$. For ease of presentation, we can reindex the vertices so that there exists r such that $\{i_0, \ldots, i_r\} = R(\sigma) \setminus S$, i.e., $\{i_{r+1}, \ldots, i_{r(\sigma)}\} = S$. Then the new simplex is

$$\tau = ((i_0, A_0), \ldots, (i_r, A_r), (i_{r+1}, R(\sigma)), \ldots, (i_{r(\sigma)}, R(\sigma)),$$
$$(i_{r(\sigma)+1}, A_{r(\sigma)+1}), \ldots, (i_d, A_d)).$$

We have $r(\tau) = r$ and $R(\tau) = R(\sigma) \setminus S$. To see that this is exactly what happens when Schlegel diagrams are extended to the entire complex, one can think of the part $\{(i_0, A_0), \ldots, (i_{r(\sigma)}, A_{r(\sigma)})\}$ as the maximal face of the corresponding cross-polytope indexed by the tuple $(-, \ldots, -)$. In the Schlegel constructions it gets replaced by simplices indexed by all possible nonempty subsets of $\{i_0, \ldots, i_{r(\sigma)+1}\}$. Then the extension of these subdivisions to the entire complex corresponds to appending this with the rest of the vertices, which is exactly what we do. The obtained d-simplices are precisely those occurring in our description of the d-simplices of X_{k+1}, where the order of the vertices is given by the construction. $\square$

Corollary 16.1.3. $X_d = \mathrm{Ch}\,\Delta^n$.

Proof. For $k = d$ we have $d - k + 2 = 2$, which means that there is at most one singleton set among $A_0, \ldots, A_d$. We see that the vertices of X_k are all the pairs (i, A) for $i \in [d], i \in A \subseteq [d]$, whereas a set of $d + 1$ vertices forms a d-simplex if and only if it can be ordered into a tuple $((i_0, A_0), \ldots, (i_d, A_d))$ satisfying $\{i_0, \ldots, i_d\} = [d]$, $|A_0| \leq \cdots \leq |A_d|$, and $A_p \subseteq A_q$ whenever $p < q$. In other words, the conditions for the sets of $d + 1$ vertices to form d-simplices translate precisely into our previous description of the simplicial complex $\mathrm{Ch}\,\Delta^n$. $\square$

Remark 16.1.4.   For $k \geq 1$, the transient simplicial complexes X_k can also be given a distributed computing interpretation. Namely, we consider all the executions where in the initial stage a certain number of processes, numbering at most $d + 1 - k$, will only perform the write operation, with the rest of processes functioning normally and performing a write with the immediate snapshot read operation. In particular, the views of those first "write-only" processes consist solely of their own names. Note that this also explains the equality $X_d = X_{d+1}$, since it does not matter whether the first process also reads or not after it wrote its name into the shared memory.

16.2 Chapter notes

The material in this chapter is adapted from Kozlov [101].

More information on Schlegel diagrams can be found in Grunbaum [74] and more on cross-polytope in Coxeter [42].

16.3 Exercises

Exercise 16.1. Let $f(d, n)$ denote the number of d-simplices in the chromatic subdivision Ch Δ^n. Give recursive formulas for the numbers $f(d, n)$. Can you estimate their asymptotics? What is the answer in the special cases $d = 0$ and $d = n$?

Exercise 16.2. Describe the links of vertices in Ch Δ^n as complexes constructed from standard chromatic subdivisions of simplices of lower dimension. How many different types of links are there?

Exercise 16.3. Describe the links of d-simplices of Ch Δ^n for arbirary d, using standard chromatic subdivisions of simplices of lower dimension.

Bibliography

[1] Abraham Ittai, Amit Yonatan, Dolev Danny. Optimal resilience asynchronous approximate agreement. In: Proceedings of the eighth international conference on principles of distributed systems, OPODIS'04. Lecture Notes in Computer Science, vol. 3544. Berlin, Heidelberg, Germany: Springer-Verlag; 2005. p. 229–239.p. 229–239.

[2] Afek Yehuda, Attiya Hagit, Dolev Danny, Gafni Eli, Merritt Michael, Shavit Nir. Atomic snapshots of shared memory. J ACM 1993;40(4):873–890.

[3] Afek Yehuda, Gafni Eli. Asynchrony from synchrony. In: Frey Davide, Raynal Michel, Sarkar Saswati, Shyamasundar Rudrapatna K, Sinha Prasun, editors. Distributed computing and networking. Lecture notes in Computer Science, vol. 7730. Berlin, Heidelberg, Germany: Springer; 2013. p. 225–239.

[4] Afek Yehuda, Gafni Eli, Rajsbaum Sergio, Raynal Michel, Travers Corentin. The k-simultaneous consensus problem. Distrib Comput 2010;22(3):185–195.

[5] Alistarh Dan, Gilbert Seth, Guerraoui Rachid, Travers Corentin. Generating fast indulgent algorithms. In: Aguilera Marcos K, Yu Haifeng, Vaidya Nitin H, Srinivasan Vikram, Choudhury Romit Roy, editors. Distributed computing and networking. Lecture notes in Computer Science, vol. 6522. Berlin, Heidelberg, Germany: Springer; 2011. p. 41–52.

[6] Anderson James H. Composite registers. Distrib Comput 1993;6(3):141–154.

[7] Armstrong MA. Basic topology (undergraduate texts in Mathematics). New York, NY, USA: Springer; 1983.

[8] Attiya Hagit, Bar-Noy Amotz, Dolev Danny. Sharing memory robustly in message-passing systems. J ACM 1995;42(1):124–142.

[9] Attiya Hagit, Bar-Noy Amotz, Dolev Danny, Peleg David, Reischuk Rüdiger. Renaming in an asynchronous environment. J ACM 1990;37(3):524–548.[ISSN: 0004-5411, http://doi.acm.org/10.1145/79147.79158, http://dx.doi.org/10.1145/79147.79158].

[10] Attiya Hagit, Borran Fatemeh, Hutle Martin, Milosevic Zarko, Schiper André. Structured derivation of semi-synchronous algorithms. In: Peleg David, editor. Distributed computing. Lecture notes in Computer Science, vol. 6950. Berlin, Heidelberg, Germany: Springer; 2011. p. 374–388.

[11] Attiya Hagit, Castañeda Armando, Herlihy Maurice, Paz Ami. Upper bound on the complexity of solving hard renaming. In: Proceedings of the 2013 ACM Symposium on principles of distributed computing, PODC '13. New York, NY, USA: ACM; 2013. p. 190–199.

[12] Attiya Hagit, Castañeda Armando. A non-topological proof for the impossibility of k-set agreement. In: Défago Xavier, Petit Franck, Villain Vincent, editors. Stabilization, safety, and security of distributed systems. Lecture notes in Computer Science, vol. 6976. Berlin, Heidelberg, Germany: Springer; 2011. p. 108–119.

[13] Attiya Hagit, Dwork Cynthia, Lynch Nancy, Stockmeyer Larry. Bounds on the time to reach agreement in the presence of timing uncertainty. J ACM 1994;41(1):122–152.[ISSN: 0004-5411, http://doi.acm.org/10.1145/174644.174649, http://dx.doi.org/10.1145/174644.174649].

[14] Attiya Hagit, Herlihy Maurice, Rachman Ophir. Atomic snapshots using lattice agreement. Distrib Comput 1995;8(3):121–132.

[15] Attiya Hagit, Paz Ami. Counting-based impossibility proofs for renaming and set agreement. In: Aguilera Marcos K, editor. Distributed computing. Lecture notes in Computer Science, vol. 7611. Berlin Heidelberg: Springer; 2012. p. 356–370.

[16] Attiya Hagit, Rajsbaum Sergio. The combinatorial structure of wait-free solvable tasks. SIAM J Comput 2002;31(4):1286–1313.

[17] Attiya Hagit, Welch Jennifer. Distributed computing fundamentals, simulations, and advanced topics. 2nd ed. Hoboken, NJ, USA: John Wiley and Sons; 2004.

[18] Biran Ofer, Moran Shlomo, Zaks Shmuel. A combinatorial characterization of the distributed tasks which are solvable in the presence of one faulty processor. In: PODC '88: Proceedings of the seventh annual ACM symposium on principles of distributed computing. New York, NY, USA: ACM; 1988. p. 263–275.

[19] Biran Ofer, Moran Shlomo, Zaks Shmuel. A combinatorial characterization of the distributed 1-solvable tasks. J Algorithms 1990;11(3):420–440.

[20] Biran Ofer, Moran Shlomo, Zaks Shmuel. Deciding 1-sovability of distributed task is np-hard. In: Möhring Rolf H., editor. Proceedings of 16th International Workshop WG '90, Berlin, Germany, June 20–22, 1990 Proceedings. Lecture Notes in Computer Science, vol. 484. London, UK: Springer-Verlag; 1991. p. 206–220.

[21] Biran Ofer, Moran Shlomo, Zaks Shmuel. Tight bounds on the round complexity of distributed 1-solvable tasks. Theor Comput Sci 1995;145(1–2):271–290.

[22] Bondy JA, Murty USR. Graph theory with applications. New York, NY, USA: Elsevier; 1976.

[23] Borowsky Elizabeth, Gafni Eli. Generalized FLP impossibility result for t-resilient asynchronous computations. In: STOC '93: Proceedings of the 25th annual ACM symposium on theory of computing. New York, NY, USA: ACM; 1993. p. 91–100.

[24] Borowsky Elizabeth, Gafni Eli. Immediate atomic snapshots and fast renaming. In: PODC '93: Proceedings of the 12th annual ACM symposium on principles of distributed computing. New York, NY, USA: ACM; 1993. p. 41–51.

[25] **Borowsky Elizabeth, Gafni Eli. The Implication of the Borowsky-Gafni simulation on the set-consensus hierarchy. Technical report, UCLA, 1993.**

[26] Borowsky Elizabeth, Gafni Eli. A simple algorithmically reasoned characterization of wait-free computations (extended abstract). In: PODC '97: Proceedings of the 16th annual ACM symposium on principles of distributed computing. New York, NY, USA: ACM; 1997. p. 189–198.

[27] Borowsky Elizabeth, Gafni Eli, Lynch Nancy, Rajsbaum Sergio. The BG distributed simulation algorithm. Distrib Comput 2001;14(3):127–146.

[28] Bracha G. Asynchronous byzantine agreement protocols. Inform Comput 1987;75(2):130–143.

[29] **Castañeda Armando, Imbs Damien, Rajsbaum Sergio, Raynal Michel. Generalized symmetry breaking tasks. Rapport de recherche PI-2007, ASAP - INRIA - IRISA, 2013.**

[30] Castañeda Armando, Imbs Damien, Rajsbaum Sergio, Raynal Michel. Renaming is weaker than set agreement but for perfect renaming: a map of sub-consensus tasks. In: Fernandez-Baca David, editor. LATIN 2012: Proceedings of the 10th Latin American symposium theoretical informatics. Lecture notes in Computer Science, vol. 7256. Berlin, Heidelberg, Germany: Springer; 2012. p. 145–156.

[31] Castañeda Armando, Rajsbaum Sergio. New combinatorial topology upper and lower bounds for renaming. In: Proceedings of the 27th ACM symposium on principles of distributed computing, PODC '08. New York, NY, USA: ACM; 2008. p. 295–304.

[32] Castañeda Armando, Rajsbaum Sergio. New combinatorial topology bounds for renaming: the lower bound. Distrib Comput 2010;22(5–6):287–301. http://dx.doi.org/10.1007/s00446-010-0108-2.

[33] Castañeda Armando, Rajsbaum Sergio. New combinatorial topology bounds for renaming: the upper bound. J ACM 2012;59(1):3:1–3:49.

[34] Castañeda Armando, Rajsbaum Sergio, Raynal Michel. The renaming problem in shared memory systems: an introduction. Comput Sci Rev 2011;5(3):229–251.

[35] Chandra Tushar, Hadzilacos Vassos, Jayanti Prasad, Toueg Sam. Generalized irreducibility of consensus and the equivalence of t-resilient and wait-free implementations of consensus. SIAM J Comput 2005;34(2):333–357.

[36] Charron-Bost Bernadette, Schiper André. The heard-of model: computing in distributed systems with benign faults. Distrib Comput 2009;22(1):49–71.

[37] Chaudhuri S. Agreement is harder than consensus: set consensus problems in totally asynchronous systems. In: Proceedings of the ninth annual ACM symosium on principles of distributed computing; 1990. p. 311–324.

[38] Chaudhuri Soma. More choices allow more faults: set consensus problems in totally asynchronous systems. Inform Comput 1993;105(1):132–158.

[39] Chaudhuri Soma, Herlihy Maurice, Lynch Nancy A, Tuttle Mark R. A tight lower bound for k-set agreement. In: Proceedings of the 34th IEEE symposium on foundations of Computer Science; 1993. p. 206–215.

[40] Chaudhuri Soma, Herlihy Maurice, Lynch Nancy A, Tuttle Mark R. Tight bounds for k-set agreement. J ACM 2000;47(5):912–943.

[41] Chaudhuri Soma, Reiners Paul. Understanding the set consensus partial order using the Borowsky-Gafni simulation (extended abstract). In: Proceedings of the 10th international workshop on distributed algorithms. London, UK: Springer-Verlag; 1996. p. 362–379.

[42] Coxeter HSM. Regular polytopes. 3rd ed. New York, NY, USA: Dover Publications; 1973.

[43] de Prisco Roberto, Malkhi Dahlia, Reiter Michael. On k-set consensus problems in asynchronous systems. IEEE Trans Parallel Distrib Syst 2001;12(1):7–21.

[44] Delporte-Gallet Carole, Fauconnier Hugues, Gafni Eli, Kuznetsov Petr. Wait-freedom with advice. In: Proceedings of the 2012 ACM symposium on principles of distributed computing, PODC '12. New York, NY, USA: ACM; 2012. p. 105–114.

[45] Delporte-Gallet Carole, Fauconnier Hugues, Guerraoui Rachid, Tielmann Andreas. The disagreement power of an adversary. In: Proceedings of the 23rd international conference on Distributed computing, DISC'09, Elche, Spain. Berlin, Heidelberg: Springer-Verlag; 2009. p. 8–21. [ISBN: 3-642-04354-2, 978-3-642-04354-3. <http://dl.acm.org/citation.cfm?id=1813164.1813173>].p. 8–21. [ISBN: 3-642-04354-2, 978-3-642-04354-3. <http://dl.acm.org/citation.cfm?id=1813164.1813173>].

[46] Delporte-Gallet Carole, Fauconnier Hugues, Guerraoui Rachid, Tielmann Andreas. The disagreement power of an adversary. Distrib Comput 2011;24(3–4):137–147.

[47] Dolev Danny, Lynch Nancy A, Pinter Shlomit S, Stark Eugene W, Weihl William E. Reaching approximate agreement in the presence of faults. J ACM 1986;33(3):499–516.

[48] Dolev Danny, Raymond Strong H. Authenticated algorithms for byzantine agreement. SIAM J Comput 1983;12(4):656–666.

[49] Dwork Cynthia, Lynch Nancy, Stockmeyer Larry. Consensus in the presence of partial synchrony. J ACM 1988;35(2):288–323.

[50] Dwork Cynthia, Moses Yoram. Knowledge and common knowledge in a byzantine environment: crash failures. Inform Comput 1990;88(2):156–186.

[51] Elrad Tzilla, Francez Nissim. Decomposition of distributed programs into communication-closed layers. Sci Comput Program 1982;2(3):155–173.

[52] Fagin Ronald, Halpern Joseph Y, Moses Yoram, Vardi Moshe Y. Reasoning about knowledge. Cambridge, MA, USA: MIT Press; 1995.

[53] Faleiro Jose M, Rajamani Sriram, Rajan Kaushik, Ramalingam G, Vaswani Kapil. Generalized lattice agreement. In: Proceedings of the 2012 ACM symposium on principles of distributed computing, PODC '12. New York, NY, USA: ACM; 2012. p. 125–134.

[54] Fan Ky. Simplicial maps from an orientable n-pseudomanifold into sm with the octahedral triangulation. J Comb Theory 1967;2(4):588–602.

[55] Fischer M, Lynch NA, Paterson MS. Impossibility of distributed commit with one faulty process. J ACM 1985;32(2).

[56] Fischer Michael J. The consensus problem in unreliable distributed systems (a brief survey). Technical Report YALEU/DCS/TR-273, Yale University, Department of Computer Science, 2000.

[57] Fischer Michael J, Lynch Nancy A. A lower bound for the time to assure interactive consistency. Inf Process Lett 1982;14(4):183–186.

[58] Fraigniaud Pierre, Rajsbaum Sergio, Travers Corentin. Locality and checkability in wait-free computing. In: Peleg David, editor. Distributed computing. Lecture notes in Computer Science, vol. 6950. Berlin, Heidelberg, Germany: Springer; 2011. p. 333–347.

[59] Fraigniaud Pierre, Rajsbaum Sergio, Travers Corentin. Locality and checkability in wait-free computing. Distrib Comput 2013;26(4):223–242.

[60] Gafni E, Guerraoui R, Pochon B. The complexity of early deciding set agreement. SIAM J Comput 2011;40(1):63–78.

[61] Gafni Eli. Round-by-round fault detectors (extended abstract): unifying synchrony and asynchrony. In: Proceedings of the 17th annual ACM symposium on principles of distributed computing, PODC '98. New York, NY, USA: ACM; 1998. p. 143–152.

[62] Gafni Eli. The extended BG-simulation and the characterization of t-resiliency. In: Proceedings of the 41st annual ACM symposium on theory of computing, STOC '09. New York, NY, USA: ACM; 2009. p. 85–92.

[63] Gafni Eli, Koutsoupias Elias. Three-processor tasks are undecidable. SIAM J Comput 1999;28(3):970–983.

[64] Gafni Eli, Kuznetsov Petr. On set consensus numbers. Distrib Comput 2011;24(3–4):149–163.

[65] Gafni Eli, Kuznetsov Petr. Relating L-resilience and wait-freedom via hitting sets. In: Aguilera Marcos K, Yu Haifeng, Vaidya Nitin H, Srinivasan Vikram, Choudhury Romit Roy, editors. Distributed Computing and Networking. Lecture notes in Computer Science, vol. 6522. Berlin, Heidelberg: Springer; 2011. p. 191–202.

[66] Gafni E, Mostfaoui A, Raynal M, Travers C. From adaptive renaming to set agreement. Theor Comput Sci 2009;410(14):1328–1335. Structural Information and Communication Complexity (SIROCCO 2007).

[67] Gafni Eli, Rajsbaum Sergio. Distributed programming with tasks. In: Proceedings of the 14th international conference on principles of distributed systems, OPODIS'10. Berlin, Heidelberg, Germany: Springer-Verlag; 2010. p. 205–218.

[68] Gafni Eli, Rajsbaum Sergio. Recursion in distributed computing. In: Dolev Shlomi, Cobb Jorge, Fischer Michael, Yung Moti, editors. Stabilization, safety, and security of distributed system. Lecture notes in Computer Science, vol. 6366. Berlin, Heidelberg, Germany: Springer; 2010. p. 362–376.

[69] Gafni Eli, Rajsbaum Sergio, Herlihy Maurice. Subconsensus Tasks Renaming Is Weaker Than Set Agreement. In: DISC; 2006. p. 329–338. http://dx.doi.org/ 10.1007/11864219_23.

[70] Gamow George, Stern Marvin. Puzzle-math. New York, NY, USA: Viking Press; 1958.

[71] **Garst PF. Cohen-Macaulay complexes and group actions. Ph.D. thesis, University of Wisconsin, 1979.**

[72] Glaser Leslie C. Geometrical combinatorial topology. vol. I. 1st ed. Van Nostrand; 1970.

[73] Jim Gray. Notes on data base operating systems. In: Operating systems, an advanced course. London, UK: Springer-Verlag; 1978. p. 393–481.

[74] Grunbaum Branko. Convex polytopes (graduate texts in Mathematics). 2nd ed. New York, Heidelberg: Springer; 2003.

[75] Havlicek J. A note on the homotopy type of wait-free atomic snapshot protocol complexes. SIAM J Comput 2004;33(5):1215–1222.

[76] Havlicek John. Computable obstructions to wait-free computability. Distrib Comput 2000;13:59–83.

[77] Henle Michael. A combinatorial introduction to topology. New York, NY, USA: Dover; 1983.

[78] Herlihy Maurice. Wait-free synchronization. ACM Trans Program Lang Syst 1991;13(1):124–149.

[79] Herlihy Maurice, Rajsbaum Sergio. Set consensus using arbitrary objects (preliminary version). In: PODC '94: Proceedings of the 13th annual ACM symposium on principles of distributed computing. New York, NY, USA: ACM; 1994. p. 324–333.

[80] Herlihy Maurice, Rajsbaum Sergio. The decidability of distributed decision tasks (extended abstract). In: STOC '97: Proceedings of the 29th annual ACM symposium on theory of computing. New York, NY, USA: ACM; 1997. p. 589–598.

[81] Herlihy Maurice, Rajsbaum Sergio. New perspectives in distributed computing (invited lecture). In: Kutyłowski Mirosław, Pacholski Leszek, Wierzbicki Tomasz, editors. Mathematical foundations of Computer Science 1999. Lecture notes in Computer Science, vol. 1672. Berlin, Heidelberg, Germany: Springer; 1999. p. 170–186.

[82] Herlihy Maurice, Rajsbaum Sergio. Algebraic spans. Math Struct Comput Sci 2000;10(4):549–573.

[83] Herlihy Maurice, Rajsbaum Sergio. A classification of wait-free loop agreement tasks. Theor Comput Sci 2003;291(1):55–77.

[84] Herlihy Maurice, Rajsbaum Sergio. The topology of shared-memory adversaries. In: Proceedings of the 29th ACM SIGACT-SIGOPS symposium on principles of distributed computing, PODC '10. New York, NY, USA: ACM; 2010. p. 105–113.

[85] Herlihy Maurice, Rajsbaum Sergio. Simulations and reductions for colorless tasks. In: Proceedings of the 2012 ACM symposium on principles of distributed computing, PODC '12. New York, NY, USA: ACM; 2012. p. 253–260.

[86] Herlihy Maurice, Rajsbaum Sergio. The topology of distributed adversaries. Distrib Comput 2013;26(3):173–192.

[87] Herlihy Maurice, Rajsbaum Sergio, Raynal Michel. Computability in distributed computing: a tutorial. SIGACT News 2012;43(3):88–110.

[88] Herlihy Maurice, Rajsbaum Sergio, Tuttle Mark. An axiomatic approach to computing the connectivity of synchronous and asynchronous systems. Electron Notes Theor Comput Sci 2009;230:79–102.

[89] Herlihy Maurice, Rajsbaum Sergio, Tuttle Mark R. Unifying synchronous and asynchronous message-passing models. In: PODC '98: Proceedings of the 17th annual ACM symposium on principles of distributed computing. New York, NY, USA: ACM; 1998. p. 133–142.

[90] Herlihy Maurice, Shavit Nir. The asynchronous computability theorem for t-resilient tasks. In: STOC '93: Proceedings of the 25th annual ACM symposium on theory of computing. New York, NY, USA: ACM; 1993. p. 111–120.

[91] Herlihy Maurice, Shavit Nir. The topological structure of asynchronous computability. J ACM 1999;46(6):858–923.

[92] Herlihy Maurice, Shavit Nir. The art of multiprocessor programming. New York, NY, USA: Morgan Kaufmann; 2008.

[93] Hoest Gunnar, Shavit Nir. Towards a topological characterization of asynchronous complexity. In: Proceedings of the 16th annual ACM symposium on principles of distributed computing, PODC '97. New York, NY, USA: ACM; 1997. p. 199–208.

[94] Hoest Gunnar, Shavit Nir. Toward a topological characterization of asynchronous complexity. SIAM J Comput 2006;36(2):457–497.

[95] Imbs Damien, Rajsbaum Sergio, Raynal Michel. The universe of symmetry breaking tasks. In: Kosowski Adrian, Yamashita Masafumi, editors. Structural information and communication complexity. Lecture notes in Computer Science, vol. 6796. Berlin, Heidelberg, Germany: Springer; 2011. p. 66–77.

[96] Imbs Damien, Raynal Michel. Visiting Gafni's reduction land: from the BG simulation to the extended BG simulation. In: Proceedings of the 11th international symposium on stabilization, safety, and security of distributed systems, SSS '09. Berlin, Heidelberg, Germany: Springer-Verlag; 2009. p. 369–383.

[97] Imbs Damien, Raynal Michel. The multiplicative power of consensus numbers. In: Proceedings of the 29th ACM SIGACT-SIGOPS symposium on principles of distributed computing, PODC '10. New York, NY, USA: ACM; 2010. p. 26–35.

[98] Junqueira Flavio, Marzullo Keith. A framework for the design of dependent-failure algorithms: research Articles. Concurr Comput: Pract Exper 2007;19(17):2255–2269.

[99] Junqueira Flavio P, Marzullo Keith. **Designing algorithms for dependent process failures. Technical report, 2003.**

[100] Kozlov Dmitry N. Combinatorial algebraic topology. Algorithms and computation in Mathematics, vol. 21. New York, Heidelberg: Springer; 2007.

[101] Kozlov Dmitry N. Chromatic subdivision of a simplicial complex. Homology, homotopy and applications 2012;14(2):197–209.

[102] Kozlov Dmitry N. **Weak symmetry breaking and abstract simplex paths. preprint, 2013.**

[103] Kuhn Fabian, Lynch Nancy, Oshman Rotem. Distributed computation in dynamic networks. In: Proceedings of the 42nd ACM symposium on theory of computing, STOC '10. New York, NY, USA: ACM; 2010. p. 513–522.

[104] Lamport Leslie. Time, clocks, and the ordering of events in a distributed system. Commun ACM 1978;21(7):558–565.

[105] Lamport Leslie. On interprocess communication, parts i and ii. Distrib Comput 1986;1(2):77–101.

[106] Lamport Leslie. The part-time parliament. ACM Trans Comput Syst 1998;16(2):133–169.

[107] Lamport Leslie, Shostak Robert, Pease Marshall. The byzantine generals problem. ACM Trans Program Lang Syst 1982;4(3):382–401.

[108] Liu Xingwu, Pu Juhua, Pan Jianzhong. A classification of degenerate loop agreement. In: Ausiello Giorgio, Karhumäki Juhani, Mauri Giancarlo, Ong Luke, editors. Fifth Ifip international conference on theoretical computer science, Tcs 2008. IFIP international federation for information processing, vol. 273. Berlin, Germany: Springer Verlag; 2008. p. 203–213. [chapter 14].

[109] Liu Xingwu, Xu Zhiwei, Pan Jianzhong. Classifying rendezvous tasks of arbitrary dimension. Theor Comput Sci 2009;410:2162–2173.

[110] Loui MC, Abu-Amara HH. Memory requirements for agreement among unreliable asynchronous processes, vol. 4. New York, NY, USA: JAI Press; 1987. p. 163–183. p. 163–183.

[111] Lubitch Ronit, Moran Shlomo. Closed schedulers: a novel technique for analyzing asynchronous protocols. Distrib Comput 1995;8(4):203–210.

[112] Malkhi Dahlia, Merritt Michael, Reiter Michael K, Taubenfeld Gadi. Objects shared by byzantine processes. Distrib Comput 2003;16(1):37–48.

[113] Matousek Jiri. Using the Borsuk-Ulam theorem: lectures on topological methods in combinatorics and geometry (universitext). New York, Heidelberg: Springer; 2007.

[114] Mendes Hammurabi, Herlihy Maurice. Multidimensional approximate agreement in byzantine asynchronous systems. In: Proceedings of the 45th annual ACM symposium on theory of computing, STOC '13. New York, NY, USA: ACM; 2013. p. 391–400.

[115] Mendes Hammurabi, Tasson Christine, Herlihy Maurice. **The topology of asynchronous byzantine colorless tasks, July 2013. preprint:** arXiv:1302.6224v3.

[116] Michailidis Dimitris. Fast set agreement in the presence of timing uncertainty. In: Proceedings of the 18th annual ACM symposium on principles of distributed computing, PODC '99. New York, NY, USA: ACM; 1999. p. 249–256.

[117] Moir Mark, Anderson James H. Wait-free algorithms for fast, long-lived renaming. Sci Comput Program 1995;25(1):1–39.

[118] Moran Shlomo, Wolfstahl Yaron. Extended impossibility results for asynchronous complete networks. Inf Process Lett 1987;26(3):145–151.

[119] Moses Yoram, Dolev Danny, Halpern Joseph Y. Cheating husbands and other stories: a case study of knowledge, action, and communication. Distrib Comput 1986;1(3):167–176.

[120] Moses Yoram, Rajsbaum Sergio. A layered analysis of consensus. SIAM J Comput 2002;31:989–1021.

[121] Mostefaoui A, Rajsbaum S, Raynal M, Travers C. The combined power of conditions and information on failures to solve asynchronous set agreement. SIAM J Comput 2008;38(4):1574–1601.

[122] Mostefaoui Achour, Rajsbaum Sergio, Raynal Michel. Conditions on input vectors for consensus solvability in asynchronous distributed systems. J ACM November 2003;50(6):922–954.

[123] Mostefaoui Achour, Rajsbaum Sergio, Raynal Michel. Synchronous condition-based consensus. Distrib Comput 2006;18:325–343.

[124] Munkres James. Elements of algebraic topology. 2nd ed. New Jersey, NJ, USA: Prentice Hall; 1984.

[125] Neda Armando C, Herlihy Maurice, Rajsbaum Sergio. An equivariance theorem with applications to renaming. In: Proceedings of the 10th latin American international conference on theoretical informatics, LATIN'12. Berlin, Heidelberg, Germany: Springer-Verlag; 2012. p. 133–144.

[126] Novikov PS. On the algorithmic unsolvability of the word problem in group theory. Trudy Mat Inst Steklov 1955;44:3–143.

[127] Rabin Michael O. Recursive unsolvability of group theoretic problems. Ann Math 1958;67(1):172+.

[128] Rajsbaum Sergio. Iterated shared memory models. In: López-Ortiz Alejandro, editor. LATIN 2010: theoretical informatics. Lecture notes in Computer Science, vol. 6034. Berlin, Heidelberg, Germany: Springer; 2010. p. 407–416.

[129] Rajsbaum Sergio, Raynal Michel. A survey on some recent advances in shared memory models. In: Kosowski Adrian, Yamashita Masafumi, editors. Structural Information and Communication Complexity. Lecture notes in Computer Science, vol. 6796. Berlin, Heidelberg, Germany: Springer; 2011. p. 17–28.

[130] **Rajsbaum Sergio, Raynal Michel, Stainer Julien. Computing in the presence of concurrent solo executions. Rapport de, recherche PI-2004, May 2013.**

[131] Rajsbaum Sergio, Raynal Michel, Travers Corentin. The iterated restricted immediate snapshot model. In: Hu Xiaodong, Wang Jie, editors. Computing and combinatorics. Lecture notes in Computer Science, vol. 5092. Berlin, Heidelberg, Germany: Springer; 2008. p. 487–497.

[132] Rajsbaum Sergio, Raynal Michel, Travers Corentin. The iterated restricted immediate snapshot model. In: Proceedings of the Computing and Combinatorics, 14th Annual International Conference, COCOON 2008, Dalian, China, June 27–29, 2008. Lecture notes in Computer Science, vol. 5092. New York, Heidelberg: Springer; 2008. p. 487–497.

[133] Raynal Michel. Fault-tolerant agreement in synchronous message-passing systems. Synthesis lectures on distributed computing theory 2010;1(1):1–189.

[134] Saks Michael, Zaharoglou Fotios. Wait-free k-set agreement is impossible: the topology of public knowledge. In: STOC '93: Proceedings of the 25th annual ACM symposium on theory of computing. New York, NY, USA: ACM; 1993. p. 101–110.

[135] Saks Michael, Zaharoglou Fotios. Wait-free k-set agreement is impossible: the topology of public knowledge. SIAM J Comput 2000;29(5):1449–1483.

[136] Santoro Nicola, Widmayer Peter. Time is not a healer. In: Monien B, Cori R, editors. STACS 89. Lecture notes in Computer Science, vol. 349. Berlin, Heidelberg, Germany: Springer; 1989. p. 304–313.

[137] Santoro Nicola, Widmayer Peter. Agreement in synchronous networks with ubiquitous faults. Theor Comput Sci 2007;384(2–3):232–249.

[138] Schmid Ulrich, Weiss Bettina, Keidar Idit. Impossibility results and lower bounds for consensus under link failures. SIAM J Comput 2009;38(5):1912–1951.

[139] Schneider Fred B. Implementing fault-tolerant services using the state machine approach: A tutorial. ACM Comput Surv 1990;22(4):299–319.

[140] Sergeraert Francis. The computability problem in algebraic topology. Adv Math 1994;104:1–29.

[141] Srikanth T, Toueg S. Simulating authenticated broadcasts to derive simple fault-tolerant algorithms. Distrib Comput 1987;2.

[142] Stillwell John. Classical topology and combinatorial group theory. 2nd ed. New York, Heidelberg: Springer; 1993.

[143] Vaidya Nitin H, Garg Vijay K. Byzantine vector consensus in complete graphs. In: Proceedings of the 2013 ACM symposium on principles of distributed computing, PODC '13. New York, NY, USA: ACM; 2013. p. 65–73.

[144] Wang Jun, Song Min. A new algorithm to solve synchronous consensus for dependent failures. In: Proceedings of the sixth international conference on parallel and distributed computing applications and technologies, PDCAT '05. Washington, DC, USA: IEEE Computer Society; 2005. p. 371–375.

[145] Yang Jiong, Neiger Gil, Gafni Eli. Structured derivations of consensus algorithms for failure detectors. In: Proceedings of the 17th annual ACM symposium on principles of distributed computing, PODC '98. New York, NY, USA: ACM; 1998. p. 297–306.

[146] Fajstrup L., Raussen M., Goubault E.. Algebraic topology and concurrency. Theor Comput Sci 2006; 357(1–3):241–278.

[147] Izmestiev I., Joswig M.. Branched coverings, triangulations, and 3-manifolds. Adv Geom 2003;3(2): 191–225.

Index

Printed and bound by CPI Group (UK) Ltd, Croydon, CR0 4YY

03/10/2024

01040322-0006